Soil Biology

Volume 38

Series Editor
Ajit Varma, Amity Institute of Microbial Technology,
Amity University Uttar Pradesh, Noida, UP, India

For further volumes:
http://www.springer.com/series/5138

Varda Kagan-Zur • Nurit Roth-Bejerano •
Yaron Sitrit • Asunción Morte
Editors

Desert Truffles

Phylogeny, Physiology, Distribution and Domestication

Springer

Editors
Varda Kagan-Zur
Nurit Roth-Bejerano
Life Sciences Department
Ben-Gurion University
Beer-Sheva, Israel

Yaron Sitrit
Jacob Blaustein Institute for Desert Research
Ben-Gurion University
Beer-Sheva, Israel

Asunción Morte
Facultad de Biología
Universidad de Murcia
Murcia, Spain

ISSN 1613-3382 ISSN 2196-4831 (electronic)
ISBN 978-3-642-40095-7 ISBN 978-3-642-40096-4 (eBook)
DOI 10.1007/978-3-642-40096-4
Springer Heidelberg New York Dordrecht London

Printed on acid-free paper

Springer is part of Springer Science+Business Media (www.springer.com)

Preface

Dozens of monographs relating to various aspects of mycorrhizal associations have been published over the last five decades or so; among them are works dealing with different edible mycorrhizal fungi, one even within this Soil Biology series, Edible Mycorrhizal Mushrooms (Zambonelli and Bonito 2012). Several volumes, written in different languages (e.g., Italian, French, English), have sought to summarize our knowledge concerning truffles or provide an overview of attempts at truffle cultivation. However, not a single one of these works deals specifically with desert truffles—the "stepbrothers" of what are termed the "true" truffles (members of the Tuberaceae family). This book is thus the first international monograph devoted to the subject of desert truffles.

Although desert truffles have been known and appreciated at least from the time of the Pharaohs, modern research into this type of truffle, together with associated international publications, was launched only in the late 1970s in Kuwait.

In contrast with the "true" truffles, all of which are members of a single family, and are phylogenetically related among the desert truffles the hypogeous form of life evolved independently several times in different families, mainly within the Pezizales order. Some genera are, therefore, phylogenetically closer to aboveground relatives than to other desert truffles.

Desert truffles have been found and described in every desert that has been explored, irrespective of the character of the habitat—cold or hot, in loamy or acidic, sandy or heavy soils. The only common denominator seems to be a limited supply of water.

Oddly enough, although in some arid areas—mainly the Mediterranean basin and the Middle East—they are known and appreciated by local inhabitants; in others they are virtually unacceptable as food and discarded when found. Wherever they are appreciated they command relatively high prices; the highest for any wild mushroom offered in the markets. As is the case for other wild commodities, yields of wild desert truffles are declining, at the same time appreciation for their nutritional value and organoleptic properties is on the rise. Moreover, interest in these truffles is increasing against the background of the search for new food and income sources for remote arid areas. The earliest reports of successful attempts at cultivation date back only some 20 years, and desert truffle cultivation is only now coming of age.

This volume offers detailed summaries of several aspects of the somewhat underdeveloped field of desert truffle research. Desert truffle taxonomy is undergoing profound changes as molecular methods of phylogenetic research come into their own. Many affiliations have been changed, a number of new genera and new species have been erected, and more changes may be expected as future research focuses on these truffles. Basic research into the mycorrhizal associations of desert truffles has revealed some interesting and unusual properties calling for further study of these underground fungi.

The book is divided into five parts: I, Phylogeny; II, Conditions favoring mycorrhiza formation; III, Distribution; IV, Fruit body attributes; and V, Cultivation. We trust that together they provide an overview of the state of the art regarding all aspects of desert truffles for the benefit of researchers, students, and members of the public with an interest in the subject.

We are grateful to Prof. Ajit Varma, editor of the series, for inviting us to edit this book and to Dr. Lindenborn of Springer, who was always available with advice as queries and problems needed to be addressed. We wish to thank all our colleagues who agreed to take part in this endeavor and helped to make this book a complete entity.

Beer-Sheva, Israel Varda Kagan-Zur
Nurit Roth-Bejerano
Yaron Sitrit
Murcia, Spain Asunción Morte

Reference

Zambonelli A, Bonito GM (eds) (2012) Edible mycorrhizal mushrooms. Soil Biology, vol 34. Springer, Heidelberg

Contents

Part III Distribution

Part IV Fruit-Body Attributes

Part I
Phylogeny

Chapter 1
Hypogeous Desert Fungi

Gabriel Moreno, Pablo Alvarado, and José Luis Manjón

1.1 Introduction

The term 'desert truffle' can be applied to multiple edible hypogeous fungi growing in arid areas throughout the world. Defining what a hypogeous fungus is, and what arid lands are, is hence critical to delimit this ecological group. Here, we choose to define 'hypogeous' fungi as those species with closed or 'sequestrate' globose fruiting bodies growing totally under soil surface or partially covered by it, their dispersal being presumably performed by animals (mainly mammals but also birds and arthropods, Fogel and Peck 1975; Maser et al. 1978; Alsheikh and Trappe 1983) actively searching for them. Occasionally, they can be partially or totally uncovered when mature, a state commonly known as semi-hypogeous. This combination of morphology and ecology is far from being a common apomorphic character of a single clade, since it is partially or totally achieved in independent branches of many unrelated families. Thus, there are hypogeous species in the former Glomeromycetes, Ascomycetes, and Basidiomycetes groups with fairly different ontogenic processes leading to a more or less similar external aspect and ecological behaviour. The term 'truffle' comes from Latin *tubera*, which was applied to the edible hypogeous fungi in the genera *Tuber* and *Terfezia* eaten by Greeks and Romans. It was the basis to coin the descriptor 'tuberoid', restricted to hypogeous Ascomycetes with asci. The idea of tuberoid species being polyphyletic was probably first developed by authors such as Malençon (1938) and later followed by Burdsall (1968) and Trappe (1979).

Similarly, Basidiomycetes (fungal species presenting basidia) were also found to present independent evolutionary lineages showing varying degrees of 'secotioid' or 'sequestrate' syndrome, as it was sometimes called (Thiers 1984; Kendrick 1992). The most evolved ones, especially those which have lost all remnants of

G. Moreno (✉) • P. Alvarado • J.L. Manjón
Dpto. Biología Vegetal, Univ. Alcalá, Alcalá de Henares, Madrid 28871, Spain
e-mail: gabriel.moreno@uah.es; pablo.alvarado@gmail.com; josel.manjon@uah.es

V. Kagan-Zur et al. (eds.), *Desert Truffles*, Soil Biology 38,
DOI 10.1007/978-3-642-40096-4_1,

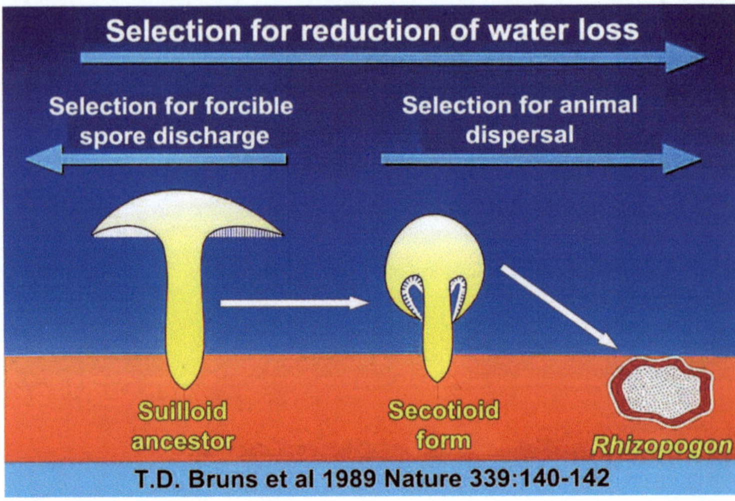

Fig. 1.1 Model for the evolution of mushrooms agaricoids, secotioids, and gastroids forms adapted from Thiers (1984) and Bruns et al. (1989)

the stipe-columella, were placed in the Gasteromycetes. They were described as 'gastroid' species and often referred to as 'false truffles'. However, some of them were soon suggested to be more related to other agaricoid lineages (Singer 1951; Heim 1971). This was first evident for the secotioid species, but the phylogenetic affiliation of some gasteroid groups was not finally unveiled until molecular tools were applied (Hibbett et al. 1997). Even nowadays sorting some of them in the field can be difficult without the aid of a microscope (Fig. 1.1).

Desert truffles are supposed to grow in arid lands. The term 'arid climate' refers to areas exposed to greater potential annual evapotranspiration than annual precipitation, which is reflected in the simple aridity index (AI) adopted by UNESCO (1979). Arid climates are hence subdivided into hyper-arid ($AI < 0.03$), arid ($0.03 < AI < 0.2$), and semi-arid ($0.2 < AI < 0.5$). Regions presenting such values usually match Köppen's B climate types (Kottek et al. 2006), such as the desertic climate Bw subtype (e.g. Arabia, Sahara, Low California, Namib, Australian outback), the continental steppe Bs climate (central Asia, American western Great Plains, Patagonia, Iran), or the milder Mediterranean Cs climate (Mediterranean Basin, California, coastal central Chile, South-African Cape coast, Australian south-western coast). For this latter climate type and some others, a fourth aridity type called dry subhumid ($0.5 < AI < 0.65$) can be recognized as an intermediate category between dry and humid climate types. In some of these climates aridity can be more pronounced in some seasons or even be present only temporally due to the highly seasonal rainfall pattern. Arid conditions can thus be found partially in the otherwise humid tropical savannas (Sahelian, Sudanian, Indian, Brazilian), corresponding to Köppen's wet and dry Aw climate. In addition, aridity may also appear because of other factors (Maliva and Missimer 2012) such as continentality and sea effects, rain shadow effects, soil type, slope, or even animal and human activity. This could affect the small-scale distribution of desert truffles especially at the boundaries of the mentioned climatic areas.

Desert truffles have been found so far in most of the geographical areas characterized by these climate types, although historically, some of them have received more attention than others. In some of these areas, hypogeous fungal species adapted to dry environments (steppes, deserts) are popularly known as 'desert truffles' (Shavit 2008; Kagan-zur and Roth-Bejerano 2008). A number of these desert truffle species can be found also in a wider range of habitats such as temperate deciduous forests, prairies, conifer forests, or even heath lands. In many cases, desert truffles establish mycorrhizal associations with specific host plants, sometimes endemics (e.g. Cistaceae in the Mediterranean). However, there are some putative exceptions to this symbiotic lifestyle, such as the genus *Carbomyces*, presumed to be ectomycorrhizal, although no suitable hosts have been identified in its natural habitat yet (Trappe 1971; Zak and Whitford 1986; Trappe and Weber 2001).

1.2 Ascomycetous 'Desert Truffles'

Ascomycetes include hypogeous truffle-like lineages with either cleistothecial (Elaphomycetales, Eurotiomycetes) or closed apothecial ascomata (Pezizales, Pezizomycetes). Among the latter group, several types of fruiting bodies were proposed by Weber et al. (1997). Here, some intermediate types are also recognized that could be useful to differentiate genera (Fig. 1.2):

1. Exothecia: external hymenium. The only hypogeous Ascomycetes presenting it are *Ruhlandiella* and *Sphaerozone* (Dissing and Korf 1980).
2. Ptychotecia: presenting an organized internal hymenium. The ascome can be:
   (a) Externally unfolded and hollow (as in *Hydnocystis*)
   (b) Externally folded and hollow (as in some *Genea*)
   (c) Externally more or less folded and filled with folded hymenium layers
      - With empty spaces between layers (as seen in *Balsamia*)
      - Completely solid without spaces between layers (as seen in *Choiromyces*)
3. Stereothecia: no organized hymenium is found in a solid gleba (e.g. *Tuber*).

Some species can present an ascome type slightly deviating from that of their genus for different reasons. Some may display truly plesiomorphic features, but others just mimic these states as a result of additional ontogenic stages. A good example is the truffle species *Tuber gennadii* (Chatin) Pat. This taxon was erected as the type species of the genus *Loculotuber* on the basis of its eye-shaped spore when young and its glebal locules, which were interpreted as a plesiomorphic feature (Álvarez et al. 1993). Molecular evidence did not support this movement (Bonito et al. 2010, 2013; Alvarado et al. 2012a), and glebal locules were reinterpreted, since they do not come from a separation between truly hymenial

Fig. 1.2 (**a**, **b**) *Ruhlandiella berolinensis* Henn., (**a**) exothecium, (**b**) detail of external hymenium and cortex. (**c**, **d**) *Genea verrucosa* Vittad, (**c**) externally folded and hollow ptychotecium. (**d**) Detail of internal hymenium layer and cortex. (**e**) *Hydnocystis piligera* Tul., unfolded and hollow ptychotecium. (**f**) *Balsamia vulgaris* Vittad, ptychotecial ascome filled with folded hymenium layers, leaving empty spaces between them. (**g**) *Choiromyces magnusii* (Mattir.) Paol., completely solid ptychotecium fored of folded hymenium layers leaving no spaces between them. (**h**) *Tuber aestivum* Vittad, stereothecial ascome, no organized hymenium is observed in a solid gleba

layers, but are left behind after the decay of the glebal hyphae surrounding the scattered asci.

Another diagnostic character, spore morphology (Fig. 1.3), is often used in Ascomycete taxonomy for lower level classification, usually for generic or specific

Fig. 1.3 (**a**) Smooth spores of *Picoa juniperi* Vittad; (**b**) verrucose spores of *Carbomyces emergens* Gilkey; (**c**) verrucose spores of *Picoa lefebvrei* (Pat.) Maire; (**d**) spiny spore of *Carbomyces gilbertsonii* N.S. Weber & Trappe; (**e**) spore presenting blunt-edged rods and cones of *Eremiomyces magnisporus* G. Moreno, P. Alvarado, Manjón & Sanz; (**f**) spore with isolated spines with a truncated tip of *Choiromyces venosus* (Fr.) Th.M. Fr.; (**g**) ellipsoid reticulated spore of *Tuber borchii* Vittad; (**h**) globose reticulated spore of *Tuber sphaerospermum* (Malençon) P. Roux, Guy García & M.C. Roux; (**i**) spore ornamented with coarse spines and short ridges of *Tuber indicum* Cooke & Massee; (**j**) spore presenting small cavities or pores of *Choiromyces magnusii* (Mattir.) Paol. Bars **a** = 10 μm, **b–g** = 5 μm, **h** = 10 μm, **i**, **j** = 5 μm

taxonomy. Spore shape and ornamentation are useful for building synoptic keys in many hypogeous groups, because of their variability between taxa. However, they should be again used cautiously since some exceptions arise. First, some degree of variability could be expected in characters such as spines length or reticulum height, even leading to confusing reticulo-spinulate intermediates that avoid a clear-cut delimitation of groups (Kovács et al. 2011). Second, eventual plesiomorphy could be present in some lineages, especially in those losing spore ornamentation and reverting to smooth ancestral states (Alvarado et al. 2012b).

The main families and genera containing desert truffles in Ascomycetes are:

Elaphomycetaceae Tul. & C. Tul. ex Paol., in Saccardo, Syll. Fung. 8: 863 (1889)
Elaphomyces Nees, Hor. Phys. Berol. (1820)

Glaziellaceae J.L. Gibson, in Gibson, Kimbrough & Benny, Mycologia 78(6): 953 (1986)
Glaziella Berk., in Warming, Vidensk. Meddel., Dansk Naturhist. Foren. Kjøbenhavn 80: 31 (1880)

Pezizaceae Dumort., Analyse des familles des plantes, avec l'indication des principaux genres qui s'y rattachent: 72 (1829)
Amylascus Trappe, Trans. Brit. Mycol. Soc. 57(1): 89 (1971)
Calongea Healy, Bonito & Trappe, Anales Jard. Bot. Madrid 66: 27 (2009)
Carbomyces Gilkey, N. Amer. Fl. 1: 27 (1954)
Delastria Tul. & C. Tul., Ann. Sci. Nat., Bot. 19: 379 (1843)
Eremiomyces Trappe & Kagan-Zur, Mycol. Res. 109(2): 244 (2005)
Fischerula Mattir., Nuovo Giorn. Bot. Ital. 34: 1348 (1928)
Hydnobolites Tul. & C. Tul., Ann. Sci. Nat., Bot. 19: 278 (1843)
Hydnotryopsis Gilkey, Univ. Calif. Publ. Bot. 6: 336 (1916)
Kalaharituber Trappe & Kagan-Zur, Mycol. Res. 109(2): 242 (2005)
Mattirolomyces E. Fisch., Nat. Pflanzenfam. 5b: 39 (1938)
Mycoclelandia Trappe & G.W. Beaton, Trans. Brit. Mycol. Soc. 83(3): 536 (1984)
Pachyphloeus Tul. & C. Tul., Giorn. Bot. Ital. 2(7–8): 60 (1845)
Ruhlandiella Henn., Hedwigia Beiblätter 42: 24 (1903)
Stouffera Kovács & Trappe, Mycologia 103(4): 836 (2011)
Terfezia (Tul. & C. Tul.) Tul. & C. Tul., Fungi hypog.: 172 (1851)
Tirmania Chatin, Truffe: 80 (1892)
Ulurua Trappe, Claridge & Kovács, in Trappe, Kovács & Claridge, Mycol. Progress 9(1): 140 (2010)

Pyronemataceae Corda [as 'Pyronemaceae'], Anl. Stud. Mycol.: 149 (1842)
Genea Vittad, Monogr. Tuberac.: 27 (1831)
Geopora Harkn., Bull. Calif. Acad. Sci. 1(3): 168 (1885)
Gilkeya M.E. Sm., Trappe & Rizzo, Mycologia 98(5): 705 (2007)
Hydnocystis Tul. & C. Tul., Giorn. Bot. Ital. 1(7–8): 59 (1845)
Picoa Vittad, Monogr. Tuberac.: 54 (1831)
Stephensia Tul. & C. Tul., Compt. Rend. Acad. Sci. Paris 21: 1433 (1845)

Tuberaceae Dumort., Comm. bot.: 69, 79 (1822)
Choiromyces Vittad, Monogr. Tuberac.: 50 (1831)
Dingleya Trappe, Mycotaxon 9(1): 331 (1979)
Labyrinthomyces Boedijn, Bull. Jard. Bot. Buitenzorg 16(2): 238 (1939)
Paradoxa Mattir., Beitr. Kryptogamenfl. Schweiz. 8(2): 32 (1935)
Reddellomyces Trappe, Castellano & Malajczuk, Austral. Syst. Bot. 5(5): 606 (1992)
Tuber P. Micheli ex F.H. Wigg., Prim. fl. holsat.: 1–112 (1780)

Incertae Sedis *Fischerula* Mattir., Nuovo Giorn. Bot. Ital. 34: 1348 (1928)

This in mind, the following could be used as a key to the most representative currently known genera of ascomycetous desert truffles, based on the abundant literature produced before on this issue (Gilkey 1939; Ceruti 1960; Burdsall 1968; Trappe 1979; Montecchi and Sarasini 2000; Hansen et al. 2001; Perry et al. 2007; Læssøe and Hansen 2007; Trappe et al. 2010; Bonito et al. 2013):

1a	Powdery gleba (dehiscent asci) when mature, thick peridium	*Elaphomyces*
1b	Powdery gleba (dehiscent asci) when mature, thin peridium	*Carbomyces*
1c	Solid gleba (asci not dehiscent) when mature	2
2a	Amyloid asci	3
2b	Non-amyloid asci	8
3a	Exothecium	*Ruhlandiella*
3b	Hollow folded ptycothecium	*Hydnoplicata*
3c	Solid ptycothecium	4
3d	Stereothecium	5
4a	Smooth spores	*Mycoclelandia*
4b	Minutely papillose spores	*Hydnotryopsis*
4c	Warted spores, paraphyses absent	*Ulurua*
4d	Spores ornamented with spines or rods	*Amylascus*
5a	4–8 ellipsoid smooth spores	*Tirmania*
5b	8 ellipsoid warty-reticulated spores	*Cazia*
5c	2–4 globose spores	6
5d	8 globose spores	7
6a	Alveolated spores	*Delastria*
6b	Warted spores	*Temperantia*
7a	Alveolated spores	*Stouffera*
7b	Spiny or verrucose spores, peridium with one layer	*Pachyphloeus*
7c	Spiny or verrucose spores, peridium with two layers	*Calongea*
8a	Unfolded hollow ptycothecium	9
8b	Irregularly folded ptycothecium	10
8c	Solid ptycothecium with internal chambers, unfolded ascoma	11

(continued)

8d	Solid ptycothecium without internal chambers, unfolded ascoma	12
8e	Stereothecium without organized hymenium	13
9a	Smooth globose yellowish ascoma, smooth globose spores	*Hydnocystis*
9b	Warted convoluted brownish ascoma, ellipsoid smooth spores	*Sepultaria*
10a	Smooth ellipsoid spores	*Geopora*
10b	Warty subglobose spores, brownish to blackish ascoma	*Genea*
10c	Warty subglobose spores, reddish ascoma	*Gilkeya*
10d	Spiny globose spores	*Genabea*
11a	Verrucose reddish peridium	*Balsamia*
11b	Lightly to moderately tomentose brownish peridium	*Labyrinthomyces*
11c	Tessellate-cracked or verrucose brownish peridium	*Dingleya*
12a	Smooth spores	*Stephensia*
12b	Verrucose spores	*Reddellomyces*
12c	Spores alveolated or covered by rods	*Choiromyces*
13a	Asci containing 1–6 spores	14
13b	Asci containing 8 spores	15
14a	Spores warted	*Fischerula*
14b	Spores mainly spiny or reticulated	*Tuber*
15a	Smooth or verrucose spores >20 μm, associated to Cistaceae, Mediterranean Basin	*Picoa*
15b	Verrucose-reticulated spores >14 μm, associated to Fabaceae	*Mattirolomyces*
15c	Verrucose spores <14 μm, Australia	*Elderia*
15d	Spores ornamented with rods, probably associated to Poaceae	*Eremiomyces*
15e	Spiny warted spores, Kalahari Desert	*Kalaharituber*
15f	Spiny warted or reticulated spores, Mediterranean Basin	*Terfezia*

1.3 Basidiomycetous 'Desert Truffles'

Unlike Ascomycetous truffles, the extreme variability of the different lineages where hypogeous fungi are found among Basidiomycetes prevents a useful synoptic key to be proposed at higher levels. Fully closed, angiocarpic or 'gastroid' habit has been developed in many independent orders and families (Fig. 1.4), although some intermediate pseudoangiocarpic or 'secotioid' development types can also be found. Gastroid species have lost the typical agaricoid external stipe, presenting just an internal columella or even lacking it as well. In turn, secotioid species usually

Fig. 1.4 (**a**) *Descomyces albus* (Berk.) Bougher & Castellano; (**b**) spores of *Descomyces albus*; (**c**) *Hydnangium carneum* Wallr.; (**d**) spores and basidia of *Hydnangium carneum* stained in ammonium Congo Red; (**e**) *Hymenogaster luteus* Vittad; (**f**) *Hysterangium cistophilum* (Tul.) Zeller & C.W. Dodge; (**g**) *Octaviania asterosperma* (Vittad) Kuntze; (**h**) *Rhizopogon buenoi* Calonge & M.P. Martín, where peridium turns violet after treating with 10 % ammonium hydroxide; (**i**) *Schenella pityophilus* (Malençon & Riousset) Estrada & Lado; (**j**) *Youngiomyces multiplex* (Thaxt.) Y.J. Yao

retain a more or less functional external stipe or stipe-columella. Both states are often simply referred to as 'sequestrate' forms, since it soon became evident that they did not represent monophyletic clades but should be combined to prioritary 'agaricoid' genera on the basis of molecular data, e.g. *Macowanites* or *Gymnomyces* in *Russula*; *Zelleromyces* in *Lactarius*; *Thaxterogaster* in *Cortinarius*; or *Cryptolepiota* into *Lepiota*. A huge body of literature has been developed on this issue (Calonge and Martín 2000, Miller et al. 2001; Peintner et al. 2001; Hibbett and Binder 2002; Moncalvo et al. 2002; Eberhardt and Verbeken 2004; Nuytinck et al. 2004). Sequestrate fruiting bodies can be displayed by fully epigeous species or else by hypogeous or semi-hypogeous fungi, these being considered 'false truffles'. The hypogeous habit does not even assure the phylogenetic independence, since it has been found to occur also in typically epigeous genera, a situation not known, so far, among Ascomycetes.

One of the most striking features of sequestrate Basidiomycetes, also found in some of their hypogeous representatives, is the afore mentioned presence of remnants of the stipe, often transformed into a columella or stipe-columella. When present, this structure shares most of its features with the epigeous stipe but appears packed, reduced, or ramified. It seems to have evolved independently in several groups, and so its phylogenetic interpretation should be taken cautiously. In fact, some researchers point to the fact that these features can even vary among specimens of the same species (Lebel and Trappe 2000), at least in some groups. The organization of fertile tissues in sequestrate basidiomycetes can also present quite diverse arrangements if compared with the typical 'agaricoid' species. The typical radial lamellate organization can be found, as well as a fully packed model where fertile tissue is regularly to irregularly folded. As in Ascomycetes, folded fertile tissue can occupy the whole basidiome, giving the aspect of a solid gleba. This can indeed have small empty pockets or be completely solid. Finally, some special arrangements of fertile tissue can be found in some groups, such as those in the Phallomycetidae (Hosaka et al. 2006, characterized by a more or less radial arrangement of the fertile tissues), producing peridioles, independent portions of packed fertile tissue resembling plant seeds.

In most groups of hypogeous basidiomycetes, ballistospores have been lost and only statimospores are developed. The colour and shape of these spores have been considered also critical features to discriminate between genera, much the same as these characters are used to define the major groups of agarics. Determining if spores are hyaline or pigmented, amyloid or non-amyloid, and finally ornamented or smooth, gives useful information about the most probable phylogenetic relationships between different species.

A small overview of the most common basidiomycetous desert truffles would include the following:

Agaricaceae Chevall., Fl. gén. env. Paris 1:121 (1826)
Cryptolepiota Kropp & Trappe, Mycologia, 104(1): 170 (2012)

Albatrellaceae Pouzar, Folia Geobot. Phytotax. 1(4):358 (1966)
Leucogaster R. Hesse, Jahrb. Wiss. Bot. 13(2):189 (1882)
Leucophleps Harkn., Proc. Calif. Acad. Sci., Ser. 3, Bot. 1:257 (1899)

Boletaceae Chevall., Fl. env. Paris 1:248 (1826)
Chamonixia Rolland, Bull. Soc. Mycol. France 15:76 (1899)
Octaviania Vittad, Monogr. Tuberac.:15 (1831)

Cortinariaceae R. Heim ex Pouzar, Ceská Mykol. 37(3):174 (1983)
Cortinarius (Pers.) Gray, Nat. Arr. Br. Pl. 1:627 (1821)
Descomyces Bougher & Castellano, Mycologia 85(2):280 (1993)
Protoglossum Massee, Grevillea 19 (92):97 (1891) (=*Cortinarius*)
Quadrispora Bougher & Castellano, Mycologia 85(2):285 (1993) (=*Cortinarius*)
Timgrovea Bougher & Castellano, Mycologia 85(2):288 (1993)

Gallaceaceae Locq. ex P.M. Kirk, in Kirk, Cannon, Minter & Stalpers, Ainsworth & Bisby's Dictionary of the Fungi, 10th edition:272 (2008)
Austrogautieria E.L. Stewart & Trappe, Mycologia 77(5):675 (1985)

Geastraceae Corda, Icon. fung. 5:25 (1842)
Radiigera Zeller, Mycologia 36(6):628 (1944)

Gomphaceae Donk, Persoonia 1(4):406 (1961)
Gautieria Vittad, Monogr. Tuberac.:25 (1831)

Hydnangiaceae Gäum. & C.W. Dodge, Comp. Morph. Fungi:485 (1928)
Hydnangium Wallr., in Dietrich, Fl. Regn. Boruss. 7:465 (1839)

Hymenogastraceae Vittad, *Monogr. Tuberac.*:11 (1831)
Hymenogaster Vittad, *Monogr. Tuberac.*:30 (1831)
Wakefieldia Corner & Hawker, Trans. Brit. Mycol. Soc. 36(2):130 (1953)

Hysterangiaceae E. Fisch., Nat. Pflanzenfamilien 1: 304 (1899)
Hysterangium Vittad, Monogr. Tuberac.: 13 (1831)

Mesophelliaceae Jülich, Biblioth. Mycol. 85: 379 (1982)
Andebbia Trappe, Castellano & Amar., *Austral. Syst. Bot.* 9(5): 808 (1996)
Castoreum Cooke & Massee, in Cooke, *Grevillea* 15(76): 100 (1887)
Chondrogaster Maire, Bull. Soc. Mycol. France 40(43): 312 (1926) [1924]
Gummiglobus Trappe, Castellano & Amar., *Austral. Syst. Bot.* 9(5): 804 (1996)
Gummivena Trappe & Bougher, *Australas. Mycol.* 21(1): 9 (2002)
Mesophellia Berk., *Trans. Linn. Soc. London* 22: 131 (1857)
Nothocastoreum G.W. Beaton, in Beaton & Weste, Trans. Brit. Mycol. Soc. 82(4): 666 (1984)

Paxillaceae Lotsy, Vortr. Bot. Stammesgesch. 1: 706 (1907)
Alpova C.W. Dodge, Ann. Missouri Bot. Gard. 18: 461 (1931)
Melanogaster Corda, in Sturm, Deut. Fl., 3(11): 1 (1831)

Pyrenogastraceae Jülich, Biblioth. Mycol. 85: 387 (1982)
Schenella T. Macbr., Mycologia 3(1): 39 (1911)

Rhizopogonaceae Gäum. & C.W. Dodge, Comp. Morph. Fungi: 468 (1928)
Rhizopogon Fr., in Fries & Nordholm, Symb. gasteromyc. 1: 5 (1817)

Russulaceae Lotsy, Vortr. bot. Stammesgesch.: 708 (1907)
Arcangeliella Cavara, Nuovo Giorn. Bot. Ital. 7: 126 (1900) (=*Lactarius*)
Cystangium Singer & A.H. Sm., Mem. Torrey Bot. Club 21: 67 (1960) (=*Russula*)
Gymnomyces Massee & Rodway, in Massee, Bull. Misc. Inform. Kew: 125 (1898) (=*Russula*)
Lactarius Pers., Tent. disp. meth. Fung.: 63 (1797)
Martellia Mattir., Malpighia 14: 78 (1900) (=*Russula*)
Russula Pers., Obs. mycol. 1: 100 (1796)
Zelleromyces Singer & A.H. Sm., Mem. Torrey Bot. Club 21(3): 18 (1960) (=*Lactarius*)

Sclerodermataceae Corda [as 'Sclerodermaceae'], Icon. fung. 5: 23 (1842)
Horakiella Castellano & Trappe, Austral. Syst. Bot. 5(5): 641 (1992)

Sclerogastraceae Locq. ex P.M. Kirk, in Kirk, Cannon, Minter & Stalpers, Ainsworth & Bisby's Dictionary of the Fungi, 10th edition: 623 (2008)
Sclerogaster R. Hesse, Hypog. Deutschl. 1: 84 (1891)

Stephanosporaceae Oberw. & E. Horak, Pl. Syst. Evol. 131: 162 (1979)
Stephanospora Pat., Bull. Soc. Mycol. France 30(3): 349 (1914)

Trappeaceae P.M. Kirk, in Kirk, Cannon, Minter & Stalpers, Ainsworth & Bisby's Dictionary of the Fungi, 10th edition: 695 (2008)
Trappea Castellano, Mycotaxon 38: 2 (1990)

Incertae Sedis *Gastrosporium* Mattir., Mem. R. Accad. Sci. Torino, Ser. 2 53: 361 (1903)

In order to build the following simplified and artificial key to these genera, some exceptions were overseen to gain some clearness. Many genera of hypogeous basidiomycetes present in most synoptic keys were excluded since their known habitat does not match the arid conditions to be considered 'desert truffles' (e.g. sequestrate *Lepiota*, sequestrate *Entoloma*, *Gallacea*, *Phallobata*, *Protubera*, *Gigasperma*). Some Cortinariaceae and Russulaceae genera were sunk under their prioritary synonyms, as many of them have been demonstrated to be just 'sequestrate' forms of typically agaricoid lineages. Epigeous or semi-hypogeous sequestrate genera as well as 'secotioid' taxa were also removed. The genus *Chamonixia* was included, despite its basidiome is here regarded as secotioid, because the external part of the stipe-columella is often broken and lost or just residual, leading to some confusion. Specialized keys for some of the included groups can be found elsewhere (Danks et al. 2010; Hosaka et al. 2006, 2008; Hosaka and Castellano 2008; Kinoshita et al. 2012; Peintner et al. 2001, 2004; Trappe et al. 1996; Trappe and Bougher 2002).

1a	Gastroid habit with fertile tissue arranged radially around a columella or its remnants	2
1b	Gastroid habit with fertile tissue irregularly folded or lamellar	5
2a	Gleba formed of brownish cells, flesh bruises blue	*Chamonixia*
2b	Gleba fleshy and greenish, spores 8–20 μm	*Hysterangium*
2c	Gleba fleshy and greenish, spores 4–6 μm	*Trappea*
2d	Gleba ramified and brownish, peridium absent, conspicuous columella	*Gautieria*
2e	Gleba ramified and brownish, peridium absent, reduced or absent columella	*Austrogautieria*
2f	Gleba arranged in peridioles	*Schenella*
2g	Gleba surrounded by protruding material and organic debris	*Chondrogaster*
2h	Gleba greenish or yellowish-orange, gummy consistence	*Sclerogaster*
2i	Gleba powdery when mature	3
3a	Gleba blackish, without mycelial cords	*Radiigera*
3b	Gleba greenish, with mycelial cords	*Gastrosporium*
3c	Gleba surrounding a sterile core, spores verrucose	*Mesophellia*
3d	Gleba surrounding a sterile core, spores ornamented with lines	*Andebbia*
3e	Brittle peridium, spores released to the air	*Nothocastoreum*
3f	Gleba with gummy tissue	4
4a	Gleba with sterile veins, peridium three-layered	*Gummivena*
4b	Columella present, peridium two-layered	*Gummiglobus*
4c	Gleba with arachnoid sterile veins, peridium two-layered	*Castoreum*
5a	Hyaline spores	6
5b	Coloured spores	10
6a	Non-amyloid spiny spores	*Hydnangium*
6b	Non-amyloid minutely ornamented or smooth spores	7
6c	Non-amyloid alveo-reticulate spores	8
6d	Amyloid spores	9
7a	Regular hymenium layers folded in a solid gleba	*Rhizopogon*
7b	Irregular fertile tissue not arranged in hymenium layers	*Alpova*
8a	Spores reticulated	*Leucogaster*
8b	Spores spiny or verrucose	*Leucophleps*
9a	Bleeding latex	*Lactarius*
9b	Not bleeding latex	*Russula*
10a	Reddish-dark spores	11
10b	Yellowish-brownish spores	12
11a	Spores smooth, blackish gleba	*Melanogaster*
11b	Spores with conical-pyramidal ornamentation	*Octaviania*

(continued)

12a	Spores smooth and ellipsoid	*Hymenogaster*
12b	Spores reticulated and subglobose, enclosed by nurse hyphae	*Horakiella*
12c	Spores with a ring at its base	*Stephanospora*
12d	Spores ornamented and fig-like	*Wakefieldia*
12e	Spores ornamented and ellipsoid	13
13a	Spores symmetrical, reticulated	*Timgrovea*
13b	Spores symmetrical, ornamentation embedded in perisporium	*Descomyces*
13c	Spores with rounded apex, ornamentation exosporial, verrucose	*Cortinarius*

1.4 Other 'Desert Truffles'

Ascomycetes and basidiomycetes constitute the subkingdom Dicarya (James et al. 2006, Hibbett et al. 2007). The remaining lineages of fungi are not monophyletic and are currently recognized as differently ranged taxa (Schübler et al. 2001, Hibbett et al. 2007). The former Zygomycetes, which included hypogeous species deserving to be considered 'desert truffles', are nowadays split between phylum Glomeromycota C. Walker & A. Schuessler and subphylum Mucoromycotina Benny.

Most species in these clades do not produce macroscopic fruiting bodies, but large soil-borne spores, and only few of them present 'sporocarps' superficially resembling those of the lineage Dicarya. Species in the Glomeromycota are mainly obligate biotroph mutualistic fungi forming the vast majority of intracellular arbuscular mycorrhizas with prothalli of Pteridophyta as well as with roots of most upper plants. These fungi have a worldwide distribution and they probably played a role in the colonization of terrestrial habitats (Dotzler et al. 2008; Simon et al. 1993; Redecker et al. 2000). They can be found in a wide range of terrestrial ecosystems, including tropical to temperate forests, grasslands, sand, dunes, and agroecosystems (Brundrett 1991), as well as extreme environments such as deserts (Dodd and Krikun 1984; McGee and Baczocha 1994). The most relevant character used to discriminate between families and genera deals with spore ontogeny and morphology. Spores are usually produced asexually at the apex of hyphae and are often termed as chlamydospores because of their resting function. Species in the Mucoromycotina can be saprotrophic or mycoparasitic but also mycorrhizal with some early lineages of terrestrial plants (Bidartondo et al. 2012). They usually form sexual spores after the fusion of gametangia.

According with the most relevant literature available (Gerdemann and Trappe 1974; Montecchi and Sarasini 2000; Oehl et al. 2011; Stürmer 2012), the scarce sporocarp-producing species in these fungal lineages can be summarized in the following genera and key:

Endogonaceae Paol., in Saccardo, Syll. fung. 8: 905 (1889)
Endogone Link, Mag. Gesell. naturf. Freunde Berlin 3: 33 (1809)

Sclerogone Warcup, Mycol. Res. 94(2): 176 (1990)
Youngiomyces Y.J. Yao, in Yao, Pegler & Young, Kew Bull. 50(2): 350 (1995)

Gigasporaceae J.B. Morton & Benny, Mycotaxon 37: 483 (1990)
Gigaspora Gerd. & Trappe, Mycol. Mem. 5: 25 (1974)

Glomeraceae Piroz. & Dalpé [as 'Glomaceae'], Symbiosis 7: 19 (1989)
Glomus Tul. & C. Tul., Giorn. Bot. Ital. 1(7–8): 63 (1845)

1a	Typically presenting zygospores	2
1b	Typically presenting chlamydospores	3
2a	With two adjacent suspensors	*Endogone*
2b	With two opposed suspensors	*Sclerogone*
2c	With 2–4 separated suspensors	*Youngiomyces*
3a	Typically presenting chlamydospores attached to the generating hypha by a bulbous vesicle	*Gigaspora*
3b	Typically presenting chlamydospores directly attached to the generating hypha	*Glomus*

Acknowledgements We wish to express our gratitude to Luis Monje and Ángel Pueblas of the Department of Drawing and Scientific Photography at the Alcalá University for their help in the digital preparation of the photographs, to Celestino Gelpi and Miguel Ángel Sanz for kindly providing some images.

Bibliography

Alsheikh AM, Trappe JM (1983) Taxonomy of *Phaeangium lefebvrei*, a desert truffle eaten by birds. Can J Mycol 61(7):1919–1925

Alvarado P, Moreno G, Manjón JL (2012a) Comparison between *Tuber gennadii* and *T. oligospermum* lineages reveals the existence of the new species *T. cistophilum* (Tuberaceae, Pezizales). Mycologia 104(4):894–910

Alvarado P, Moreno G, Manjón JL (2012b) A new *Tuber* without spore ornamentation. *Tuber melosporum* comb. nov. Bol Soc Mycol Madrid 36:23–28

Álvarez IF, Parladé J, Trappe JM (1993) *Loculotuber gennadii* gen. et comb. nov. and *Tuber multimaculatum* sp. nov. Mycologia 84:926–929

Bidartondo MI, Read DJ, Trappe JM, Merckx V, Ligrone R, Duckett JG (2012) The dawn of symbiosis between plants and fungi. Biol Lett 7:574–577

Bonito G, Smith ME, Nowak M, Healy RA, Guevara G, Cázares E, Kinoshita A et al (2013) Historical biogeography and diversification of truffles in the Tuberaceae and their newly identified Southern Hemisphere sister lineage. PLoS ONE 8(1):e52765

Bonito G, Trappe JM, Rawlinson P, Vilgalys R (2010) Improved resolution of major clades within *Tuber* and taxonomy of species within the *Tuber gibbosum* complex. Mycologia 102:1042–1057

Brundrett MC (1991) Mycorrhizas in natural ecosystems. Academic Press, London. pp. 171–313

Bruns TD, Fogel R, White TJ, Palmer JD (1989) Accelerated evolution of a false-truffle from a mushroom ancestor. Nature 339:140–142

Burdsall HH Jr (1968) A revision of the genus *Hydnocystis* (Tuberales) and of the hypogeous species of *Geopora* (Pezizales). Mycologia 60:496–525

Calonge FD, Martín MP (2000) Morphological and molecular data on the taxonomy of *Gymnomyces*, *Martellia* and *Zelleromyces* (Russulales). Mycotaxon 76:9–15

Ceruti A (1960) Elaphomycetales et Tuberales. In: Bresadola G (ed) Iconographiae mycologicae 28: text unnumbered, 47 plates. Trento, Italy

Danks M, Lebel T, Vernes K (2010) 'Cort short on a mountaintop'—eight new species of sequestrate Cortinarius from sub-alpine Australia and affinities to sections within the genus. Persoonia 24:106–126

Dissing H, Korf RP (1980) Preliminary studies in the genera *Ruhlandiella*, *Sphaerosoma*, and *Sphaerozone* (order Pezizales). Mycotaxon 12:287–306

Dodd JC, Krikun J (1984) Observations on endogonaceous spores in the Negev Desert. Trans Br Mycol Soc 82:536–540

Dotzler N, Walker C, Krings M, Haas H, Taylor TN, Agerer R (2008) Acaulosporoid glomeromzycotan spores with a germination shield from the 400 million-year-old Rhynie Chert. Mycol Res 112:1107–1114

Eberhardt U, Verbeken A (2004) Sequestrate *Lactarius* species from tropical Africa: *L. angiocarpus* sp. nov. and *L. dolichocaulis* comb. nov. Mycol Res 108(9):1042–1052

Fogel R, Peck SB (1975) Ecological studies of hypogeous fungi. I. Coleoptera associated with sporocarps. Mycologia 67(4):741–747

Gerdemann JW, Trappe JM (1974) The Endogonaceae in the Pacific Northwest. Mycol Mem 5:1–76

Gilkey HM (1939) Tuberales of North America. Oregon State Monogr 1:1–63

Hansen K, Læssøe T, Pfister DH (2001) Phylogenetics of the Pezizaceae, with an emphasis on *Peziza*. Mycologia 93(5):958–990

Heim R (1971) The interrelationships between the Agaricales and gasteromycetes. In: Petersen RH (ed) Evolution in the higher basidiomycetes. University of Tennessee, Knoxville, Tennessee, pp 505–534

Hibbett DS, Pine EM, Langer E, Langer G, Donoghue MJ (1997) Evolution of gilled mushrooms and puffballs inferred from ribosomal DNA sequences. Proc Natl Acad Sci U S A 94:12002–12006

Hibbett DS, Binder M (2002) Evolution of complex fruitingbody morphologies in homobasidiomycetes. Proc Roy Soc Lond B 269:1963–1969

Hibbett DS, Binder M, Bischoff JF, Blackwell M, Cannon PF, Eriksson OE, Huhndorf S et al (2007) A higher-level phylogenetic classification of the fungi. Mycol Res 111:509–547

Hosaka K, Bates ST, Beever RE, Castellano MA, Colgan W III, Domínguez LS, Nouhra ER et al (2006) Molecular phylogenetics of the gomphoid-phalloid fungi with an establishment of the new subclass Phallomycetidae and two new orders. Mycologia 98(6):949–959

Hosaka K, Castellano MA (2008) Molecular phylogenetics of Geastrales with special emphasis on the position of Sclerogaster. Bull Natl Mus Sci Ser B 34(4):161–173

Hosaka K, Castellano MA, Spatafora JW (2008) Biogeography of the Hysterangiales (Phallomycetidae, Basidiomycota). Mycol Res 112:448–462

James TY, Kauff F, Schoch CL, Matheny PB, Hofstetter V, Cox C, Celio G et al (2006) Reconstructing the early evolution of the fungi using a six gene phylogeny. Nature 443:818–822

Kagan-zur V, Roth-Bejerano N (2008) Desert truffles. Fungi 1(3):32–37

Kendrick B (1992) The fifth kingdom. Focus Information Group, Newburyport, MA

Kinoshita A, Sasaki H, Nara K (2012) Multiple origins of sequestrate basidiomes within Entoloma inferred from molecular phylogenetic analyses. Fungal Biol 116:1250–1262

Kottek M, Grieser J, Beck C, Rudolf B, Rubel F (2006) World map of the Köppen-Geiger climate classification updated. Meteorol Zeitschr 15(3):259–263

Kovács GM, Trappe JM, Alsheikh AM, Hansen K, Healy RA, Vagi P (2011) *Terfezia* disappears from the American truffle mycota as two new genera and *Mattirolomyces* species emerge. Mycologia 103(4):831–840

Lebel T, Trappe JM (2000) Type studies of sequestrate Russulales. I. Generic type species. Mycologia 92(8):1188–1205

Læssøe T, Hansen K (2007) Truffle trouble: what happened to the Tuberales? Mycol Res 111:1075–1099

Malençon G (1938) Les truffes européennes. Historique, morphogenie, organographie, classification, culture. Revue Mycologique 3. Mémoirs Hors Série 1:1–92

Maliva RG, Missimer TM (2012) Arid lands water evaluation and management. Springer, series: environmental science and engineering series. Springer, Berlin

Maser C, Trappe JM, Nussbaum R (1978) Fungal-small mammal interrelationships with emphasis on Oregon coniferous forests. Ecology 59:799–809

McGee PA, Baczocha N (1994) Sporocarpic Endogonales and Glomales in the scats of *Rattus* and *Perameles*. Mycol Res 98(2):246–249

Miller SL, McClean TM, Walker J, Buyck B (2001) A molecular phylogeny of the Russulales including agaricoid, gasteroid and pleurotoid taxa. Mycologia 93:344–354

Moncalvo JM, Vilgalys R, Redhead SA, Johnson JE, James TY, Aime MC, Hofstetter V et al (2002) One hundred and seventeen clades of euagarics. Mol Phylogenetic Evol 23:357–400

Montecchi A, Sarasini M (2000) Funghi ipogei d'Europa, AMB Fondazione Centro Studi Micologici. Trento, Italy

Nuytinck J, Verbeken A, Delarue S, Walleyn R (2004) Systematics of European sequestrate lactarioid Russulaceae with spiny spore ornamentation. Belgian J Bot 136(2):145–153

Oehl F, da Silva GA, Goto BT, Sieverding E (2011) Glomeromycota: three new genera and glomoid species reorganized. Mycotaxon 116:75–120

Peintner U, Bougher NL, Castellano MA, Moncalvo JM, Moser MM, Trappe JM, Vilgalys R (2001) Multiple origins of sequestrate fungi related to *Cortinarius* (Cortinariaceae). Am J Bot 88:2168–2179

Peintner U, Moncalvo JM, Vilgalys R (2004) Toward a better understanding of the infrageneric relationships in *Cortinarius* (Agaricales, Basidiomycota). Mycologia 96(5):1042–1058

Perry B, Hansen K, Pfister DH (2007) A phylogenetic overview of the family Pyronemataceae (Ascomycota, Pezizales). Mycol Res 111:549–571

Redecker D, Morton JB, Bruns TD (2000) Molecular phylogeny of the arbuscular mycorrhizal fungi *Glomus sinuosum* and *Sclerocystis coremioides*. Mycologia 92:282–285

Schübler A, Schwarzott D, Walker C (2001) A new fungal phylum, the Glomeromycota: phylogeny and evolution. Mycol Res 105(12):1413–1421

Shavit E (2008) Truffles roasting in the evening fires. Pages from the history of desert truffles. Fungi 1(3):18–23

Simon L, Bousquet J, Levesque RC, Lalonde M (1993) Origin and diversification of endomycorrhizal fungi and coincidence with vascular land plants. Nature 363:67–69

Singer R (1951) *Thaxterogaster*—a new link between Gasteromycetes and Agaricales. Mycologia 43:215–288

Stürmer SL (2012) History of the taxonomy and systematics of arbuscular mycorrhizal fungi belonging to the phylum Glomeromycota. Mycorrhiza 22(4):247–58

Thiers HD (1984) The secotiod syndrome. Mycologia 76(1):1–8

Trappe JM (1971) A synopsis of the Carbomycetaceae and Terfeziaceae (Tuberales). Trans Br Mycol Soc 57:85–92

Trappe JM (1979) The orders, families, and genera of hypogeous Ascomycotina (truffles and their relatives). Mycotaxon 9:297–340

Trappe JM, Weber NS (2001) North American desert truffles: the genus *Carbomyces* (Ascomycota, Carbomycetaceae). Harvard Papers in Botany 6:209–214

Trappe JM, Bougher NL (2002) Australasian sequestrate (truffle-like) fungi, XI, *Gummivena potorooi* gen. & sp. nov. (Basidiomycota, Mesophelliaceae), with a key to the 'gummy' genera and species of the Mesophelliaceae. Australasian Mycol 21(1):9–11

Trappe JM, Castellano AM, Amaranthus MP (1996) Australasian truffle-like fungi, VIII, *Gummiglobus* and *Andebbia* gen. nov. (Basidiomycotina, Mesophelliaceae) and a supplement to the nomenclatural bibliography of Basidiomycotina. Australasian Syst Bot 9:803–811

Trappe JM, Kovács GM, Claridge AW (2010) Comparative taxonomy of desert truffles of the Australian outback and the African Kalahari. Mycol Progress 9:131–143

United Nations Educational, Scientific and Cultural Organization (UNESCO) (1979) Map of the world distribution of arid regions: map at scale 1:25,000,000 with explanatory note. MAB Technical Notes 7, UNESCO, Paris

Weber NS, Trappe JM, Denison WC (1997) Studies on western American Pezizales, Collecting and describing ascomata—macroscopic features. Mycotaxon 61:153–176

Zak J, Whitford WG (1986) The occurrence of a hypogeous ascomycete in the northern Chihuahuan desert. Mycologia 78:840–841

Chapter 2
Nomenclatural History and Genealogies of Desert Truffles

Gábor M. Kovács and James M. Trappe

2.1 Introduction

The term "desert truffle" is neither a phylogenetic nor a taxonomic term, and as those fungi have played an important role in the life of the native people of deserts, they have had from ancient times—and still do—a common knowledge of truffles (see Chap. 15 by Shavit as well as other chapters in this volume). We are unaware of any definition of a truffle, especially above the species rank, for which desert habitat was used as a taxonomic character.

To discuss the phylogenetic history and the genealogies of desert truffles, we first need to define what fungi are considered to be desert truffles. Although this whole book focuses on these fungi, and generally those who study them understand what they are, it is not easy to state an operative definition. First we confront the problem of how to define a truffle. A wide definition considers all fungi with hypogeous or partly hypogeous, macroscopic sporocarps as truffles. Using a narrow definition, we could consider only hypogeous ascomycetes as truffles once belonging to the order Tuberales (Læsøe and Hansen 2007). Sometimes in everyday or commercial contexts only fungi belonging to the genus *Tuber* are considered as truffles.

Second, we must face the definition of a desert, or at least the question of the common features of deserts. Truffles are diverse, especially when defined sensu

G.M. Kovács (✉)
Department of Plant Anatomy, Institute of Biology, Eötvös Loránd University,
Pázmány Péter sétány 1/C, 1117 Budapest, Hungary
e-mail: gmkovacs@caesar.elte.hu

J.M. Trappe
Department of Forest Ecosystems and Society, Oregon State University, Corvallis,
OR 97331-5752, USA

Forestry Sciences Laboratory, USDA Forest Service, Pacific Northwest Research Station,
Corvallis, OR 97331, USA
e-mail: trappej@onid.orst.edu

V. Kagan-Zur et al. (eds.), *Desert Truffles*, Soil Biology 38,
DOI 10.1007/978-3-642-40096-4_2, © Springer-Verlag Berlin Heidelberg 2014

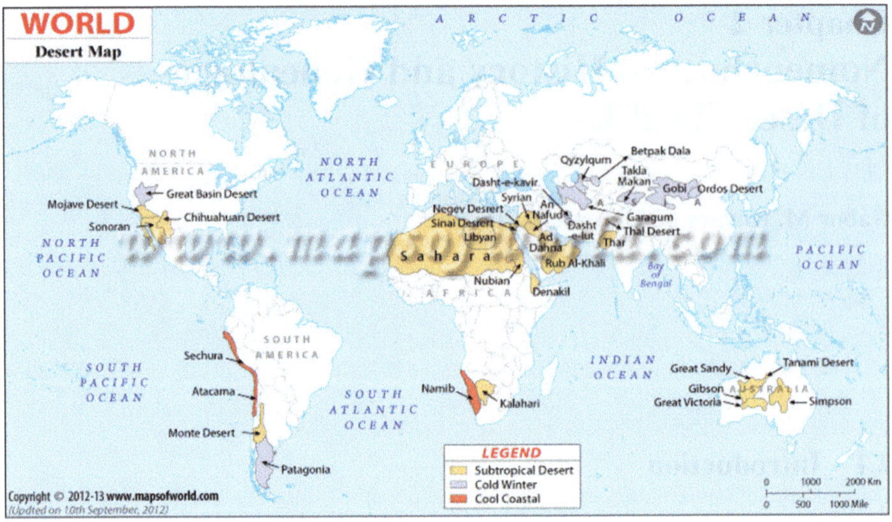

Fig. 2.1 Desert of the world. http://www.mapsofworld.com/world-desert-map.htm. Duly purchased from, and approved by MapsofWorld for the present edition. May 26, 2013

lato, but deserts are diverse as well. According to Ward (2008), “Deserts are defined by their aridity, yet differ enormously in their abiotic characteristics. The variation among deserts is probably greater than for any other biome, largely because deserts are so widely spaced on the planet and have arisen for very different reasons.” Excluding many small areas, more than 20 main deserts (Ward 2008) have been listed worldwide (Fig. 2.1).

For these reasons, we delimit desert truffles to taxa generally considered as desert truffles, belonging to these genera (genera with all species known only from deserts are in boldface: ***Carbomyces*, *Elderia*, *Eremiomyces*, *Kalaharituber*,** *Mattirolomyces*, ***Mycoclelandia***, *Picoa*, ***Stouffera***, *Terfezia*, *Tirmania*, and ***Ulurua***). Although the distinction of semiarid and xeric areas from deserts is often ambiguous, we do not consider hypogeous fungi of the former areas as desert truffles for purposes of this chapter.

Here we focus on taxonomic and phylogenetic studies of desert truffles. We review the taxonomic/nomenclatural history of two genera: (1) *Terfezia*, which, despite recent revisions, is still the most speciated and probably most widely known desert truffle genus and (2) *Mattirolomyces*, a recently expanded genus containing desert truffles as well as species of less extreme environments. We aimed to complete a detailed overview of all published studies that present information on phylogenies of desert truffles.

2.2 Taxonomy and Nomenclature

2.2.1 *Terfezia: The Most Speciated Desert Truffle Genus*

2.2.1.1 The Pre-molecular Era

The genus *Terfezia* (Tul. & Tul.) Tul. & Tul. (Ascomycotina) was raised by Tulasne and Tulasne (1851) from the rank of subsection within the genus *Choiromyces* they had described earlier (Tulasne and Tulasne 1845). They included three species in *Choiromyces* subsection *Terfezia*: *C. leonis* Tul. & C. Tul., *C. leptodermus* Tul. & C. Tul., and *C. olbiensis* Tul. & C. Tul., which they had described in 1844. In 1851 the Tulasne brothers added two more species: *T. berberidiodora* Tul. & C. Tul. and *T. oligosperma* Tul. & C. Tul., both of which were later transferred to other genera. Nevertheless, Moris (1829) was the first to describe a *Terfezia* species (as *Tuber arenarium* Moris) from Sardinia, later recombined as *Terfezia arenaria* (Moris) Trappe. Zobel (1854) erected a new genus, *Tulasneinia*, to accommodate *T. leonis* and *T. olbiensis*.

Late in the nineteenth century colonial officers and botanists exploring Asia and Africa sent truffle specimens bought in local markets or collected in the field to mycologists in Europe, where most species were described and named (Chatin 1891a, b, c, d, e, 1892a, b, c, 1893, 1894a, b, 1895a, b, c, 1896a, b, 1897; Maire 1907; Patouillard 1894a, b, 1899). Many names have proven to be synonyms of earlier described species (Alsheikh 1994). Malençon (1973) described *T. eremita* Malençon from Mauritania and discussed some ecological aspects of *Terfezia* species in North Africa.

The information on early history, taxonomy, ecology, and chemical composition of desert truffles from North Africa, southwest Asia, and southern Europe was summarized in Chatin's (1892a) book "La Truffe" (recent information is summed up in the various chapters of this volume).

Meanwhile, other newly described *Terfezia* species from around the world have been transferred to other genera when data permitted or remain of uncertain status pending further investigation. From Italy Mattirolo (1887) described *Choiromyces terfezioides* Mattir., which Fischer (1897) transferred to *Terfezia* and then to a new genus, *Mattirolomyces* (Fischer 1938). Trappe (1971) subsequently reduced *Mattirolomyces* to a subgenus in *Terfezia* (see below). Ellis (1887) reported *T. leonis* from southern Louisiana; Harkness (1899) later determined this to be a new species, which he named *T. spinosa* Harkn. Spegazzini (1887) described *Tuber argentinum* var. *pampeanum* Speg. from Argentina, but it was found to be synonymous with *Terfezia longii* Gilkey (1947), described from New Mexico. Hennings (1897) described *T. pfeilii* Henn. from southwest Africa, and Heim (1934) described *T. decaryi* R. Heim from Madagascar. Marasas and Trappe (1973) described *T. austroafricana* Trappe & Marasas from southwest Africa and noted that Pole-Evans' (1918) *T. claveryi* Chatin from the Kalahari Desert was *T. pfeilii*. From Japan, Imai (1933) described *T. gigantea* S. Imai, later also found in the USA

(Gilkey 1947, 1954; Trappe and Sundberg 1977), while Boedijn (1939) described *T. indica* Boedijn from Sumatra as the first collection from Southeast Asia.

Fischer (1897, 1938), Bataille (1921), and Mattirolo (1922) provided keys for the known species in Africa, Asia, and Europe. Gilkey (1947) provided a key for the North American species. Ceruti (1960) presented Latin descriptions and watercolor paintings of most North African and European *Terfezia* species. Alsheikh (1994) monographed *Terfezia* worldwide with keys and descriptions. Note, however, that these treatments preceded phylogenetic analytical techniques, so many include species now placed in other genera.

Fischer (1897) first proposed family status for the Terfeziaceae and included it in the Tuberales. Trappe (1971) emended the family concept and later transferred it to the order Pezizales after abandoning the polyphyletic order Tuberales (Trappe 1979).

2.2.1.2 Molecular Phylogenetic Revisions: Decreasing *Terfezia* Species Richness, Increasing Genetic and Morphological Uniformity

Trappe was correct considering Tuberales as polyphyletic (see the comprehensive review by Læsøe and Hansen 2007), but some decisions regarding desert truffles have been falsified by later molecular phylogenetic analyses (see below in detail). The "Terfeziaceae" was not supported by molecular phylogenetic analyses; the species are considered as belonging to Pezizaceae (Læsøe and Hansen 2007), but "Terfeziaceae" is still sporadically used as a fungal family. The molecular taxonomic revisions have increased the geographic uniformity of the genus *Terfezia*, only species from the Mediterranean region and the Middle East being proven to belong to the *Terfezia* s.str. Generic assignments of species thought to represent *Terfezia* in North America (Kovács et al. 2008, 2011a), South Africa (Ferdman et al. 2005; Trappe et al. 2010a, b), and Australia (Trappe et al. 2010a, b) had to be revised. Also, the very first molecular phylogenetic analyses (see below) revealed the genus *Mattirolomyces* to be distinct from *Terfezia*. The molecular taxonomic revisions also revealed an intraspecific diversity of *Terfezia* species and diverse species complexes (Díez et al. 2002; Aviram et al. 2004; Ferdman et al. 2009; Sbissi et al. 2011; Kovács et al. 2011b; and see Chap. 3 by Bordallo and Rodrígez).

2.2.2 Mattirolomyces*: An Expanding Genus Containing Desert Truffles*

Mattirolomyces is not a typical desert truffle genus. Our recent revisions of *Terfezia* species and truffles collected in widely separated desert regions have revealed that *Mattirolomyces* is more diverse geographically and taxonomically than thought before, so we present a short history of the genus here. Mattirolo originally described *Choiromyces terfezioides* from Piemonte, Northern Italy (Mattirolo 1887).

Fischer (1938) concluded that the species represented a distinct genus, which he named *Mattirolomyces* for "the Mattirolo fungus" with the sole species *M. terfezioides* (Mattir.) E. Fisch. Besides its most frequent habitat of mixed *Robinia pseudoacacia* forests on sandy soils in the Danube-Tiscia interfluves, it occurs in other habitats such as farm and urban environments, but not deserts or semiarid areas.

Trappe (1971) moved *Mattirolomyces* into *Terfezia* and changed its rank to a subgenus. Later molecular phylogenetic analyses proved *Mattirolomyces* to be a distinct genus (Percudani et al. 1999; Díez et al. 2002; Hansen et al. 2001; Læsøe and Hansen 2007) within the Pezizaceae. Healy (2003) described a second *Mattirolomyces* species, *M. tiffanyae* Healy, from a non-desert habitat in Iowa, but molecular phylogenetic analysis showed it does not belong to *Mattirolomyces* but represents a new genus (*Temperantia*) within Pezizaceae (Kovács et al. 2011a). True desert truffles first appeared in *Mattirolomyces* when hypogeous ascomycetes of the Australian Outback and the Kalahari were revised by use of molecular phylogenetic methods (Trappe et al. 2010a, b). *Mattirolomyces mulpu* Kovács, Trappe & Claridge was described as a new species and *M. austroafricanus* (Marasas & Trappe) Kovács, Trappe & Claridge as a new combination when the generic position of *Terfezia austroafricana* was corrected by phylogenetic analysis. The taxon richness of the genus increased by two more species when North American *Terfezia* species were analyzed (Kovács et al. 2011a). Although *M. mexicanus* Kovács, Trappe & Alsheikh probably has an arid location and *M. spinosus* (Harkn.) Kovács, Trappe & Alsheikh does not, the habitat collection data for either are inadequate, so they cannot be confirmed as desert truffles. *Mattirolomyces* had been a monotypic taxon for decades. Now it includes five species, and we are aware of one more (Kovács et al. unpublished data) from four continents (or five, if we consider the Beijing urban collection of *M. terfezioides* as well) and both Northern and Southern Hemispheres. Considering the habitats of the species, *Mattirolomyces* represents the widest geographic and climatic range of truffle genera that include desert truffles.

2.3 How Many Lineages Are Out There? Phylogenies of Desert Truffles Worldwide

The first molecular phylogenetic study mentioning desert truffles was by O'Donnell et al. (1997), a landmark paper on phylogeny of truffles and morels. Partial sequences of the small (18S, SSU) and large (28S, LSU) subunits of the nuclear ribosomal RNA gene (nrDNA) were analyzed. The paper reveals the misplacement of several taxa, shows the traditional concept of Tuberales to be wrong, and proposes that the hypogeous life-form has evolved several times within Pezizales. *Redellomyces donkii* (Malençon) Trappe, Castellano & Malajczuk was included in the analyses, but it is still unclear if *Redellomyces westralensis* (G.W. Beaton & Malajczuk) Trappe, Castellano & Malajczuk should be considered a desert truffle (see below and Chap. 14 by Claridge et al.). The phylogenetic position of *Picoa* is suggested by O'Donnell et al. (1997): "Preliminary 28S rDNA sequence data from the type species of *Picoa*, *P. juniperi* Vittad. (O'Donnell et al. unpublished data), suggests that it is more closely related to *Otidea* (Pers.) Bonord. than the taxa sampled in this study." The comprehensive review by Læsøe and Hansen (2007) also cited this unpublished information. O'Donnell et al. (1997) were the first to support Trappe's (1979) hypothesis that transitions from epigeous to hypogeous life-forms happened independently within the Pezizales, and those characters are homoplasic.

In 1999 two papers included desert truffles in the phylogenetic analyses. The first desert truffle sequence was analyzed and published in Percudani et al. (1999) and also used by Norman and Egger (1999). Percudani et al. (1999) sequenced the nrDNA SSU of nine hypogeous fungi and the ITS of three of them. The nine species included *Terfezia arenaria* and *Mattirolomyces terfezioides* (as *Terfezia terfezioides*), but only SSU sequences were obtained from those two taxa. They extended their dataset with sequences published by O'Donnell et al. (1997). In these analyses both *M. terfezioides* and *T. arenaria* grouped unambiguously in the Pezizaceae. The result "strongly supports the emendation of Pezizaceae to include the hypogeous genera *Pachyphloeus* and *Terfezia*" (Percudani et al. 1999). *Terfezia arenaria* and *M. terfezioides* had distant positions in the phylogeny, and it was the first molecular analysis supporting the distinct nature of *Mattirolomyces* after its emendation into *Terfezia*. Epigeous taxa were underrepresented and only a few taxa of Pezizaceae were included in the analyses. *Terfezia arenaria* was a sister species of *Cazia flexiascus* Trappe, and their clade got a 92 % ML bootstrap support. The authors discussed in detail the phylogenetic resolution power of the nrDNA SSU sequences and considered the gene as "capable of resolving phylogenetic relationships." Norman and Egger (1999) used the same two SSU sequences, but their study considerably expanded the taxon sampling from Pezizaceae, especially from *Peziza*. Although they analyzed SSU, ITS, and LSU sequences, these two taxa were represented only in the SSU dataset. In the analyses *T. arenaria* also grouped with *Cazia* with high reliability but low bootstrap value. *Mattirolomyces terfezioides* branched into a distant lineage of the phylogeny and grouped with the *Peziza vesiculosa* group with moderate support. Norman and Egger (1999) strengthened the hypothesis that hypogeous life-forms had originated several times within the Pezizales. They also emphasized the problems of the concept of Pezizaceae and the distinct nature of the genus *Mattirolomyces*, stating that "*Terfezia* is paraphyletic, and that morphological similarities between *T. arenaria* and *T. terfezioides* are a result of convergence" (Norman and Egger 1999). The genus *Terfezia* was polyphyletic instead of paraphyletic in their analyses (Norman and Egger 1999).

Hansen and her colleagues (Hansen et al. 2001) published a comprehensive phylogenetic study with a massive taxon sampling of Pezizaceae with a special emphasis on *Peziza*. They focused on phylogenetic relations and reevaluation of morphological characters used in the systematics of the group. In the detailed introduction and overview of the genera they discussed general features of the taxa and previous molecular phylogenetic results on *Terfezia* and *Tirmania*. They amplified and sequenced a segment containing the D1–D2 loop region of the nrDNA LSU gene. The choice of this segment strongly influenced later studies of desert truffles aiming to analyze a well-established reliable dataset. *Tirmania pinoyi and T. nivea* but no *Terfezia* species were involved in their sophisticated phylogenetic analyses. *Tirmania* grouped into "group VI" similarly to *Cazia*; *Cazia flexiascus* was a sister group of *Terfezia arenaria* in previous studies (see above). Hansen et al. (2001) considered *Terfezia arenaria* also belonging to "group VI," along with various other hypogeous taxa. They discuss the excipular structure of stereothecia of *Cazia*, *Terfezia*, and *Tirmania*. Based on Norman and Egger's (1996) results, Hansen et al. (2001) considered *Mattirolomyces terfezioides* to be "supported within, or as a sister lineage to, group IV based on nSSU sequences."

Díez et al. (2002) presented the first molecular phylogenetic examination of intrageneric relations of desert truffles. The nrDNA ITS of 38 specimens of two *Tirmania* and four *Terfezia* spp. plus *Mattirolomyces terfezioides* were studied by use of RFLP; in all, 18 species were sequenced. They also confirmed the separation of *Mattirolomyces terfezioides* from *Terfezia* species. They concluded, "Phylogenetic analyses indicated a close genetic relationship between *Tirmania* and *Terfezia*. They may have arisen from a single evolutionary lineage of pezizalean fungi that developed the hypogeous habit as an adaptation to heat and drought in Mediterranean ecosystems." However, as noted by Læsøe and Hansen (2007), the limited taxon sampling from Pezizaceae makes both the relationship of the two taxa and the evolutionary hypothesis a bit uncertain. Díez et al. (2002) could not satisfactorily align two *Peziza* ITS sequences, so they excluded those from the analyses; they interpreted this as proof for the monophyly of the *Terfezia*/*Tirmania* clade. The most important findings of the study concerned host relations and edaphic preferences of *Terfezia* spp. and the prediction of intraspecific genetic variations of *Terfezia* species.

Hansen et al. (2005) revisited their earlier phylogenetic analysis of *Peziza* and the Pezizaceae (Hansen et al. 2001). The new dataset contained 69 species of the Pezizales, and in addition to the nrDNA LSU segment, RNA polymerase II (RPB2) and the β-tubulin gene sequences were sequenced and involved in the analyses. Besides the two *Tirmania* species that were studied previously, three *Terfezia* species (*T. arenaria*, *T. boudieri*, *T. claveryi*) were also included. However, none of the three *Terfezia* spp. were represented by all three loci: from *T. arenaria* only the β-tubulin were analyzed, from *T. boudieri* Chatin the nrDNA LSU and the β-tubulin, and from *T. claveryi* the nrDNA LSU and RPB2 sequences. Fourteen "fine-scale lineages" of the Pezizaceae were identified during the sophisticated phylogenetic analyses of the dataset, and the topology was identical with the previous LSU nrDNA-based results. Nevertheless, the relative branching orders and the relations of the lineages could not be completely resolved. Both *Terfezia* and *Tirmania* nested into the *Peziza depressa-Ruhlandiella* lineage, which gained strong support in the combined analyses of the three loci. *Terfezia* showed close relationship with *Peziza* spp. Their phylogenetic analyses of the Pezizales showed no novelty about the desert truffles (only one *Terfezia* sequence was included in the dataset). Hansen et al. (2005) hypothesized that "the *P. depressa-Ruhlandiella* lineage could be a mycorrhizal lineage." Comparing the three loci they found RPB2 as "the most informative single gene region based on resolution and clade support." However, the LSU sequences were almost as useful and much easier to amplify. This plus the reliable, massive dataset published for the Pezizaceae (Hansen et al. 2001, 2005) and the Pezizales (Hansen and Pfister 2006) made the LSU D1–D2 region a useful locus to study generic positions of desert truffles.

Two new genera, *Kalaharituber* and *Eremiomyces,* were erected when *Terfezia pfeilii* Henn. and *Choiromyces echinulatus* Trappe & Marasas were subjected to molecular taxonomic analysis (Ferdman et al. 2005). The D1–D2 LSU and ITS regions were amplified and analyzed together with sequences of *Terfezia, Tirmania, Mattirolomyces, Tuber,* and *Choiromyces* species, so the sampling focused on hypogeous taxa related to the original generic positions of the species. In spite of the narrow taxon sampling of the lineages, the relative branching order of the lineages unambiguously showed the wrong generic assignation of the two species. *Eremiomyces echinulatus* (Trappe & Marasas) Trappe & Kagan-Zur was distant from *Choiromyces* s. str. and grouped into *Pezizaceae* instead of Tuberaceae. *Kalaharituber pfeilii* (Henn.) Trappe & Kagan-Zur separated from *Terfezia* species, but its branching did not reveal any information about its family position, although it was closer to Pezizaceae than to Tuberaceae. However, the molecular results and the morphological features convinced the authors to erect new genera, and later analyses (Læsøe and Hansen 2007) confirmed that the lineages were distinct even when more representatives of the Pezizaceae and Pezizales were involved in the phylogenetic datasets.

Læsøe and Hansen (2007) published a comprehensive paper on the former Tuberales genera now assigned to the Pezizales. The LSU sequences of almost 200 species, 55 hypogeous and 139 epigeous, were analyzed. Several truffle genera containing desert truffles were involved in the phylogenetic analyses: *Terfezia, Tirmania, Mattirolomyces, Kalaharituber,* and *Eremiomyces.* Inclusion of the latter two, both new genera, confirmed their inclusion in the Pezizaceae: *Kalaharituber* formed a lineage with *Iodowynnea,* whereas the position of *Eremiomyces* was ambiguous within the family. The others generally nested into the same lineages as in previous analyses of the LSU sequences. Their detailed presentation of the different truffle genera includes *Picoa* and *Carbomyces.* Concerning the phylogenetic position of *Picoa,* Læsøe and Hansen (2007) cite O'Donnell et al. (1997) as close to *Otidea. Carbomyces* was also mentioned as receiving ongoing molecular taxonomic study by K. Hansen.

Kovács et al. (2008) removed the Japanese/North American *Terfezia gigantea* S. Imai from both *Terfezia* and the Pezizaceae on the basis of molecular phylogenetic results and morphology. Although SSU, ITS, and LSU sequences were gained, only the SSU phylogeny of representatives of five families of the Pezizales was presented. They erected the new genus *Imaia* with its single species, *I. gigantea* (S. Imai) Trappe & Kovács. It is not a desert truffle, but results of the revision together with previous molecular phylogenetic results of *Mattirolomyces terfezioides* and *Kalaharituber pfeilii* showed that the genus *Terfezia* sensu lato contained numerous, well-known desert truffles, was not monophyletic, and needed revision.

Sbissi et al. (2010) presented phylogenetic results on *Picoa juniperi* Vittad. and *P. lefebvrei* (Pat.) Maire. In addition to a morphological comparison of the two species, they sequenced the ITS and LSU region of 12 specimens, as the sequencing of one LSU failed, two *P. lefebvrei* and nine *P. juniperi* specimens were involved in the final phylogenetic analyses of the LSU and ITS dataset, including sequences from public databases. They stated: "Ribosomal DNA analyses have enabled the genus *Picoa* to be assigned to the Pyronemataceae and to confirm that *Picoa* is closely related to *Geopora* (Tedersoo et al. 2010)." Although the phylogenetic trees presented show only the topology and the supports of the branches/clades, it was clear that *Geopora* is a closer relative of *Picoa* than *Otidea* was. However, the unpublished results of O'Donnell et al. (1997) that *Picoa* is related to *Otidea* were true, considering the taxon sampling in their analysis.

In comparative taxonomy of the desert truffles of the Australian Outback and the Kalahari, six Australian and three Kalahari taxa were analyzed and redescribed: eight of the nine belonged to the Pezizaceae (Trappe et al. 2010a, b). The SSU, ITS, and LSU segments were amplified and sequenced when possible and the partial LSU sequences (see above) used to analyze the phylogenetic position of the taxa within the Pezizaceae. *Mycoclelandia* [represented by *M. arenacea* (Trappe) Trappe & G.W. Beaton and *M. bulundari* (G.W. Beaton) Trappe & G.W. Beaton)] and *Ulurua nonparaphysata* Trappe, Claridge & Kovács (a new genus and species described in the paper) nested into the *P. depressa-Ruhlandiella* lineage sensu (Hansen et al. 2005). *Elderia arenivaga* (Cooke & Massee) McLennan formed a well-supported clade with *Mattirolomyces,* and both branched with the *Peziza* s. str. lineage. It was the first paper evidencing that *Mattirolomyces* is not monotypic. Two desert truffles were found to belong to the genus: *M. mulpu* described as a new species from Australia while *M. austroafricanus* was published as a new combination for *Terfezia austroafricana*. The other desert truffle taxa, which have been analyzed in previous phylogenetic works, branched at the same position as previously on the phylogram of the family. Based on the phylogenetic analyses presented, if only the so-called fine-scale lineages sensu Hansen et al. (2005) of the family are considered, four desert truffle lineages appeared in the phylogeny of the Pezizaceae. The paper deals with *Redellomyces westraliensis* belonging to the Tuberaceae—but the authors emphasized the species "was found in a swale along a streambank."

When American *Terfezia* species and *Mattirolomyces tiffanyae,* now *Temperantia tiffanyae* (Healy) K. Hansen, Healy & Kovács., were revised (Kovács et al. 2011a), the SSU, ITS, and LSU sequences were obtained, but only the phylogeny calculated from the LSU sequence dataset was presented (Fig. 2.2). All desert truffle taxa mentioned above were involved in the phylogenetic analyses, and their branching was as in previous analyses (see above). Species and collections previously considered as *Terfezia* (*T. spinosa*, *"T. mexicana,"* *T. longii*) were assigned to different genera. *Mattirolomyces spinosus* and *M. mexicanus* expanded the genus *Mattirolomyces*, the former being represented by one collection that

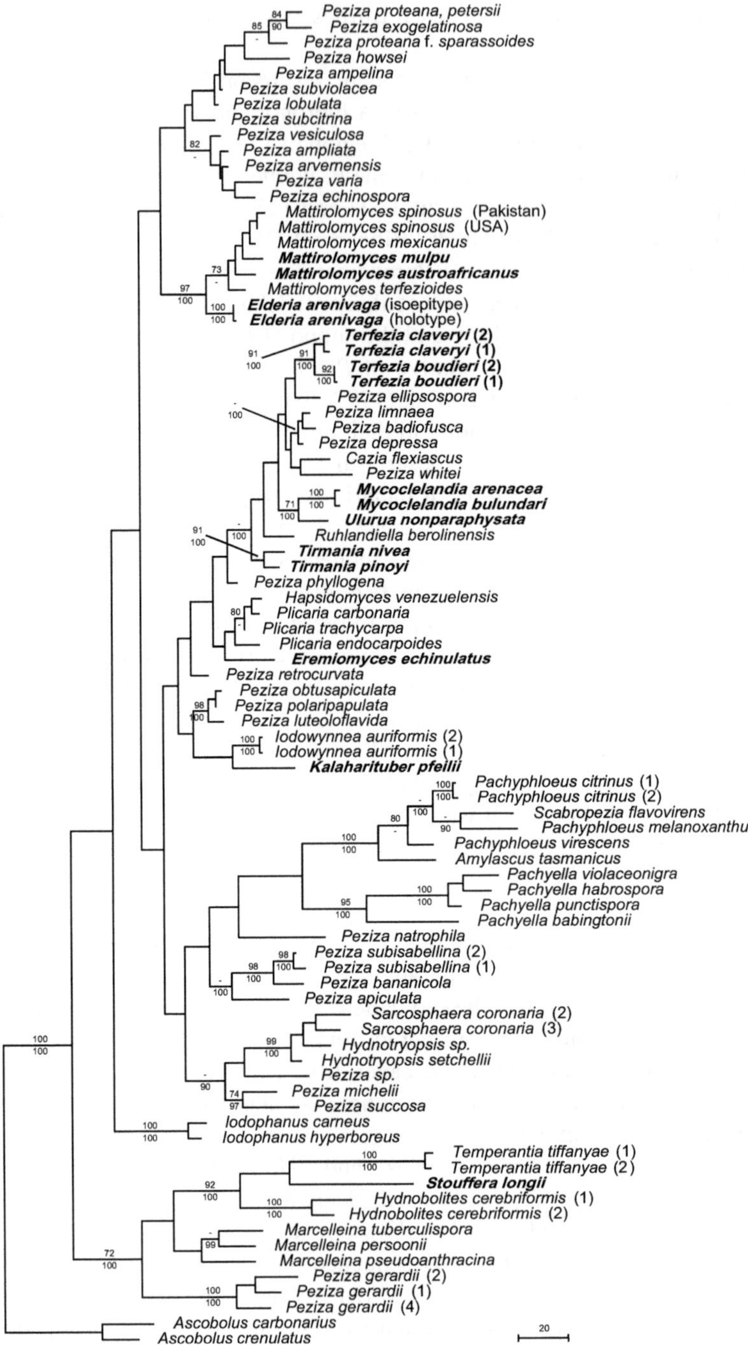

Fig. 2.2 One of 375 most parsimonious phylogenetic trees inferred from a dataset of partial LSU sequences showing positions desert truffle genera within the Pezizaceae. *Ascobolus crenulatus* served as outgroup. Desert truffles are shown in *boldface*. Parsimony bootstrap (PB) values are shown *above* the branches, while the Bayesian posterior probabilities (PP) are *below*. Bootstrap

might have been collected in an Arizona desert (see Trappe et al. Chap. 8, this volume). *Stouffera* was described as a new genus for *T. longii*. *Stouffera longii* (Gilkey) Kovács & Trappe represented a distinct lineage at a basal branch of the Pezizaceae forming a clade with *Hydnobolites* and *Temperantia tiffanyae* (Fig. 2.2). However, as all those taxa represented quite long branches, this group might result from long-branch attraction. The desert truffle genus *Terfezia* disappeared from the American truffle mycota (Kovács et al. 2011a).

Sbissi et al (2011) presented the intraspecies variability of *Terfezia boudieri* from southern Tunisian locations based on RFLP analyses of the ITS region of 163 samples. Thirty ITS were sequenced and analyzed; the phylogram showed only topology. Apparently *T. boudieri* has considerable intraspecific variation. A high intraspecific and intrasporocarpic ITS variability was detected when Spanish *Terfezia* collections were studied (Kovács et al. 2011b). Kovács et al. (2011b) described a new *Terfezia* species and suggested that probably the lineages of the *T. olbiensis* species complex represent several morphospecies described previously but considered as synonyms. These two papers (Sbissi et al 2011; Kovács et al. 2011b) focus more on intrageneric, intraspecific variations and intrageneric species relations of desert truffles and not the phylogenetic position of desert truffles in higher taxonomic groups. Intraspecies/intrahyphal genetic variation was reported from *T. boudieri* as well (Aviram et al. 2004; Ferdman et al. 2009).

When the first report of *Picoa lefebvrei* from Iran was published (Ammarellou et al. 2011), both ITS and LSU regions were sequenced from an Iranian (*P. lefebvrei*) and a Spanish sample (*P. juniperi*), and the partial LSU sequences were analyzed together with species from Pyronemataceae gained from public databases. Although the phylogenetic tree published shows no information about distances, its broad taxon sampling of the family enables more precise interpretation of the phylogenetic position of *Picoa*. The genus nested into the *Geopora-Tricharina* clade of the family and showed a close relationship to *Geopora*.

Carbomyces represents an enigmatic desert truffle genus (see Trappe et al. Chap. 8, this volume), and the first molecular phylogenetic results about these truffles were published only recently (Moreno et al. 2012). They did not mention that they are first to do so, but no phylogenetic information about *Carbomyces* had been published before. Moreno et al. (2012) obtained ITS and LSU sequences only from *C. emergens* Gilkey. When the LSU sequence was analyzed together with others from the family Pezizaceae, *C. emergens* formed a clade with the *Kalaharituber-Iodowynnea* lineage but with low support. Moreno et al. (2012) write that their results are "in accordance with data obtained by Karen Hansen (pers. comm.)." The LSU analyses confirmed by BLAST of the ITS sequences "linked *C. emergens* to *Terfezia* and *Kalaharituber* (84 % and 83 % identity in 68 % and 65 % coverage, respectively)." It shows the ambiguous branching of *Carbomyces* in accordance with the blast result of the LSU sequence. *Carbomyces* is similarly distant from all Pezizaceae represented in the

Fig. 2.2 (continued) values below 70 % and PP below 95 % are not shown (modified from Kovács et al. 2011a; with kind permission of Mycologia)

analyses. As the family position of *Carbomyces* position is not completely clear, it is important to analyze a broader taxon sampling of Pezizales.

Hansen et al. (2013) published a comprehensive multilocus phylogeny of the Pyronemataceae with analyses of different life history traits and ancestral states. Although no *Picoa* species was included in the analyses, the taxa overlap with the aforementioned studies (Sbissi et al. 2010; Ammarellou et al. 2011) enabling one to assume that *Picoa* belongs to the *Scutellinia-Trichophaea* lineage, and together with that group, most probably to the *Geopora-Tricharina-"Pustularia"* clade. Thus, biological evolutionary aspects of fungi have been explored through phylogenetic analyses in which lineages with desert truffles were represented.

Basidiomycetes have been overrepresented in diversity studies of ectomycorrhizal fungi before the advent of molecular diversity screening methods. Tedersoo et al. (2006) revealed that several lineages of the Pezizales contain EM fungi, identified by use of ITS and LSU sequences. Although the EM sampling area did not represent deserts or even semiarid regions, the results revealed ectomycorrhizal strategy in *Terfezia-Peziza depressa* clade of Pezizaceae and the *Geopora-Tricharina* lineage of Pyronemataceae where the desert truffles *Terfezia* (s.str.), *Tirmannia*, and *Picoa* belong, respectively.

In a literature review plus meta-analyses, Tedersoo et al. (2010) discussed the possible ectomycorrhizal lifestyle and biogeography of fungal phylogenetic lineages. EcM were identified in the "/geopora" lineage and "/terfezia-peziza depressa" lineages. Based on missing EcM hosts in desert habitats, Tedersoo et al. (2010) doubted the (ecto)mycorrhizal strategy of *Carbomyces*, *Eremiomyces*, and *Kalaharituber*; the endomycorrhizal interactions of the latter were proven previously (Kagan-Zur et al. 1999 and see Chap. 5 by Roth-Bejerano et al.). They also discussed another desert truffle lineage: "*Mattirolomyces* and an Australian desert truffle *Elderia* are probably nested within *Peziza* s. str., a clade with no known EcM species." Kovács et al. (2003, 2007, 2011c) have questioned the mycorrhizal strategy of *M. terfezioides* as a result of their studies on that species.

Healy et al. (2013) presented a surprising discovery of widespread presence of mitotic spore mats within EcM Pezizales. The ITS and LSU regions were used for phylogenetic analyses of datasets containing several desert truffle taxa as well. The sampling areas of the mats were not arid regions. No desert truffle was conspecific with mitosporic spore mat samples, but the "/terfezia-peziza depressa" lineage contained several examples. In the "News and views" part of the same issue of the journal, Tedersoo et al. (2013) presented interesting results on "endophytic" (i.e., mostly only sequences gained from plant tissues) and "endolichenic" lifestyle within Pyronemataceae. No *Picoa* sequences were included, but these results reveal the diversity of distribution and lifestyle of pezizalean taxa.

2.4 Considerations for Future Research

What can all these results show? Trappe and Claridge (2005) hypothesized that the extraordinary richness of hypogeous fungal taxa in drought-prone Australia is due in part to the selection pressure imposed by dry weather. Wet weather that initiates sporocarp formation is often interrupted by warm, drying weather there. Epigeous fungi often desiccate before maturing spores, whereas fungi protected from desiccation belowground have a better chance to complete their life cycle. Claridge et al. (Chap. 14 in this volume) detail the dependence of desert truffles on rainfall events. An important question is how complicated might be the adaptation of truffles to the extreme biotic conditions of deserts. One cannot consider this question independently from the evolution of hypogeous ascomata. Several phylogenetic studies on Pezizales presented above discussed the problem of the evolution of truffles and hypothesized that the hypogeous nature of fruiting bodies might be under the control of a not very complex genetic toolbox, and that is why hypogeous fungi could appear in several phylogenetic lineages. And the desert truffles spread through these lineages. Considering the fine-scale lineages of Pezizaceae sensu Hansen et al. (2005), adaptation of truffles to desert environments has evolved in at least four lineages (Fig. 2.2). This is a very conservative estimate, as we do not consider the potential within-lineage appearance of this adaptation, and we regard the clade formed by the long branches of the *Carbomyces-Kalaharituber-Iodowynnea* as one lineage. Besides these adaptations in Pezizaceae, at least one desert truffle lineage evolved in Pyronemataceae where *Picoa* resides. Based on this multiple adaptation we might hypothesize that, similarly to the evolution of the hypogeous fruiting body, adaptation of truffles to desert environments might not have a very complicated genetic background.

The advent of the next generation sequencing technique has enabled the fast and cost-effective sequencing of complete genomes of filamentous fungi. Several comparative genomic analyses revealed genomic and functional diversities in biological strategies such as wood decaying (Eastwood et al. 2011; Floudas et al. 2012) or ectomycorrhiza forming (Martin et al. 2010). *Terfezia boudieri* is on the list of fungi whose genome has recently been sequenced as planned (Plett and Martin 2011) but is still being annotated (Sitrit et al. personal communication). The 1,000 fungal genome initiative (Grigoriev et al. 2011) aims to sequence at least two genomes from each family of fungi; as we write this chapter, *T. boudieri* is the sole desert truffle whose complete genome has been sequenced. We agree with a conclusion of Sterflinger et al. (2012) in their review of fungi of hot and cold deserts: "Due to their enormous stress tolerance, desert fungi could also be a promising source for new biotechnological and medical adaptations…." Comparative genomics may reveal the key components of how desert truffles evolved the adaptation to their extreme environments.

Acknowledgements The authors are grateful to Abdulmagid M. Alsheikh for sharing his important results on the taxonomic history of *Terfezia*. Gábor M. Kovács was supported by the Hungarian Research Fund (OTKA K72776) and János Bolyai Scholarship (Hungarian Academy of Sciences).

References

Alsheikh AM (1994) Taxonomy and mycorrhizal ecology of the desert truffles in the genus *Terfezia*. Ph.D. dissertation. Oregon State University, Corvallis

Ammarellou A, Smith ME, Tajick MA, Trappe JM (2011) The phylogenetic placement of *Picoa*, with a first report on *Picoa lefebvrei* (Pat.) Maire (=*Phaeangium lefebvrei*) from Iran. Int J Environ Res 5:509–514

Aviram S, Roth-Bejerano N, Kagan-Zur V (2004) Two ITS forms co-inhabiting a single genet of an isolate of *Terfezia boudieri* (Ascomycotina), a desert truffle. A van Leeuw J Microb 85:169–174. doi:10.1023/B:ANTO.0000020149.89753.7a

Bataille F (1921) Flore analytique et descriptive des Tuberoïdées de l'Europe et de l'Afrique du Nord. Bull Soc Mycol France 37:155–207

Boedijn KB (1939) The Tuberales of Netherlands Indies. Bull Jard Bot Buitenzorg Ser 3 16:236–244

Ceruti A (1960) Elaphomycetales et Tuberales. Iconographia mycologica. J Bresadola 28(suppl 2), Trento

Chatin A (1891a) Contribution a 'l'histoire botanique de la truffe. Deuxiéme note: Terfâs ou truffes d'Afrique (et d'Arabie), genres *Terfezia* et *Tirmania*. Compt Rend Acad Sci, Paris, sér 3. 112:136–141

Chatin A (1891b) Contribution a 'l'histoire botanique de la truffe, kammé de Damas (*Terfezia claveryi*); troisiéme note. Compt Rend Acad Sci, Paris, sér 3, 113:381–384

Chatin A (1891) Contribution a 'l'histoire botanique de la truffe (quatriéme note).- Kamés de Bagdad (*Terfezia hafizi* et *Terfezia metaxasi*) et de Smyrne (*Terfezia leonis*). Compt Rend Acad Sci, Paris, sér 3. 113:530–534

Chatin A (1891d) Contribution a 'l'histoire naturelle de la truff e - Parallèle entre les terfaz ou kamés (*Terfezia*, *Tirmania*) d'Afrique et d'Asie, et les truffes d'Europe. Compt Rend Acad Sci, Paris, sér 3. 113:582–586

Chatin A (1891e) Contribution a' l'histoire naturelle de la truffe. II Terfas ou truffes d'Afrique et d'Arabie, genres *Terfezia* et *Tirmania*. Bull Soc Bot France 38:54–64

Chatin A (1892a) La Truffe. J. B. Baillie're et fils, Paris

Chatin A (1892b) Contribution nouvelle á l'histoire chimique de la Truffe. Paralléle entre les Terfâs ou Kamés d'Afrique et d'Asie et les Truffes d'Europe, sous les rapport de la composition chimique des terres et des tubercules. Compt Rend Acad Sci, Paris, sér 3. 114:46–53

Chatin A (1892c) Nouvelle contribution á l'histoire botanique de la truffe: Kamés de Bagdad (*Terfezia hafizi*, *T. metaxasi*) et de Smyrne (*T. leonis*)—Paralléle entre les terfaz ou kamés d'Afrique et d'Asie et les truffes de France. Bull Soc Bot France 39:10–20

Chatin A (1893) Sur une truffe du Caucase, La Touboulane. Bull Soc Bot France 40:301–304

Chatin A (1894a) Truffe (Terfâs) de Tunisie et de Tripoli. Compt Rend Acad Sci, Paris, sér 3, 119:485–487

Chatin A (1894b) Truffe (Domalan) de Smyrne. Compt Rend Acad Sci, Paris, sér 3, 119:523–526

Chatin A (1895a) Terfas du Maroc et de Sardaigne. Bull Soc Bot France 42:489–493

Chatin A (1895b) Truffes (terfas) de Chypre, de Smyrne et de La Calle. Bull Soc Bot France, sér 3, 42:549–551

Chatin A (1895c) Truffe (*Terfezia hanotauxii*) de Téhéran. Bull Soc Bot France 43:619–623

Chatin A (1896a) Un terfas d'Espagne et trois nouveaux terfas du Maroc. Bull Soc Bot France 43:397–399

Chatin A (1896b) Truffes (terfaz) de Gréce, *Terfezia gennadii*. Bull Soc Bot France 43:611–617

Chatin A (1897) Un nouveau terfas (*Terfezia aphroditis*) de l'ile de Chypre. Bull Soc Bot France, sér 3, 44:290–292

Díez J, Manjón JL, Martin F (2002) Molecular phylogeny of the mycorrhizal desert truffles (*Terfezia* and *Tirmania*), host specificity and edaphic tolerance. Mycologia 94:247–259

Eastwood DC, Floudas D, Binder M, Majcherczyk A, Schneider P, Aerts A, Asiegbu FO, Baker SE, Barry K, Bendiksby M, Blumentritt M, Coutinho PM, Cullen D, de Vries RP, Gathman A,

Goodell B, Henrissat B, Ihrmark K, Kauserud H, Kohler A, LaButti K, Lapidus A, Lavin JL, Lee YH, Lindquist E, Lilly W, Lucas S, Morin E, Murat C, Oguiza JA, Park J, Pisabarro AG, Riley R, Rosling A, Salamov A, Schmidt O, Schmutz J, Skrede I, Stenlid J, Wiebenga A, Xie X, Kües U, Hibbett DS, Hoffmeister D, Högberg N, Martin F, Grigoriev IV, Watkinson SC (2011) The plant cell wall-decomposing machinery underlies the functional diversity of forest fungi. Science 333:762–765. doi:10.1126/science.1205411

Ellis JB (1887) *Terfezia leonis*, Tul.—*Tuber niveum* (Desf.). J Mycol 3:10

Ferdman Y, Aviram S, Roth-Bejerano N, Trappe JM, Kagan-Zur V (2005) Phylogenetic studies of *Terfezia pfeilii* and *Choiromyces echinulatus* (Pezizales) support new genera for southern African truffles: *Kalaharituber* and *Eremiomyces*. Mycol Res 109:237–245. doi:10.1017/S0953756204001789

Ferdman Y, Sitrit Y, Li Y-F, Roth-Bejerano N, Kagan-Zur V (2009) Cryptic species in the *Terfezia boudieri* complex. A van Leeuw J Microb 95:351–362. doi:10.1007/s10482-009-9321-z

Fischer E (1897) Ascomycetin: Tuberaceen und Hemiascenn. Rabenhorst's Kryptogamen Flora von Deutschland 1(5):1–131

Fischer E (1938) Klasse Ascomycetes, Reihe Euascales. Unterreihe VIII. Tuberineae. In: Engler A, Harms H (eds) Natürlichen Pflanzenfamilien V. Wilhelm Engelmann, Leipzig

Floudas D, Binder M, Riley R, Barry K, Blanchette RA, Henrissat B, Martínez AT, Otillar R, Spatafora JW, Yadav JS, Aerts A, Benoit I, Boyd A, Carlson A, Copeland A, Coutinho PM, de Vries RP, Ferreira P, Findley K, Foster B, Gaskell J, Glotzer D, Górecki P, Heitman J, Hesse C, Hori C, Igarashi K, Jurgens JA, Kallen N, Kersten P, Kohler A, Kües U, Kumar TK, Kuo A, LaButti K, Larrondo LF, Lindquist E, Ling A, Lombard V, Lucas S, Lundell T, Martin R, McLaughlin DJ, Morgenstern I, Morin E, Murat C, Nagy LG, Nolan M, Ohm RA, Patyshakuliyeva A, Rokas A, Ruiz-Dueñas FJ, Sabat G, Salamov A, Samejima M, Schmutz J, Slot JC, St John F, Stenlid J, Sun H, Sun S, Syed K, Tsang A, Wiebenga A, Young D, Pisabarro A, Eastwood DC, Martin F, Cullen D, Grigoriev IV, Hibbett DS (2012) The Paleozoic origin of enzymatic lignin decomposition reconstructed from 31 fungal genomes. Science 336:1715–1719. doi:10.1126/science.1221748

Gilkey HM (1947) New or otherwise noteworthy species of Tuberales. Mycologia 36:441–452

Gilkey HM (1954) North American flora: Tuberales. Contr New York Bot Garden 2:1–36

Grigoriev IV, Cullen D, Goodwin SB, Hibbett D, Jeffries TW, Kubicek CP, Kuske C, Magnuson JK, Martin F, Spatafora JW, Tsang A, Baker SE (2011) Fueling the future with fungal genomics. Mycology 2:192–209. doi:10.1080/21501203.2011.584577

Hansen K, Pfister DN (2006) Systematics of the Pezizomycetes—the operculate discomycetes. Mycologia 98:1029–1040. doi:10.3852/mycologia.98.6.1029

Hansen K, Læsøe T, Pfister DH (2001) Phylogenetics of the Pezizaceae, with an emphasis on *Peziza*. Mycologia 93:958–990

Hansen K, Lobuglio KF, Pfister DH (2005) Evolutionary relationships of the cup-fungus genus *Peziza* and Pezizaceae inferred multiple nuclear genes: RPB2, β-tubulin, and LSU rDNA. Mol Phylogenet Evol 36:1–23. doi:10.1016/j.ympev.2005.03.010

Hansen K, Perry BA, Dranginis AW, Pfister DH (2013) A phylogeny of the highly diverse cup-fungus family Pyronemataceae (Pezizomycetes, Ascomycota) clarifies relationships and evolution of selected life history traits. Mol Phylogenet Evol 67:311–35. doi:10.1016/j.ympev.2013.01.014

Harkness HW (1899) Californian hypogeous fungi. Proc Calif Acad Sci, ser 3 1(8):241–292

Healy RA (2003) *Mattirolomyces tiffanyae*, a new truffle from Iowa, with ultrastructural evidence for its classification in the Pezizaceae. Mycologia 95:765–772

Healy RA, Smith ME, Bonito GM, Pfister DH, Ge ZW, Guevara GG, Williams G, Stafford K, Kumar L, Lee T, Hobart C, Trappe J, Vilgalys R, McLaughlin DJ (2013) High diversity and widespread occurrence of mitotic spore mats in ectomycorrhizal Pezizales. Mol Ecol 22:1717–1732. doi:10.1111/mec.12135

Heim R (1934) Observations sur la flore mycologique Malagache. Ann Cryptog Exotique 7(5):5–10

Hennings P (1897) Fungi Camerunenses I. Bot Jahrb Syst Pflanzengeschichte Pflanzengeogr 22:72–111

Imai S (1933) On two new species of Tuberaceae. Proc Imp Acad Jpn 9:182–184

Kagan-Zur V, Kuang J, Tabak S, Taylor FW, Roth-Bejerano N (1999) Potential verification of a host plant for the desert truffle *Terfezia pfeilii* by molecular methods. Mycol Res 103:1270–1274. doi:10.1017/S0953756299008448

Kovács GM, Vágvölgyi C, Oberwinkler F (2003) *In vitro* interaction of the truffle *Terfezia terfezioides* with *Robinia pseudoacacia* and *Helianthemum ovatum*. Folia Microbiol 48:369–378

Kovács GM, Jakucs E, Bagi I (2007) Identification of host plants and description of sclerotia of the truffle *Mattirolomyces terfezioides*. Mycol Progr 6:19–26. doi:10.1007/s11557-006-0520-y

Kovács GM, Trappe JM, Alsheikh AM, Bóka K, Elliott TF (2008) *Imaia*, a new truffle genus to accommodate *Terfezia gigantea*. Mycologia 100:930–939. doi:10.3852/08-023

Kovács GM, Trappe JM, Alsheikh AM, Hansen K, Healy RA, Vági P (2011a) *Terfezia* disappears from the American truffle mycota as two new genera and *Mattirolomyces* species emerge. Mycologia 103:831–840. doi:10.3852/10-273

Kovács GM, Balázs TK, Calonge FD, Martín MP (2011b) The diversity of *Terfezia* desert truffles: new species and a highly variable species complex with intrasporocarpic nrDNA ITS heterogeneity. Mycologia 103:841–853. doi:10.3852/10-312

Kovács GM, Balázs TK, Calonge FD, Martín MP (2011c) First record of the truffle *Mattirolomyces terfezioides* from the Iberian Peninsula—the most western and southern locality of the species in Europe. Mycotaxon 103:841–853

Læsøe T, Hansen K (2007) Truffle trouble: what happened to the Tuberales? Mycol Res 111:1075–1099. doi:10.1016/j.mycres.2007.08.004

Maire R (1907) Contribution á l'étude de la flore mycologique de l'Afrique du Nord. (1906). Bull Soc Bot France 53:180–215

Malençon G (1973) Champignons hypogés du Nord de l'Afrique—I Ascomycetes. Persoonia 7:261–288

Marasas WFO, Trappe JM (1973) Notes on southern African Tuberales. Bothalia 11:139–141

Martin F, Kohler A, Murat C, Balestrini R, Coutinho PM, Jaillon O, Montanini B, Morin E, Noel B, Percudani R, Porcel B, Rubini A, Amicucci A, Amselem J, Anthouard V, Arcioni S, Artiguenave F, Aury JM, Ballario P, Bolchi A, Brenna A, Brun A, Buée M, Cantarel B, Chevalier G, Couloux A, Da Silva C, Denoeud F, Duplessis S, Ghignone S, Hilselberger B, Iotti M, Marçais B, Mello A, Miranda M, Pacioni G, Quesneville H, Riccioni C, Ruotolo R, Splivallo R, Stocchi V, Tisserant E, Viscomi AR, Zambonelli A, Zampieri E, Henrissat B, Lebrun MH, Paolocci F, Bonfante P, Ottonello S, Wincker P (2010) Périgord black truffle genome uncovers evolutionary origins and mechanisms of symbiosis. Nature 464:1033–1038. doi:10.1038/nature08867

Mattirolo O (1887) Illustrazione di tre nuove specie di Tuberacee Italiane. Mem della Reale Accad Sci Torino (Seconda serie) 38:377–393

Mattirolo O (1922) Osservazioni sopra due ipogei della Cirenaica e considerazioni intorno ai generi *Tirmania* e *Terfezia*. Reale Accademia Nazionale dei Lincei 5:545–569

Moreno G, Lizárraga M, Esqueda M, Galán R, Alvarado P (2012) New records of little-known species of *Carbomyces* (Carbomycetaceae, Ascomycota). Mycotaxon 120:89–98. doi:10.5248/120.89

Moris JH (1829) Stirpium Sardoanum Elenchus. Fasc 3. Bessone, Torino

Norman JE, Egger KN (1999) Molecular phylogeny analysis of *Peziza* and related genera. Mycologia 91:820–829

O'Donnell KO, Cigelnik E, Weber NS, Trappe JM (1997) Phylogenetic relationships among ascomycetous truffles and the true and false morels inferred from 18S and 28S ribosomal DNA sequence analysis. Mycologia 89:48–65

Patouillard N (1894a) Les Terféz de la Tunisie. J Bot (Morot) 8(8):153–156

Patouillard N (1894b) Les Terféz de la Tunisie (2eme note). J Bot (Morot) 8(10):181–182

Patouillard N (1899) Champignons du Nord de l'Afrique. Bull Soc Mycol France 15:54–59
Percudani R, Trevisi A, Zambonelli A, Ottonello S (1999) Molecular phylogeny of truffles (Pezizales: Terfeziaceae, Tuberaceae) derived from nuclear rDNA sequence analysis. Mol Phylogenet Evol 13:169–180. doi:10.1006/mpev.1999.0638
Plett JM, Martin F (2011) Blurred boundaries: lifestyle lessons from ectomycorrhizal fungal genomes. Trends Genet 27:14–22. doi:10.1016/j.tig.2010.10.005
Pole-Evans IB (1918) Note on the genus *Terfezia*; a truffle from the Kalahari. Trans Roy Soc South Afr 7:117–118
Sbissi I, Neffati M, Boudabous A, Murat C, Gtari M (2010) Phylogenetic affiliation of the desert truffles *Picoa juniperi* and *Picoa lefebvrei*. A van Leeuw J Microb. doi:10.1007/s10482-010-9456-y
Sbissi I, Ghodhbane-Gtari F, Neffati M, Ouzari H, Boudabous A, Gtari M (2011) Diversity of the desert truffle *Terfezia boudieri* Chatin. in southern Tunisia. Can J Microbiol 57:599–605. doi:10.1139/W11-040
Spegazzini C (1887) Las trufas Argentinas. Anal Soc Ci Argent 24:120–127
Sterflinger K, Tesei D, Zakharova K (2012) Fungi in hot and cold deserts with particular reference to microcolonial fungi. Fung Ecol 5:453–462. doi:10.1016/j.funeco.2011.12.007
Tedersoo L, Hansen K, Perry BA, Kjøller R (2006) Molecular and morphological diversity of pezizalean ectomycorrhiza. New Phytol 170:581–96. doi:10.1111/j.1469-8137.2006.01678.x
Tedersoo L, May TW, Smith ME (2010) Ectomycorrhizal lifestyle in fungi: global diversity, distribution, and evolution of phylogenetic lineages. Mycorrhiza 20:217–63. doi:10.1007/s00572-009-0274-x
Tedersoo L, Arnold AE, Hansen K (2013) Novel aspects in the life cycle and biotrophic interactions in Pezizomycetes (Ascomycota, Fungi). Mol Ecol 22:1488–1493
Trappe JM (1971) A synopsis of the Carbomycetaceae and Terfeziaceae (Tuberales). Trans Br Mycol Soc 57:85–92
Trappe JM (1979) The orders, families, and genera of hypogeous Ascomycotina (truffles and their relatives). Mycotaxon 9:297–340
Trappe JM, Claridge AW (2005) Hypogeous fungi: evolution of reproductive and dispersal strategies through interactions with animals and mycorrhizal plants. In: Dighton J, White JF, Oudemans P (eds) The fungal community: its organization and role in the ecosystem, 3rd edn. Taylor & Francis, Boca Raton, FL
Trappe JM, Sundberg WJ (1977) *Terfezia gigantea* (Tuberales) in North America. Mycologia 69:433–437
Trappe JM, Kovács GM, Claridge AW (2010a) Comparative taxonomy of desert truffles of the Australian Outback and African Kalahari. Mycol Prog 9:131–143. doi:10.1007/s11557-009-0612-6
Trappe JM, Kovács GM, Claridge AW (2010b) Validation of the new combination *Mattirolomyces austroafricanus*. Mycol Prog 9:145. doi:10.1007/s11557-009-0631-3
Tulasne L-R, Tulasne C (1844) Fungi nonnulli hypogaei, nov. v. minus cogniti. Giorn Bot Ital 2:55–63
Tulasne L-R, Tulasne C (1845) De gen. *Choiromyces* et *Picoa* e Tuberacearum familia. Ann Sci Nat Bot, sér. 3, 3:348–353
Tulasne L-R, Tulasne C (1851) Fungi hypogaei. Friedrich Klincksieck, Paris
Ward D (2008) The biology of deserts. Oxford University Press, Oxford
Zobel JB (1854) Iconum Fungorum Hucusque Cognitorum, vol. 6. In: Corda ACJ (ed.) Abbildungen von Pilzen und Schwaemmen. Pragae

Chapter 3
Cryptic and New Species

Juan-Julián Bordallo and Antonio Rodríguez

3.1 Introduction

The common vision from last century's microbiology literature, as proposed by Baas-Becking "Everything is everywhere, but the environment selects" (Bass Becking 1934), suggests that endemic species should be rare, and the distinction between different populations of the same species should be minimal. This idea was supported by the common spread of fungi spores by the wind, foreseeing a widespread distribution of the species, and from their minimal morphologic differentiation. An example of all this, referring to the species that interest us, is the spore diversity that Mattirolo (1922), using methods available at the time, described within the same *Terfezia* species, without doubt pressed by the similarity between ascomata. Nevertheless, it would seem that these hypotheses were incorrect for fungi (Taylor et al. 2006). Such hypotheses have conditioned that in numerous occasions only what was already expected would be seen (Burnett 2003). This explains why some authors are surprised by unexpected fungal diversity in nature (Buée et al. 2009; Egger 2006). The idea of such a limited diversity was false, as suggested by Peay et al. (2007), and an underrated amount of fungal species are being discovered in all environments (Blackwell 2011; Hawksworth 1997; Hawksworth and Rossman 1997).

The spectacular increase in the number of fungal species is being widely reported throughout the literature (Hawksworth 2001, 2010). Some authors give clues to where and how to look for the "lost" species (Hyde 2001). We think that perhaps the same sites should be visited and, with help of molecular and biogeographic methods, complete the old studies.

J.-J. Bordallo (✉) • A. Rodríguez
Departamento de Biología Vegetal (Botánica), Facultad de Biología, Campus de Excelencia Internacional Regional "Campus Mare Nostrum", Universidad de Murcia, Murcia 30100, Spain
e-mail: juanjulian.bordallo@um.es

V. Kagan-Zur et al. (eds.), *Desert Truffles*, Soil Biology 38,
DOI 10.1007/978-3-642-40096-4_3, © Springer-Verlag Berlin Heidelberg 2014

The great transformation generally taking place in taxonomy (Padial et al. 2010; Ebach et al. 2011; Wiens 2007) and in particular the studies on lichen-forming fungi (Crespo and Pérez-Ortega 2010; Leavitt et al. 2011; Grube and Kroken 2000) and of pathogenic fungi (Telle et al. 2011) are expanding over the taxonomy of mycorrhiza-forming fungi (Ferdman et al. 2009; Bordallo et al. 2013).

This transformation consists of specifying what new criteria and characters, linked with traditional ones, are useful for defining the demarcation of species (Petersen and Hughes 2009). Massive DNA sequencing and its application to molecular phylogeny favour changes to the concepts of cryptic, interrelated and boundary species. Conclusively, the criteria when considering different species are changing. Some authors propose amendments when considering new species, by suggesting different nomenclature based on the increasing number of fungal DNA sequences and the continuous enumeration of newly discovered *molecular operational taxonomic units* (MOTUs) (Hibbett et al. 2011). Other authors emphasise the problems in the nomenclature regarding species arising from sequencing of environmentally extracted DNA as compared to DNA sequencing from ascocarps (Taylor 2011; Nilsson et al. 2011).

3.2 The New and Cryptic *Terfezia* Species

Recently, some new desert truffle species have been reported (Kovăcs et al. 2011; Bordallo et al. 2013) in the Claveryi and Fanfani groups. A morphological characterisation is described below for each species, and a phylogenetic relationship among them has been established (Fig. 3.1) using the Maximum Parsimony method and the sister species *T. claveryi* (Fig. 3.4c, d) and *T. arenaria* (Fig. 3.4e, f).

3.2.1 Claveryi Group

3.2.1.1 *Terfezia canariensis* Bordallo & Ant. Rodr (Bordallo et al. 2012) (Fig. 3.2a, b)

T. canariensis has been historically mistaken for *T. claveryi*. Both species of *Terfezia* share prosenchymatous peridium and reticulate spores, but *T. canariensis* spores (mean values: 21–23 μm diam, median = 22 μm) are bigger than *T. claveryi* spores (mean values: 18–21 μm diam, median = 20 μm), and *T. canariensis* spores have a well-developed, complete reticulum, more clear than reticulum of *T. claveryi* spores, often reduced to isolated warts and ridges.

The *T. canariensis* species develop in calcareous, clayey or sandy soils, associated with *Helianthemum canariense*, February through April.

T. canariensis ascomata are hypogeous to partially emergent at maturity, 2–8 (–10) cm in size, subglobose, turbinate, obpyriform, often with short, hemispherical

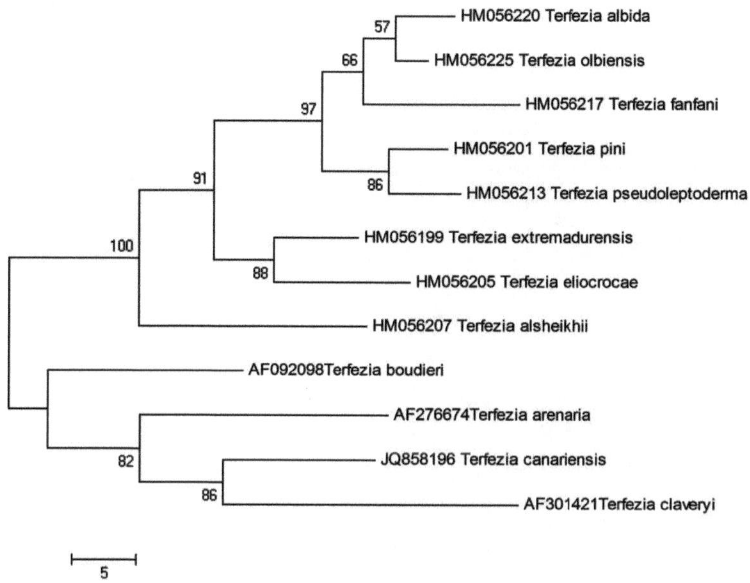

Fig. 3.1 Evolutionary relationship of 12 taxa. The evolutionary history was inferred using the Maximum Parsimony method. The bootstrap consensus tree inferred from 500 replicates is taken to represent the evolutionary history of the taxa analysed. Branches corresponding to partitions reproduced in less than 50 % bootstrap replicates are collapsed. The percentages of replicate trees in which the associated taxa clustered together in the bootstrap test (500 replicates) are shown next to the branches. The MP tree was obtained using the Close-Neighbour-Interchange algorithm with search level 3 in which the initial trees were obtained with the random addition of sequences (10 replicates). The tree is drawn to scale with branch lengths calculated using the average pathway method and are in the units of the number of changes over the whole sequence. All positions containing gaps and missing data were eliminated from the dataset (Complete Deletion option). There were a total of 499 positions in the final dataset, out of which 58 were parsimony informative. Phylogenetic analyses were conducted in MEGA4

to obconic base, very light brown at first, becoming reddish brown, blackening with age, initially smooth, but often furrowed, associated with rapid growth. Peridium 1–1.5(–3) mm thick, composed of hyphae, 8–20 μm broad with walls ±1 μm thick. Gleba solid, fleshy, succulent, whitish with pale orange pink pockets at first, maturing to pink meat pockets of fertile tissue separated by whitish pink, sterile veins, inconspicuous at maturity, sometimes with yellowish brown spots. Nonamyloid, subglobose to ellipsoid, pyriform, sessile or short-stipitate asci, 60–90 × 50–70 μm, with 6–8 irregularly disposed spores, randomly arranged in fertile pockets. Globose ascospores, (20–)21–23(–25) μm diam (median = 22 μm) including ornament, hyaline, smooth and uniguttulate at first, then yellow and ornamented with a well-developed, small-meshed reticulum, polygonal meshes variable in form and size, 0.5–1 μm thick, 1 μm tall. This reticulum is conspicuous from very early stages until complete maturity.

Fig. 3.2 *T. canariensis*: (**a**) Ascomata and (**b**) ascospores. *T. extremadurensis*: (**c**) ascomata and (**d**) ascospores. *T. eliocrocae*: (**e**) ascomata and (**f**) ascospores stained with acid fuchsine. *T. alsheikhii*: (**g**) ascomata and (**h**) ascospores

3.2.2 *Fanfani Group*

3.2.2.1 *Terfezia extremadurensis* Mohedano, Ant. Rodr. & Bordallo (Bordallo et al. 2013) (Fig. 3.2c, d)

T. extremadurensis differs from other spiny spored *Terfezia* species in its Tuber-like glebal morphology, with meandering veins of the gleba not completely surrounding the fertile tissue and not forming pockets, as all the other *Terfezia* species. Microscopically, *T. extremadurensis* has the largest spores of the spiny spored *Terfezia* species, average spore size 22–26 μm including ornament, with the thickest spines, up to 3 μm wide at the base. *T. fanfani*, described below, shares the same habitat, associated with *T. guttata*, but *T. extremadurensis* lacks reddish tones and the gleba is different. Microscopically, *T. fanfani* has smaller spores, average spore size 19–22 μm, with thin spines, 1 μm wide at the base.

Hypogeous ascomata are partially emergent at maturity, 2–5 cm in size, subglobose, sometimes furrowed and nodulose, often cracked, often with a small basal depression, rarely with pseudostipe, cream colour at first, becoming brown, black spots on the sun-exposed parts or where handled, smooth. Peridium 300–600 μm thick, well-defined, concolorous with surface in cross section, pseudoparenchymatous, composed of subglobose cells, hyalines and thin walled in the innermost layers, yellowish and with thicker walls in the outermost layers. Gleba solid, fleshy, succulent, whitish at first, soon becoming salmon pink, darkening with age, greenish grey at maturity, marbled with thin, white, meandering veins, sometimes arising from the base and inconspicuous in very mature specimens. Frequently, it presents small holes indicating mycophagous activity. Odour is faint, not distinctive. Nonamyloid, subglobose to ovate, sessile or short-stipitate asci, 60–80 × 50–65 μm, walls 1–2 μm thick, with 6–8 irregularly disposed spores, randomly arranged in the gleba. Globose ascospores, (21–)22–26(–27) μm diam (median = 24 μm) including ornament (16–)17–19(–20) μm (median = 18 μm) without ornament, hyaline, smooth and uniguttulate at first, by maturity yellow and ornamented with conical, blunt, thick spines, sometimes truncated, sometimes finger-like, often joined at the base, 3–4(–5) μm long, 1–3 μm wide at the base.

3.2.2.2 *Terfezia eliocrocae* Bordallo, Morte & Honrubia (Bordallo et al. 2013) (Fig. 3.2e, f)

The species *T. eliocrocae* grows in grasslands of calcareous soils in the presence of *H. almeriense* and *H. violaceum* (Cistaceae). Hypogeous, then the ascomata cracks the soil to rise to the surface. A distinctive character is the inner milky white colour that persists in contact with air.

Ascomata more or less crushed shapes, unsmooth, embossed, exceeding 5 cm in size. Cream colour, blackish aspect when ripe with a rough, cracked texture. Sticky mycelial remains and pseudo-stipe. Thin peridium, well delimited and inseparable from the gleba; from cream to blackish-brown colour; partially rough when ripe. A pseudoparenchymatous structure, up to 200 μm thick, round cells of 30 × 15 μm in size. Hyphal extensions can be observed on its external layer. The gleba has intense white colour, which does not darken during maturation or when in contact with air. Fertile and sterile tissue of a milky white colour; small islets of fertile tissue and surrounded by very thin and ramified sterile veins. Subglobose, ovoid or pyriform asci elongated, sessile, containing eight ascospores spherical with a well-developed reticulum, regular, 17–19 μm in size including ornamentation. Initially hyaline and smooth, when ripe, they acquire a yellow colour and ornamentation with a very marked reticulum (up to 1 μm height).

3.2.2.3 *Terfezia alsheikhii* Kovács, M.P. Martín & Calonge (Kovăcs et al. 2011) (Fig. 3.2g, h)

The species grows in sandy acidic soils associated with Cistaceae plants. Gregarious specimens have been found by raking under *C. monspeliensis* L. on the border of a forest track.

Regular, globose ascomata, 0.5–1.5 cm in size, smooth, firm. No strong odour detected. Smooth peridium, thin and inseparable from the gleba. Brown-ochre colour. A pseudoparenchymatous structure throughout, with thick-walled cells and up to 50 μm in diameter. Gleba is formed by big pinkish islets of fertile tissue, varying in size and surrounded by a scarce sterile tissue of white colour, stained by brown-reddish zones when in contact with air. Subglobose, ovoid, pyriform asci, 60–100 × 40–70 μm in size, sessile or with a short and thick peduncle. Most contain eight ascospores spherical and reticulated, 15–18 μm in size including ornamentation. Initially hyaline, they acquire a yellow colour when mature. Decorated by an irregular, well-developed reticulum with thickish net (nearly 1 μm). Warts appear in very mature spores with roundish to flat shapes and up to 2 μm long and 2 μm wide.

3.2.2.4 *Terfezia pini* Bordallo, Ant. Rodr. & Mohedano (Bordallo et al. 2013) (Fig. 3.3a, b)

The most distinctive characters of this *Terfezia* species are that it is not associated with Cistaceae plants and the ornamentation of its spores that consists of long spines joined on their base, forming the pseudo-reticulum. The species grows mainly in sandy pine forests from November to May. It is often found at a depth of 3–5 cm and is rarely observed on the surface but usually under mosses.

Globose ascomata, round, regular, less than 2 cm in size, smooth. Frequently, it presents residual mycelium at the base. Peridium smooth, slightly tomentous, difficult to separate from the gleba, thin and not clearly delimited, 200–400 μm thick. It is partially covered by a whitish film. Initial cream colour becomes ochre and in cross section grey. Pseudoparenchymatous, thick-walled cells up to 40 μm in

Fig. 3.3 *T. pini*: (**a**) Ascomata and (**b**) ascospores. *T. pseudoleptoderma*: (**c**) ascomata and (**d**) ascospores. *T. fanfani*: (**e**) ascomata and (**f**) ascospores. *T. albida*: (**g**) ascomata and (**h**) ascospores

diameter, pigmented in the outer part. Gleba is initially whitish. When mature, the fertile tissue forms round islets of various sizes, initially pale pinkish that become greenish brown and greyish when the gleba is very mature. It is always surrounded by a sterile tissue of white veins. Ovoid, ellipsoid, subglobose, sessile asci, 60–90 × 45–60 μm in size and walls 1 μm thick; they contain 6–8 ascospores spherical with spines, 20–23(–25) μm in size including ornamentation. Initially, they are hyaline and smooth, with a big central drop. When mature, they acquire a yellowish-ochre colour and are decorated by cylindrical spines with a round tip, 3–4 (–5) μm long and 1 μm wide. Spines are joined by their basal part in the shape of crests that often form a pseudo-reticulum.

3.2.2.5 *Terfezia pseudoleptoderma* Bordallo, Ant. Rodr. & Mohedano (Bordallo et al. 2013) (Fig. 3.3c, d)

The species grows bordering pine and oak forests in the presence of plants of Cistaceae family.

Globose ascomata, round, quite regular, up to 2 cm in size, smooth when immature and partially rough at maturity. Initially it presents a cream colour that becomes reddish brown when maturing. Smooth peridium or slightly rough, inseparable from the gleba; cream colour when in contact with light, becoming darker when in contact with air. Gleba is initially whitish. Its fertile tissue forming islets becomes translucent greyish blue, while veins or the sterile tissue remain white in colour. Asci from globose to ovoid; containing eight ascospores spherical, 19–23 μm in size including ornamentation of separated, blunt spines and asymmetric base, from 2 to 5 μm long and up to 1 μm at their base. Initially hyaline, they acquire a yellow colour when maturing.

3.2.2.6 *Terfezia albida* Ant. Rodr., Mohedano & Bordallo (Bordallo et al. 2013) (Fig. 3.3g, h)

T. albida differs from other spiny spored *Terfezia* species in the larger average size of ascomata, white colour of peridium and spermatic odour. *T. albida* is the only spiny spored *Terfezia* species associated with *Helianthemum* spp. in alkaline soils. *T. eliocrocae*, described below, and *T. claveryi* share habitat with *T. albida*, but they have reticulate spores. Grows in southeast of the Iberian Peninsula, limited to arid and semiarid areas, in calcareous, alkaline soils, associated with *Helianthemum* spp., from late April to mid-May.

Hypogeous ascomata to partially emergent at maturity, 2–4 cm across, 3–4 cm high, subglobose to turbinate, pulvinate, often with tapered, sterile base, white at first, becoming light cream, often black spots on the sun-exposed parts or where handled, greenish with age on injured areas, smooth. Peridium 200–500 μm thick, poorly delimited, white in cross section, pseudoparenchymatous, composed of subglobose cells, of variable size, hyalines and thin walled in the innermost layers, yellowish and with thicker walls in the outermost layers. Gleba solid, fleshy, succulent, white at first, maturing to greyish green pockets of fertile tissue separated by whitish, sometimes with pink spots, sterile veins. Spermatic odour, more remarkable in young specimens. Asci nonamyloid, subglobose to ovate, sessile or short stipitate, 70–85 (–90) × 55–70 μm, walls 1 μm thick, with 6–8 irregularly disposed spores, randomly arranged in the gleba. Globose ascospores, (18–) 19–22

(–23) μm diam (median = 20 μm) including ornament, 14–17(–18) μm (median = 16 μm) without ornament, hyaline, smooth and uniguttulate at first, by maturity yellow ochre and ornamented with conical, blunt, straight spines, sometimes cylindrical and curved, sometimes truncated, separate, 2–3 μm long, 1–2 μm wide at the base, and sometimes connected to form a pseudo-reticulum.

3.2.3 Boudieri Complex

Molecular studies on evolution of the taxon *T. boudieri* have been made by Ferdman et al (2009). These authors studied the presence of cryptic species within the *T. boudieri* complex. For this purpose SNPs markers were used, demonstrating the existence of three consistent genotypes, correlated with a few selected physiological characteristics.

3.3 Other *Terfezia* Species Phylogenetically Close to Claveryi, Fanfani and Boudieri Groups

3.3.1 Terfezia olbiensis Tul. & C. Tul (Fig. 3.4a, b)

The species grows in limestone and clayey pine and oak woodlands without *Helianthemum* spp. from mid-March to mid-April. It is the first *Terfezia* species to produce ascocarps in limestone soils. Initially hypogeous, it later rises to the surface, where it remains firm. They usually end their life cycle parasitised underground by larvae or eaten by rabbits, probably because they appear early in the year, when there is greater humidity and less sunlight. Distinctive smell, as if rotten; less flavour than other edible *Terfezia* species. The most distinguished characteristic is the reduced length of its spines, always shorter than 2.5 μm.

Ascomata globose, round, regular, rarely presents pseudostipe, 2–5 cm in size. Smooth peridium, inseparable from the gleba, thin, 300–500 μm thick and not clearly delimited. It is initially cream and changes to brown. It frequently presents black maculae in the zones exposed to sun, which increase when manipulated. Its cross section is white. Pseudoparenchymatous structure formed by more or less rounded hyaline cells of different sizes with thin walls. It acquires a yellow colour and a prismatic shape towards the periphery. Gleba initially white, when maturing, the fertile tissue creates small grey islets that become greenish grey, always surrounded by white sterile tissue with salmon highlights. From ellipsoidal to ovoid asci, citrus shaped, 60–90 × 50–60 μm in size and walls 1–2 μm thick. Most are sessile but some present a short, thick peduncle. They contain eight spores. When immature, they present a dextrinoid content. Spherical and spiny

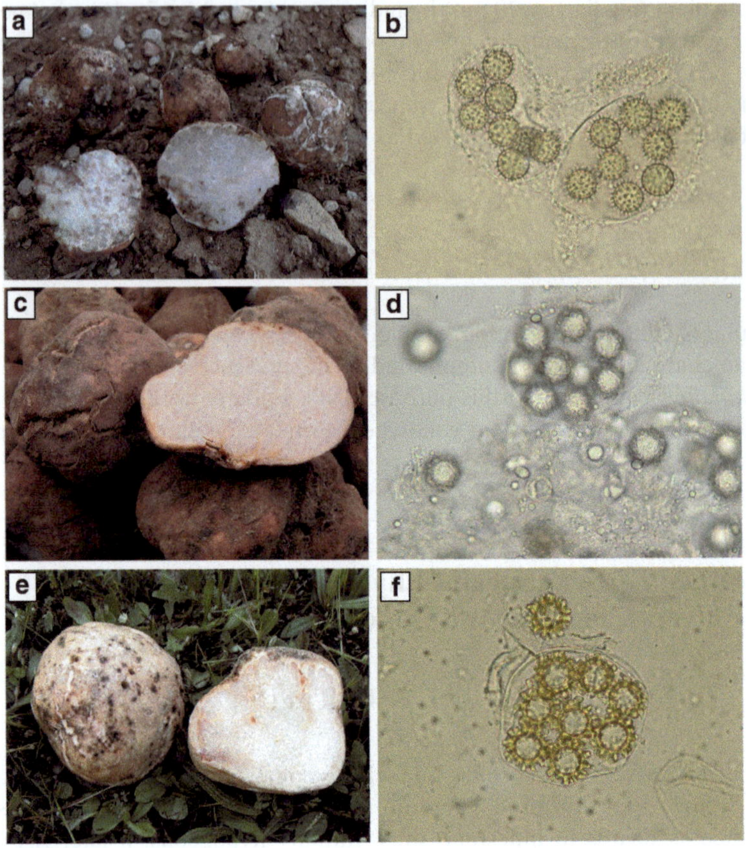

Fig. 3.4 *T. olbiensis*: (**a**) Ascomata and (**b**) ascospores. *T. claveryi*: (**c**) ascomata and (**d**) ascospores. *T. arenaria*: (**e**) ascomata and (**f**) ascospores

ascospores, 15–19 μm in diameter including ornaments, initially hyaline and smooth, with a great central drop. When mature, they acquire an ochre yellow colour and are decorated with pointy thin conical spines (some truncated) of 1–2 (–2.5) μm in length and a base of 1 μm; these spines are sparse and not joined through the base.

3.3.2 Terfezia fanfani Mattir. (Fig. 3.3e, f)

The species grows in acidic grasslands associated with *T. guttata* from the end of March to the end of April. It appears before *T. arenaria* and usually prefers saltier and less deep soils. Initially hypogeous but later it rises to the surface; firm and odourless. It shares both habitat and fructification season with *T. arenaria*. The most distinguishing characteristics are its reddish colour, the fact that it can be found growing together with *T. guttata* and its spores decorated with long and isolated spines.

Globose ascomata, round, regular, occasionally lobed, usually with mycelial remains, 2–5 cm in size, smooth or slightly rough. Smooth peridium, slightly rough, inseparable from the gleba, thin, 200–700 μm thick. Initially white but soon changes to reddish brown and darkens as it matures, it presents black maculae in the zones exposed to the sun and its cross section is white. Pseudo-parenchymatous structure is formed by rounded and prismatic cells. Gleba initially white, when maturing, the fertile tissue creates islets which are firstly pale pink, then olive green to become blackish grey when very mature. It is always surrounded by white sterile tissue. Subglobose, elongated and ovoid asci, 70–80 × 55–70 μm in size with 1 μm thick walls. The majority are sessile, containing eight ascospores spherical with spines, 19–23 (–25) μm in diameter including ornaments. Hyaline and smooth initially, with a great central drop. When mature, they acquire an ochre yellow colour and are decorated with thin elongated conic spines with a sharp point end (2–) 3–4 (–5) μm long and a base of 1 μm; these spines are not joined through the base.

3.4 Conclusions

Molecular analyses consider a single locus, the ITS location, when the phylogenetic results (ITS) coincide with phenotypic or ecological data (Grube and Kroken 2000). However, when phylogenetic results (ITS) do not coincide with phenotypic or ecological characteristics, additional loci or genes have to be used (Ferdman et al. 2009). Molecular phylogenies support the morphological conclusions of traditional taxonomy in some cases but not in others. An increase in species being considered is found across contemporary studies. Whereas morphology indicates taxon variety, molecular phylogeny indicates an independent lineage and therefore a different phylogenetic species (Crespo and Lumbsch 2010). In previous decades, the criteria for separating and identifying groups of species were limited to morphological, anatomic and occasionally chemical characters. Nowadays taxonomic studies are supported by molecular genetics methods (Del-Prado et al. 2010; Horton and Bruns 2001).

We think that in desert truffles (mycorrhizal fungi) there is a strong endemic component due to their hypogeal fructification (limited dispersion) and their symbiotic nature (host co-dispersion and necessary microhabitat). The preference-specificity factor of the host is manifested as an important factor in mycorrhizal fungi (Carriconde et al. 2008; Burnett 2003; Murat et al 2004). Regarding that in mycorrhizal fungi, biological interactions with their host are part of the environment that conditions their evolution in different species (Molina and Trappe 1982).

From this we can extract the working hypothesis of the host's importance in specialisation (Ishida et al. 2007), allowing the host to condition the distribution of fungal species (Murat et al. 2004; Tedersoo et al. 2008; Jumpponen et al. 2004). Grubisha et al. (2007) show how nearby mycorrhizal fungal populations can have significant genetic differences. If we consider the persistence of different genomes of sympatric form due to the impossibility of genetic material exchange, this would favour the criteria for species separation (de Queiroz 2007).

The difficulty of sampling desert truffles implies their discovery only in specific locations. This allows the hypothesis that a thorough study of the same and other collection zones and during different seasons of the year would consequently favour the discovery of new species (Claridge et al. 2000a, b; Henkel et al. 2012).

Taking into account the different definitions of species (Taylor et al. 2000), it is necessary to consider that to determine cryptic species with morphologic criteria or molecular genetics data would produce dissimilar results (Carriconde et al. 2008; Bickford et al. 2007). A particular case showing this within the desert truffles is the *Terfezia boudieri*, a group of cryptic species widely studied (Aviram et al. 2004; Ferdman et al. 2009).

The convergence of morphological characters in cryptic species may be caused due to their peculiar ecology or due to the fact that not always speciation (genetic change) is accompanied by morphological change. Because of this, it is expected that when molecular methods are used different species are found within the morphologically indistinguishable fungi (cryptic species) (Bickford et al. 2007).

We propose to make headway in this change of paradigm on mycology taxonomy (Lumbsch and Leavitt 2011; Beheregaray and Caccone 2007). The application of novel molecular methods to hypogeous fungal group, on which desert truffles can be found, allows the discovery of new species. We think the finding of undiscovered species within the genera of desert truffles will rise throughout the coming years (own studies in progress), similar to those carried out on genera like Tuber (Bonito et al. 2010).

Furthermore, we think that only subtle and precise tools will make it possible to delve and reveal the extreme complexity of nature (Beheregaray and Caccone 2007).

We would like to acknowledge from these pages the labour, often carried selflessly and passionately, which many amateurs and mycological societies carry out in order to document the biodiversity of these exceptional desert fungi. Recognition and joint efforts between academics and amateurs would without a doubt benefit their knowledge (Hawksworth 2003).

Acknowledgements This work has been supported by project CGL2011-29816 (Spanish Ministry of Science and Innovation, Spain). Authors thank Dr. Asun Morte for her help and inexhaustible patience in the writing of this chapter.

References

Aviram S, Roth-Bejerano N, Kagan-Zur V (2004) Two ITS forms co-inhabiting a single genet of an isolate of *Terfezia boudieri* (Ascomycotina), a desert truffle. Ant Leeu Int J 85:169–174

Bass Becking LGM (1934) Geobiologie of Inleiding Tot de Milieukunde. Van Stockum & Zoon, The Hague, The Netherlands

Beheregaray LB, Caccone A (2007) Cryptic biodiversity in a changing world. J Biol 6:9

Bickford D, Lohman DJ, Sohdi NS, Ng PKL, Meier R, Winker K, Ingram KK, Das I (2007) Cryptic species as a window on diversity and conservation. Trends Ecol Evol 22(3):148–155

Blackwell M (2011) The fungi: 1, 2, 3 … 5.1 million species? Am J Bot 98:426–438

Bonito G, Gryganskyi A, Trappe J, Vilgalys R (2010) A global meta- analysis of tuber its rDNA sequences: species diversity, host specificity, and long-distance dispersal. Mol Ecol 19:4994–5008. doi:10.1111/j.1365-294X.2010.04855.x

Bordallo JJ, Rodriguez A, Honrubia M, Morte A (2012) *Terfezia canariensis sp. nov.* una nueva especie de trufa encontrada en las Islas Canarias. J Cantarela 56:1–8

Bordallo JJ, Rodriguez A, Muñoz-Mohedano JM, M.Suz L, Honrubia M, Morte A (2013) Morphological and molecular characterization of five new *Terfezia* species from the Iberian Peninsula. Mycotaxon (in press)

Buée M, Reich M, Murat C, Morin E, Nilsson RH, Uroz S, Martin F (2009) 454 pyrosequencing analyses of forest soils reveal an unexpectedly high fungal diversity. New Phytol 184:449–456

Burnett J (2003) Fungal population and species. Oxford University Press, Oxford

Carriconde F, Gardes M, Jargeat P, Heilmann-Clausen J, Mouhamadou B, Gryta H (2008) Population evidence of cryptic species and geographical structure in the cosmopolitan ectomycorrhizal fungus *Tricholoma scalpturatum*. Microb Ecol 56:513–524

Claridge AW, Cork SJ, Trappe JM (2000a) Diversity and habitat relationships of hypogeous fungi. I. Study design, sampling techniques and general survey results. Biodivers Conserv 9:151–173

Claridge AW, Barry SC, Cork SJ, Trappe JM (2000b) Diversity and habitat relationships of hypogeous fungi. II. Factors influencing the occurrence and number of taxa. Biodivers Conserv 9:175–199

Crespo A, Lumbsch HT (2010) Cryptic species in lichen-forming fungi. Int Mycol Assoc 1 (2):167–170

Crespo A, Pérez-Ortega S (2010) Cryptic species and species pairs in lichens: a discussion on the relationship between molecular phylogenies and morphological characters. Anales del Jardín Botanico de Madrid 66:71–81

de Queiroz K (2007) Species concepts and species delimitation. Syst Biol 56:879–886

Del-Prado R, Cubas P, Lumbsch HT, Divakar P, Blanco O, Amo De Paz G, Molina MC, Crespo A (2010) Genetic distances within and among species in monophyletic lineages of Parmeliaceae (Ascomycota) as a tool for taxon delimitation. Mol Phylogenet Evol 56:1329–1337

Ebach MC, Valdecasas AG, Wheeler QD (2011) Impediments to taxonomy and users of taxonomy: accessibility and impact evaluation. Cladistics 27:550–557

Egger KN (2006) The surprising diversity of ascomycetous mycorrhizas. New Phytol 170:421–423

Ferdman Y, Sitrit Y, Li Y-F, Roth-Bejerano N, Kagan-Zur V (2009) Cryptic species in the Terfezia boudieri complex. A van Leeuw J Microb 95:351–362

Grube M, Kroken S (2000) Molecular approaches and the concept of species and species complexes in lichenized fungi. Graphis 104:1284–1294

Grubisha LC, Bergemann SE, Bruns TD (2007) Host islands within the California Northern Channel Islands create fine-scale genetic structure in two sympatric species of the symbiotic ectomycorrhizal fungus *Rhizopogon*. Mol Ecol 16:1811–1822

Hawksworth DL (1997) The fascination of fungi: exploring fungal diversity. Mycologist 11 (1):18–22

Hawksworth DL (2001) The magnitude of fungal diversity: the 1.5 million species estimate revisited. Mycol Res 12:1422–1432

Hawksworth DL (2003) Monitoring and safeguarding fungal resources worldwide: the need for an international collaborative MycoAction Plan. Fung Diver 13:29–45

Hawksworth DL (2010) Cryptic speciation: how common is it and how should it be handled taxonomically? Rimc9.info/prog_sig3_detail_hawksworth.htm

Hawksworth DL, Rossman AY (1997) Where are all the undescribed fungi? Phytopathology 87:888–891

Henkel TW, Aime MC, Chin MML, Miller SL, Vilgalys R, Smith ME (2012) Ectomycorrhizal fungal sporocarp diversity and discovery of new taxa in Dicymbe monodominant forests of the Guiana Shield. Biodivers Conserv 21:2195–2220. doi:10.1007/s10531-011-0166-1

Hibbett DS, Ohman A, Glotzer D, Nuhn M, Kirk P, Nilsson RH (2011) Progress in molecular and morphological taxon discovery in *Fungi* and options for formal classification of environmental sequences. Fung Biol Rev 25:38–47

Horton TR, Bruns TD (2001) The molecular revolution in ecto-mycorrhizal ecology: peeking into the black-box. Mol Ecol 10:1855–1871

Hyde KD (2001) Where are the missing fungi? Mycol Res 105:1409–1412

Ishida TA, Nara K, Hogetsu T (2007) Host effects on ectomycorrhizal fungal communities: insight from eight host species in mixed conifer-broadleaf forests. New Phytol 174:430–440

Jumpponen AM, Claridge AW, Trappe JM, Lebel T, Claridge DL (2004) Ecological relationships among hypogeous fungi and trees: inferences from association analysis integrated with habitat modelling. Mycologia 96:510–525

Kovăcs G, Calonge D, Martín MP (2011) The diversity of *Terfezia* desert truffles: new species and a highly variable species complex with intrasporocarpic nrDNA ITS heterogeneity. Mycologia 103:841–853. doi:10.3852/10-312

Leavitt SD, Johnson L, St. Clair LL (2011) Species delimitation and evolution in morphologically and chemically diverse communities of the lichen-forming genus *Xanthoparmelia* (Parmeliaceae, Ascomycota) in western North America. Am J Bot 98:175–188

Lumbsch HT, Leavitt SD (2011) Goodbye morphology? A paradigm shift in the delimitation of species in lichenized fungi. Fung Diver 50:59–72

Mattirolo O (1922) Osservazioni sopra due ipogei dell Cirenaica e considerazioni intorno ai generi Tirmania e Terfezia. Reale Accarlemiu Nuzionule dei Lincei 5:545–569

Molina R, Trappe J (1982) Patterns of ectomycorrhizal host specificity and potential among Pacific Northwest conifers and fungi. Forest Sci 28:423–458

Murat C, Diez LP, Delaruelle C, Dupré C, Chevalier G, Bonfante P, Martin F (2004) Polymorphism at the ribosomal DNA ITS and its relation to postglacial recolonization of the Perigord truffle Tuber melanosporum. New Phytol 164:401–411

Nilsson RH, Ryberg M, Sjokvist E, Abarenkov K (2011) Rethinking taxon sampling in the light of environmental sequencing. Cladistics 27:197–203

Padial JM, Miralles A, Riva IDL, Vences M (2010) The integrative future of taxonomy. Front Zool 7:16

Peay K, Bruns T, Kennedy P, Bergemann S, Garbelotto M (2007) A strong species-area relationship for eukaryotic soil microbes: island size matters for ectomycorrhizal fungi. Ecol Lett 10:470–480

Petersen RH, Hughes KW (2009) Names, names, names: when nomenclature meets molecules. McIlvainea 18(1):22–28

Taylor JW (2011) One fungus = one name: DNA and fungal nomenclature twenty years after PCR. International Mycological Association 2(2):113–120

Taylor JW, Jacobson DJ, Kroken S, Kasuga T, Geiser DM, Hibbett DS, Fisher MC (2000) Phylogenetic species recognition and species concepts in fungi. Fungal Genet Biol 31:21–32

Taylor JW, Turner E, Townsend JP, Dettman JR, Jacobson D (2006) Eukaryotic microbes, species recognition and the geographic limits of species: examples from the kingdom Fungi. Phil Trans R Soc B Biol Sci 361:1947–1963

Tedersoo L, Jairus T, Horton BM, Abarenkov K, Suvi T, Saar I, Koljalg U (2008) Strong host preference of ectomycorrhizal fungi in a Tasmanian wet sclerophyll forest as revealed by DNA barcoding and taxon-specific primers. New Phytol 180:479–490

Telle S, Shivas RG, Ryley MJ, Thines M (2011) Molecular phylogenetic analysis of Peronosclerospora (Oomycetes) reveals cryptic species and genetically distinct species parasitic to maize. Eur J Plant Pathol 130:521–528. doi:10.1007/s10658-011-9772-8

Wiens JJ (2007) Species delimitation: new approaches for discovering diversity. Syst Biol 56 (6):875–878

Part II
Conditions Favoring Mycorrhiza Formation

Chapter 4
Soil Properties

Eleonora Bonifacio and Asunción Morte

4.1 Introduction

Little is known about the properties of the soils of desert truffles and even the term desert soils is probably not the best to encompass the wide range of properties they show. Desert truffles are reported to occur in Europe, Africa, the Middle East, China and Australia, although with different species. In many cases they do occur in sandy soils, matching therefore the common idea of deserts: an endless, sandy undulating plain, with dunes in the foreground. In Europe and often in Northern Africa, however, the areas where desert truffles are found belong to the Mediterranean type of climate; sandy soils are not the rule anymore, and a much wider range of soil properties is present. Drylands and Mediterranean areas share a low amount of annual precipitation, but they differ in rainfall distribution over the year. As a consequence, these two climatic areas differ in the length of the plant growing season: drylands have a growing season of 0–103 days, while the typical growing season in the Mediterranean ranges from 67 to 250 days (Watson 1992).

Climatic factors are among the most important drivers of desert truffle occurrence (e.g. Kagan-Zur and Roth-Bejerano 2008; Trappe et al. 2008a, b), and indeed the effect of sudden rainfall after a dry period is perceived to be one of the most important factors in the appearance of truffles (Mandeel and Al-Laith 2007). Climate is, however, also one of the factors of soil formation; its importance was recognized by Jenny (1941) in his state function that is still at present the basis to understand world soil distribution (e.g. Schaetzl and Anderson 2005). Because of the great effect water availability has on soil development, desert truffles may therefore occur both in very poorly developed soils, those that have inherited

E. Bonifacio (✉)
Università di Torino, DISAFA, via L. da Vinci 44, 10095 Grugliasco (TO), Italy
e-mail: eleonora.bonifacio@unito.it

A. Morte
Departamento de Biología Vegetal, Universidad de Murcia, Campus de Espinardo, 30100 Murcia, Spain

V. Kagan-Zur et al. (eds.), *Desert Truffles*, Soil Biology 38,
DOI 10.1007/978-3-642-40096-4_4, © Springer-Verlag Berlin Heidelberg 2014

most of their characteristics from the parent material, as the typical dryland ones, and on the contrary may grow in soils whose characteristics are the result of alternating relatively moist and dry periods.

4.2 Soils of Drylands

Desert soils are the soils associated with the climatic area of the deserts in Köppen-Geiger classification, an area, indicated by the B letter, where annual precipitation is always below a certain dryness threshold, which in turn depends on precipitation distribution (Kottek et al. 2006). Desert truffles occur in the Bwh subtype, characterized by annual temperature above 18 °C. The soil moisture regime is generally aridic in these areas indicating therefore that the soil moisture control section is dry for more than half of the time when the air temperature is above 5 °C and is moist for less than 90 consecutive days when soil temperature is above 8 °C (Soil Survey Staff 2010). This lack of water is not only important for plant growth, motivating the scarce vegetation of deserts and drylands, but also for soil development. Poorly degradable quartz is abundant in the soil parent material. Weathering, thus, occurs only to a very limited extent, and leaching of the cations released by the dissolution of minerals is only partial. As a consequence, the soil solution easily becomes oversaturated for even the most soluble salts that can therefore precipitate giving rise to some of the most typical features of the soils of these areas. These soils may contain carbonates and gypsum but also other minerals such as halite, mirabilite and thenardite which are sodium and magnesium chlorides and sulphates that are not normally found in other environments (Driessen et al. 2001). Depending, thus, on the amounts of rainfall and on the presence of sources of ions for salt precipitation, soils may range from Arenosols (sandy soils), to other more specific types such as Calcisols (containing pedogenic carbonates), Gypsisols (with gypsum) and Solonchaks (precipitation of other salts, more soluble than gypsum). Calcisols, Gypsisols and Solonchaks are very often classified within the Aridisol order of the USDA Soil Taxonomy. In rocky deserts, Arenosols are not present and the most common soil types are Regosols, the typical group of soils with almost no development, or Leptosols when soil thickness is very limited.

According to the WRB definition (ISSS 1998) Arenosols have more than 70 % sand (Ø 0.05–2 mm) and less than 15 % clay; they broadly correspond to the Psamment suborder of the USDA Soil Taxonomy. Arenosols are the most extensive soils in the world (Driessen et al. 2001) and are often associated with areas of sand dunes. Pedogenesis cannot occur until the dune is colonized by vegetation and held in place, but, even when shifting has stopped, inputs of organic matter are limited and mineralization proceeds rapidly, reducing the incorporation of organic material in the topsoil horizon. A proper structure cannot develop, and Arenosols have a sandy A horizon with single grain or poorly developed crumb structure over one or more C horizon. Most of the soil pores are large and they easily lose water. Because of the paucity in clay and organic matter, these soils also have small pools of

Table 4.1 General properties of desert truffles soils (A horizon) from Namibia and Botswana

	n	Mean	Standard deviation	Minimum	Maximum
pH	30	6.0	0.7	5.1	7.6
Organic matter (g kg^{-1})	21	6.24	4.99	2.92	22.4
C/N	19	10.6	3.5	6.5	23
Clay (%)	30	2.5	1.1	0.3	5.0
Silt (%)	30	1.8	1.0	0.1	3.8
Sand (%)	30	95.8	1.5	92.7	98.8
$CaCO_3$ (g kg^{-1})	30	3.3	6.7	0.0	31.0

nutrients. On one hand nitrogen is limited because it is strongly linked to the organic matter biogeochemical cycle; on the other Ca, K and Mg are either not available as still included in the mineral structures or poorly retained due to the lack of sorbing surfaces. Cation exchange capacity is indeed extremely low in these soils and 1–3 $cmol_c$ kg^{-1} is the rule (Santoni et al. 2001). Phosphorus becomes a limiting factor later on with soil development (Wardle et al. 2004) and is therefore rarely limiting in Arenosols. Even if the total P concentrations are low when compared to other soil types, the available P pool may form from 1 to 8 % of total P, and the lowest availability is caused by the presence of carbonates (Bonifacio et al. 1998b).

In Southern Africa, one of the area where desert truffles are common (Trappe et al. 2008a), Arenosols are widely spread and cover 169 million ha, mainly in Angola and Botswana but also in Zimbabwe, South Africa and Mozambique (Hartemink and Huting 2008). Differences in chemical properties among the Arenosols of Southern Africa were found by Hartemink and Huting (2008): the average sand contents are lower in Mozambique than in Namibia and Botswana, and, consequently the former show a higher amount of C and nutrients; in addition, Arenosols in Angola and Zimbabwe have a lower pH, below 6, than those located in the other countries (pH between 6 and 7).

Arenosols of Botswana and Namibia are well known for the occurrence of desert truffles (Taylor et al. 1995), and the chemical properties of some truffle sites in these two countries are reported in Table 4.1. In addition to the data obtained by Taylor et al. (1995) from several places around the Kalahari Desert, the characteristics of top A horizons (about 10 cm on the average) of soils taken in the areas of Serowe and Shakawe (Botswana) and Kanovlei and Hamoye (Namibia) are reported. Soil properties do not differ from those of other sandy soils of the same areas (Jamagne 1990; 1998a; Bonifacio et al. 2000) and well match the expected properties of Arenosols: they always contain more than 90 % sand and have a very low concentration of organic matter, which is easily lost by mineralization, as indicated by the low C to N ratio. The pH is more variable, from 5 up to more than 7, due to the presence of carbonates in some of the areas covered by the lithostratigraphic unit of Kalahari bed (Carney et al. 1994). The pH does not seem therefore to represent a limiting factor for the development of desert truffles in these countries.

Table 4.2 Nutrient concentration in the A horizons of desert truffles soils (A horizon) from Namibia and Botswana

	n	Mean	Standard deviation	Minimum	Maximum
Total N (g kg^{-1})	19	0.37	0.26	0.10	1.10
Available P (mg kg^{-1})	18	1.8	1.9	0.4	6.4
Exchangeable Ca (mg kg^{-1})	21	267	233	66	912
Exchangeable Mg (mg kg^{-1})	21	41.0	29.9	9.6	110.4
Exchangeable K (mg kg^{-1})	21	53	40	8	183
Exchangeable Na (mg kg^{-1})	21	33.2	18.1	2.3	85.1

Nutrients necessary for plant development are rather low in these soils: total N ranged from 0.1 to 1 g kg^{-1}, available P (Olsen P) is below 2 mg kg^{-1}, available Mg and K are on the average 50 and 40 mg kg^{-1}, respectively, while Ca is higher, about 200 mg kg^{-1} (Table 4.2). N is assumed to be the main factor limiting plant growth in desert environments while the amount of P is often sufficient (Day and Ludeke 1993). This situation is rather normal in poorly developed soils (Wardle et al. 2004): P is readily available to plants and organisms thanks to the dissolution of apatite minerals present in the soil parent material; organic compounds form instead the main pool of soil N, thus the building up of N reservoirs is closely linked to vegetation development. Mycorrhizae are able to solubilize P from minerals (e.g. Berthelin et al. 1991) and are extremely efficient in taking up P from the soil solution (Plassard and Dell 2010); therefore the amounts of P which is available to plants (Olsen P) may not be the best indicator of P-fertility for desert truffle soils. In Botswana, available P concentration in soils was found to vary from North to South along the Kalahari Transect, i.e. with decreasing rainfall, ranging from about 5 mg kg^{-1} in Pandamatenga and Tshane to 10–15 in Ghanzi (O'Halloran et al. 2010). The data reported in Table 4.2 show that even lower amounts are common. Total P concentration is extremely low as well: total P was measured only in two of these soil samples, but it was 32 and 46 mg kg^{-1}, thus in good agreement with the scarce P availability. Bohrer et al (2003) found a significant depletion of available P from some Botswana soils upon seedling inoculation with VAM and explained their results with an increased ability of the VAM-infected roots to absorb P from the available pool. The low P availability of desert truffle soils might therefore also represent a consequence of mycorrhiza presence, but to the best of our knowledge, no specific comparison between soils has been carried out in the field.

4.3 Soils of Mediterranean Environments and of the Middle East Regions

Areas with a Mediterranean type of climate take their name from the Mediterranean sea geographical region but are not restricted to Southern Europe, Northern Africa and the Middle East as they do occur also in America (e.g. California and Chile) and

Southern Australia (Kottek et al. 2006). All these areas have a Cs type of climate according to Köppen-Geiger classification, meaning they have a warm temperate climate with dry summers. A warm climate is defined from the mean temperature of the coldest month, which should be above −3 °C but below 18 °C, while the lowest monthly summer precipitation is below 40 mm. Precipitation may however be abundant during the winter half-year as the maximum monthly winter precipitation is at least three times more than the summer monthly minimum (Kottek et al. 2006). As a consequence, soils in Mediterranean environments may be rather well developed as water is available for cation leaching and pedogenic processes at a time of the year when competition for water by plants is lower and, particularly, when evaporation from the soil surface occurs to a lesser extent. These are some of the characteristics that define the xeric soil moisture regime: the moisture control section is dry in all parts for 45 or more consecutive days in the 4 months following the summer solstice and moist in all parts for 45 or more consecutive days in the 4 months following the winter solstice (Soil survey Staff 2010). Depending therefore on the absolute amounts of water, precipitation of carbonate and gypsum may occur at some depth in the profile or may not be present at all. As a consequence, Calcisols and Gypsisols may be present, but Luvisols and sometimes Alisols occur as well. In these two soil types, leaching was sufficient to decrease the cation concentration below the threshold for clay flocculation, and therefore clay translocation at depth has occurred. The clay enriched deep horizon roughly corresponds to the lower limit of cation leaching; Alisols differ from the most common Luvisols as they are more acidic and have a lower base saturation. These two soil groups approximately correspond to the Ultisol and Alfisol orders of the USDA Soil Taxonomy, and the clay enriched horizon must have at least 20 % more clay than the clay-depleted one (Soil Survey Staff 2010). Soil distribution in Mediterranean environments is however not only shaped by climate, as other pedogenic factors, such as time and relief, need to be taken into account as well. Time is very important as Mediterranean environments are generally young from the geological point of view and the soils located on the relatively older surfaces often reflect previous climatic conditions. In addition, the highly corrugated surfaces typical of some Mediterranean areas, combined with the frequent occurrence of heavy rainfall on dry soils, lead to soil rejuvenation, and Cambisols, Regosols and Leptosols are frequently found. Arenosols are also present, mainly along the coasts, although in some cases sandy soils develop on dunes that originated during the Pleistocene glaciations, as, for example, the soil described by Catoni et al. (2012) in North-western Italy.

Soil characteristics in the Middle East region may for practical reasons be considered as intermediate between those of Mediterranean environments and of drylands, with a gradient of increasing aridity going from the coasts inland. Aridic moisture regime is indeed the prevailing one in the Middle East (Khademi and Mermut 2003), and several suborders of the USDA Soil Taxonomy Aridisol order are found. In Israel, for example, Haplargids, Calciorthids, Camborthids and even Salorthids are common (http://cals.arizona.edu/OALS/soils/israel/israel.html), as well as in Iran (Mojiri et al. 2011). In Saudi Arabia soils are sandy, with rather

Table 4.3 General characteristics of desert truffle soils from the Mediterranean (MED) and the Middle East (MEAST) areas

		n	Mean	Standard deviation	Minimum	Maximum
pH	MED	16	7.1	1.1	5.0	8.6
	MEAST	5	8.3	0.6	7.6	9.0
Organic matter (g kg^{-1})	MED	16	15.3	8.1	4.1	34.5
	MEAST	5	10.3	10.2	1.2	25.9
Clay (%)	MED	14	16.0	12.4	2.6	43.4
	MEAST	2	3.8	0.2	3.7	4.0
Silt (%)	MED	14	25.5	13.6	4.4	43.4
	MEAST	2	3.2	1.9	1.9	4.6
Sand (%)	MED	14	58.4	21.1	29.6	92.1
	MEAST	2	32.9	1.7	91.7	94.1
$CaCO_3$ (g kg^{-1})	MED	14	222.0	260.1	0.0	691.0
	MEAST	2	28.0	2.8	26.0	30.0

Table compiled using the data reported by Hashem and Al-Obaid (1996), Ammarellou et al. 2007; Akyuz et al. (2012), Slama et al. 2010, in addition to Authors' data

large amounts of coarse fragments and often contain calcite and gypsum (Ehlen and Ponrder Henley 1991).

Occurrence of desert truffles has been reported in almost all countries around the Mediterranean basin and in the Middle East, and Diez et al. (2002) tried to group them not only from the phylogenetic point of view but also based on some climatic and soil characteristics of the area of origin. They found that several species were taken in both acid and basic soils and suggested that their tolerance to soil conditions might just reflect the wide edaphic tolerance of the *Helianthemum* hosts. Desert truffles also occur in Italy where, however, they are of very little commercial interest due to the presence of the highly valued *Tuber* species. The same should hold also for France: Fortas and Chevalier (1992) reported the characteristics of a "Terfezia soil" taken close to Les Baux de Provence in Southern France, but again the interest for desert truffles is extremely low. Only in a few cases, however, reports on desert truffle occurrence are accompanied by a complete characterization of soil properties, more frequently soil is defined very broadly as an acid or basic sandy soil. Indeed the variation in desert truffle soils of the Mediterranean area and the Middle East is much greater than a change in pH, as shown in Table 4.3. The few samples from the Middle East are extremely sandy Arenosols, similar to those of other drylands, and show a pH that may be as high as 9, but the range of properties of the Mediterranean desert truffle soils is much wider. The pH ranges from 5 to more than 8, clay content may be as high as 40 %, and carbonates may be absent or instead form almost 70 % of the fine earth fraction (Table 4.3). The samples taken in Italy ($n = 2$) correspond to non-calcareous Arenosols, the two samples from Tunisia (Slama et al. 2010) have a high pH (i.e. 7.1 and 7.7), and one of them is defined as gypsiferous, suggesting therefore a high electrical conductivity, but from the reported data it is unclear if it is a Gypsisol; those from Morocco (Bonifacio et al. 1998b) comprise a non-calcareous

Table 4.4 Nutrient concentration in the A horizons of desert truffles soils from Mediterranean areas

	n	Mean	Standard deviation	Minimum	Maximum
Total N (g kg^{-1})	10	1.32	0.56	0.70	2.20
Total P (mg kg^{-1})	6	232	205	50	575
Available P (mg kg^{-1})	10	23.7	23.0	0	76
Exchangeable Ca (mg kg^{-1})	12	3,049	4,228	128	14,924
Exchangeable Mg (mg kg^{-1})	13	84.4	57.2	20.4	198.0
Exchangeable K (mg kg^{-1})	15	156.1	120.9	18.8	429.0
Exchangeable Na (mg kg^{-1})	13	131.9	112.2	25.3	306.0

Arenosol and a moderately calcareous soil (e.g. 6 % $CaCO_3$) with about 20 % clay and relatively high concentration of organic matter (26 g kg^{-1}). The situation is extremely variable also when nutrient contents are taken into account (Table 4.4).

In general, element availability in the soils of the Mediterranean areas is much higher than in Arenosols; thanks to the larger presence of clay and organic matter, which both provide sites for cation sorption, exchangeable cations are two to three times more concentrated than in Arenosols. The abundance of available Ca (ten times more on the average than in Southern African desert truffle soils) is linked to the presence of carbonates that add soluble Ca to the exchangeable pool. Indirectly, the presence of carbonates may be linked also to the total P pool; in moderately developed soils, P is still present in primary minerals such as apatites (Ca-phosphates), while if Calcisols have developed and pedogenic Ca carbonates have formed, these latter may act as a sink for phosphate sorption and precipitation (e.g. Buckingham et al. 2010). P availability in desert truffle soils from Mediterranean environments is around 20 mg kg^{-1} on the average (Table 4.4), a concentration which is considered to be sufficient even for agricultural crops (Matar et al. 1992). P deficiency deeply affects plant development, and in vitro studies (Fortas and Chevalier 1992) have shown that inoculation with *Terfezia* has a positive influence on the growth of *Helianthemum guttatum*; on the other hand, they did not find any significant effect of P availability on mycorrhization.

A more systematic survey of desert truffle soils is available for Spanish sites of different regions: Fig. 4.1 indicates the location of the sampling sites, together with the dominant soil type, taken from the soil maps of Andalusia (Mudarra et al. 1988) and of Spain (Instituto Nacional de Edafologia y Agrobiologia C.S.I.C. 1966). The wide variability in chemical properties is confirmed also in these cases. Even within a relatively more homogenous geographical area, such as Southern and Southwestern Spain, the pH ranges from about 6 to more than 8, depending on the presence of pedogenic carbonates and on the type of parent material: calcareous or metamorphic rocks with low carbonate contents. As a consequence, carbonates in soil may vary between 34 g kg^{-1} to about 700 g kg^{-1} (Table 4.5). The variations in carbonate contents and in pH affect species occurrence: Morte et al. (2009) evidenced in fact that distribution of several species of desert truffles belonging to the *Terfezia* genus in Spain are related to soil pH; *T. claveryi* and *T. boudieri* are the

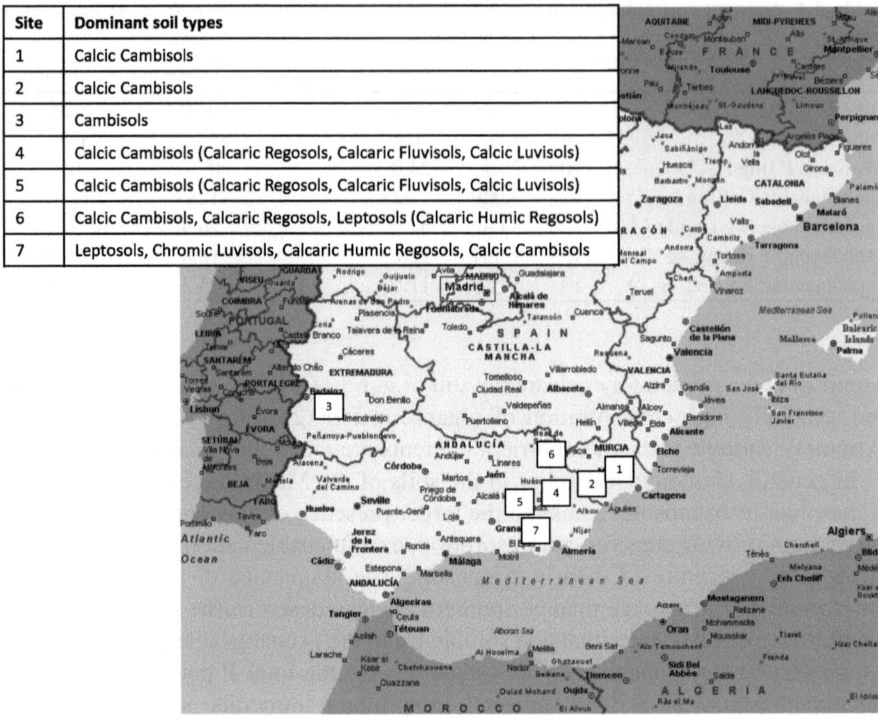

Fig. 4.1 Location of some desert truffle sites in Spain and dominant soil types in the areas

Table 4.5 Soil types and selected chemical properties from desert truffle sites in Southern Spain

Site[a]	Location	pH	$CaCO_3$ (g kg^{-1})	OM[b] (g kg^{-1})	C/N	Olsen P (mg kg^{-1})	Sand (%)	Clay (%)
1	Corvera	7.6	691	12.0	7	0	35.2	29.7
2	Zarzadilla de Totana	7.2	440	7.0	6	0	43.3	13.5
3	Merida	6.2	34	11.0	8	38.7	49.6	10.5
4	Baza	7.6	204	6.0	–	4.9	63.3	11.2
5	Guadix	7.8	643	21.8	8	45.5	29.6	39.0
6	Puebla de Don Fadrique	7.8	535	22.5	7	29.6	37.0	40.1
7	Fuente Alta	8.6	391	19.0	5	22.9	68.8	10.4

[a]Numbers refer to Fig. 4.1; [b]OM: organic matter

most common species in calcareous soils, while *T. arenaria* and *T. leptoderma* prefer acidic environments. The contents of organic matter in desert truffle soils of Spanish environments reflect the abundance of vegetation and the outputs of litter induced by erosion. No differences in organic matter transformations are instead visible, as shown by the C to N ratio, which is always very low (Table 4.5) because of the high mineralization rate typical of Mediterranean environments (e.g. Garcia and Hernandez 1996; Kurganova et al. 2012). Texture varies from clay to sandy in sample 6 and 7, respectively, and the highest sand contents are in the poorly developed Regosols of the area around Granada (Fig. 4.1). The clay-richest soils (sample 5 and 6) are also those with the highest concentration of organic matter; the negative effects that large clay contents have on soil aeration may therefore be partially compensated by the improvement of soil structure induced by organic matter (Bronick and Lal 2005). The lowest P availability is found where the concentration of organic matter is the lowest, suggesting an effect of the organic pool and, if only the carbonate-rich samples are considered; P availability is indeed highly correlated to organic matter ($r_P = 0.886$).

4.4 Conclusions

The soils of desert truffles show a remarkable variability that reflects the climatic conditions in which they form. At a very general level therefore desert truffles (or their hosts) seem to be able to adapt to a wide range of soil pH, edaphic conditions and texture. In arid environments, poor fertility conditions linked to the sandy texture and low inputs of organic matter are the rule, but even in these relatively homogeneous soils, desert truffles are found both in acidic and in basic situations. They seem also well adapted to soils with high electrical conductivity, such as those containing gypsum and more soluble salts. In Mediterranean environments, soil variability is enhanced, and desert truffles are able to grow both in well-aerated sandy soils and heavy clay-rich ones, with or without carbonates, at high or relatively low pH, in poor fertility conditions or instead where the availability of nutrients is more than acceptable. The occurrence of desert truffles in a wide range of situations may be related to the possibility of having several hosts or to the presence of several species of commercial interest within the *Terfezia* genus.

Acknowledgements This work was partially supported by the projects CGL2011-29816 (Ministerio de Educación y Ciencia, Spain) and 08812/PI/08 (Fundación Séneca, Región de Murcia, Spain).

References

Akyuz M, Kirbag S, Kursat M (2012) Ecological aspects of the arid and semi-arid truffle in Turkey: evaluation of soil characteristics, morphology, distribution and mycorrhizal relationships. Turk J Bot 36:386–391

Ammarellou A, Saremi H, Gucin F (2007) Evaluation of morphology, citology and mycorrhizal relationships of desert truffles (*Terfeziaboudieri*) in Iran. Pak J Biol Sci 10:1486–1490

Berthelin J, Leyval C, Laheurte F, De-Giudici P (1991) Involvement of roots and rhizosphere microflora in the chemical weathering of soil minerals. In: Atkinson D (ed) Plant root growth. An ecological perspective. British Ecological Society Special Publication, London
Bohrer G, Kagan-Zur V, Roth-Bejerano N, Ward D, Beck G, Bonifacio E (2003) Effects of different Kalahari-desert VA mycorrhizal communities on mineral acquisition and depletion from the soil by host plants. J Arid Envir 55:193–208
Bonifacio E, Mateke SM, Taylor FW, Zanini E (1998) Native soils of some wild fruit trees in the veld of Eastern Botswana. Proceedings of 16th World Congress of Soil Science, Montpellier, 20–26 August 1998, http://natres.psu.ac.th/Link/SoilCongress/bdd/symp27/2112-t.pdf. Accessed 5 Nov 2012
Bonifacio E, Scalenghe R, Barberis E, Zanini E (1998b) Phosphorus status in some natural desert truffle stands. Commun Soil Sci Plant Anal 29:87–96
Bonifacio E, Santoni S, Zanini E (2000) Soil properties required by some Southern Africa fruit trees as assessed by discriminant analysis. Arid Soil Res Rehab 14:253–263
Bronick CJ, Lal R (2005) Soil structure and management: a review. Geoderma 12:3–32
Buckingham SE, Neff J, Titiz-Maybach B, Reynolds RL (2010) Chemical and textural controls on phosphorus mobility in drylands of southeastern Utah. Biogeochemistry 100:105–120
Carney JN, Aldiss DT, Lock NP (1994) The geology of Botswana. Bulletin no. 37, Geological Survey Department, Lobatse, Botswana
Catoni M, Falsone G, Bonifacio E (2012) Assessing the origin of carbonates in a complex soil with a suite of analytical methods. Geoderma 175–176:47–57
Day AD, Ludeke KL (1993) Plant nutrients in desert environments. Springer, Berlin
Diez J, Manjon JL, Martin F (2002) Molecular phylogeny of the mycorrhizal desert truffles (*Terfezia* and *Tirmania*), host specificity and edaphic tolerance. Mycologia 94:247–259
Driessen P, Deckers J, Spaargaren O, Nachtergaele F (2001) Lecture notes on the major soils of the world. World Soil Resources Reports no. 94. http://www.fao.org/docrep/003/Y1899E/Y1899E00.HTM. Accessed 5 Nov 2012
Ehlen J, Ponrder Henley J (1991) A comparison of soils from Twentynine Palms, California and Saudi Arabia. US Army Engineer Topographic Laboratories Technical report ETL-0583. http://www.dtic.mil/cgi-bin/GetTRDoc?AD=ADA238681. Accessed 5 Nov 2012
Fortas Z, Chevalier G (1992) Effet des conditions de culture sur la mycorhization de l'*Helianthemum guttatum* par trois espèces de terfez des genres *Terfezia* et *Tirmania* d'Algérie. Can J Bot 70:2453–2460
Garcia C, Hernandez T (1996) Organic matter in bare soils of the Mediterranean region with a semiarid climate. Arid Soil Res Rehab 10:31–41
Hartemink AE, Huting J (2008) Land cover, extent, and properties of arenosols in Southern Africa. Arid Land Res Manag 22:134–147
Hashem AR, Al-Obaid AM (1996) Mineral composition of soil and wild desert truffles in Saudi Arabia. J King Saud Univ 8:5–10
Institito Nacional de Edafologia y Agrobiologia del CSIC (1966) Mapa de suelos de Espana. http://eusoils.jrc.ec.europa.eu/EuDASM/ES/sp_x1.jpg. Accessed 5 Nov 2012
ISSS Working Group RB (1998) World reference base for soil resources: introduction. In: Deckers JA, Nachtergaele FO, Spaargarren OC (eds) World reference base for soil resources, 1st edn. Acco, Leuven, International Soil Science Society (ISSS), International Soil Reference and Information Centre (ISRIC) and Food and Agriculture Organization of the United Nations (FAO)
Jamagne P (1990) Soils and land suitability of North West Ngamiland. FAO/UNDP/Government of Botswana Soil Mapping and Advisory Services Project AG: BOT/85/011. Field Document 26. Ministry of Agriculture, Gaborone
Jenny H (1941) Factors of soil formation. McGraw-Hill, New York, NY
Khademi H, Mermut AR (2003) Micromorphology and classification of Argids and associated gypsiferous aridisols from central Iran. Catena 54:439–455
Kagan-Zur V, Roth-Bejerano N (2008) Desert Truffles. Fungi 1:32–37

Kottek M, Grieser J, Beck C, Rudolf B, Rubel F (2006) World map of the Koeppen-Geiger classification updated. Meteorol Z 15:259–263

Kurganova IN, de Gerenyu VOL, Lancho JFG, Oehm CT (2012) Evaluation of the rates of soil organic matter mineralization in forest ecosystems of temperate continental, Mediterranean and tropical monsoon climates. Eura Soil Sci 45:68–79

Mandeel QA, Al-Laith AAA (2007) Ethnomycological aspects of the desert truffle among native Bahraini and non-Bahraini peoples of the Kingdom of Bahrain. J Ethnopharmacol 110:118–129

Matar A, Torrent J, Ryan J (1992) Soil and fertilizer phosphorus and crop responses in the dryland Mediterranean zone. Adv Soil Sci 18:81–146

Mojiri A, Honarjoo N, Jalalian A (2011) Comparison between keys to soil taxonomy and WRB to classification of soils in Segzi Plain (Iran). J Appl Sci 11:579–583

Morte A, Zamora M, Gutierrez A, Honrubia M (2009) Desert truffle cultivation in semiarid Mediterranean areas. In: Azcon-Aguilar C et al (eds) Mycorrhizas: functional processes and ecological impact. Springer, New York, NY

Mudarra JL, Barahona E, Banos C, Iriarte A, Santos F (1988) Mapa de suelos de Andalucia. http://eusoils.jrc.ec.europa.eu/EuDASM/ES/port_x24.jpg. Accessed 5 Nov 2012

O'Halloran LR, Shugart HH, Wang L, Caylor KK, Ringrose S, Kgope B (2010) Nutrient limitations on aboveground grass production in four savanna types along the Kalahari Transect. J Arid Environ 74:284–290

Plassard C, Dell B (2010) Phosphorus nutrition of mycorrhizal trees. Tree Physiol 30:1129–1139

Santoni S, Bonifacio E, Zanini E (2001) Indophenol blue colorimetric method for measuring cation exchange capacity in sandy soils. Commun Soil Sci Plant Anal 32:2519–2530

Schaetzl R, Anderson S (2005) Soils: genesis and geomorphology. Cambridge University Press, New York, NY

Slama A, Fortas Z, Boudabous A, Neffati M (2010) Cultivation of an edible desert truffle (*Terfeziaboudieri* Chatin). Afr J Microbiol Res 4:2350–2356

Soil Survey Staff (2010) Keys to soil taxonomy, 11th edn. USDA-Natural Resources Conservation Service, Washington, DC

Taylor FW, Thamage DM, Roth-Bejerano N, Kagan-Zur V (1995) Notes on the Kalahari desert truffle, *Terfeziapfeilii*. Mycol Res 7:874–878

Trappe JM, Claridge AW, Arora D, Smit WA (2008a) Desert truffles of the African Kalahari: ecology, ethnomycology, and taxonomy. Econ Bot 62:521–529

Trappe JM, Claridge AW, Claridge DL, Liddle L (2008b) Desert truffles of the Australian outback: ecology, ethnomycology, and taxonomy. Econ Bot 62:497–506

Wardle DA, Walker LR, Bardgett RD (2004) Ecosystem properties and forest decline in contrasting long-term chronosequences. Science 305:509–513

Watson A (1992) Desert soils. In: Martini IP, Chesworth W (eds) Weathering soils and paleosoils. Developments in earth surface processes. Elsevier, Amsterdam

Chapter 5
Types of Mycorrhizal Association

Nurit Roth-Bejerano, Alfonso Navarro-Ródenas, and Almudena Gutiérrez

5.1 Introduction

A mycorrhiza is a symbiotic, generally mutualistic association between a fungus and the roots of a vascular plant (Kirk et al. 2001) in which each benefits from the other: the plant supplies the fungus with photosynthetic products such as sugars, and the fungus supplies the plant with minerals and water (Smith and Read 2008). In such associations the fungal hyphae may develop either intracellularly inside the plant cortex cells, as in arbuscular mycorrhizas, or intercellularly around the cortex cells, as in ectomycorrhizas (Gutiérrez et al. 2003). The fungi that produce arbuscular mycorrhizas belong to the division Glomeromycota, while ectomycorrhiza-forming fungi may belong either to the Ascomycota or to the Basidiomycota. Truffles are mycorrhizal fungi belonging to the Ascomycota (Smith and Read 2008).

It has been established that ectomycorrhizal fungi are capable of producing endomycorrhizas with underdeveloped intracellular hyphal structures, as well as ectendomycorrhizas with both intra- and intercellular hyphae, in response to particular mineral levels (Fortas and Chevalier 1992; Gutiérrez et al. 2003); the same phenomenon has been observed in certain fungal mutants (Gea et al. 1994; Gay et al. 1994).

Desert truffles, as the name indicates, are mycorrhizal fungi that grow almost exclusively in semiarid and arid zones (Kagan-Zur 2001). They are now classed in

N. Roth-Bejerano (✉)
Department of Life Sciences, Ben-Gurion University of the Negev, Beer Sheva 84105, Israel
e-mail: nuritrb@bgu.ac.il

A. Navarro-Ródenas
Servicio de Experimentación Agroforestal, Edificio CAID, Universidad de Murcia, Campus de Espinardo, 30100 Murcia, Spain

A. Gutiérrez
Departamento de Biología Vegetál, Facultad de Biología, Universidad de Murcia, Campus de Espinardo, 30100 Murcia, Spain

V. Kagan-Zur et al. (eds.), *Desert Truffles*, Soil Biology 38,
DOI 10.1007/978-3-642-40096-4_5,

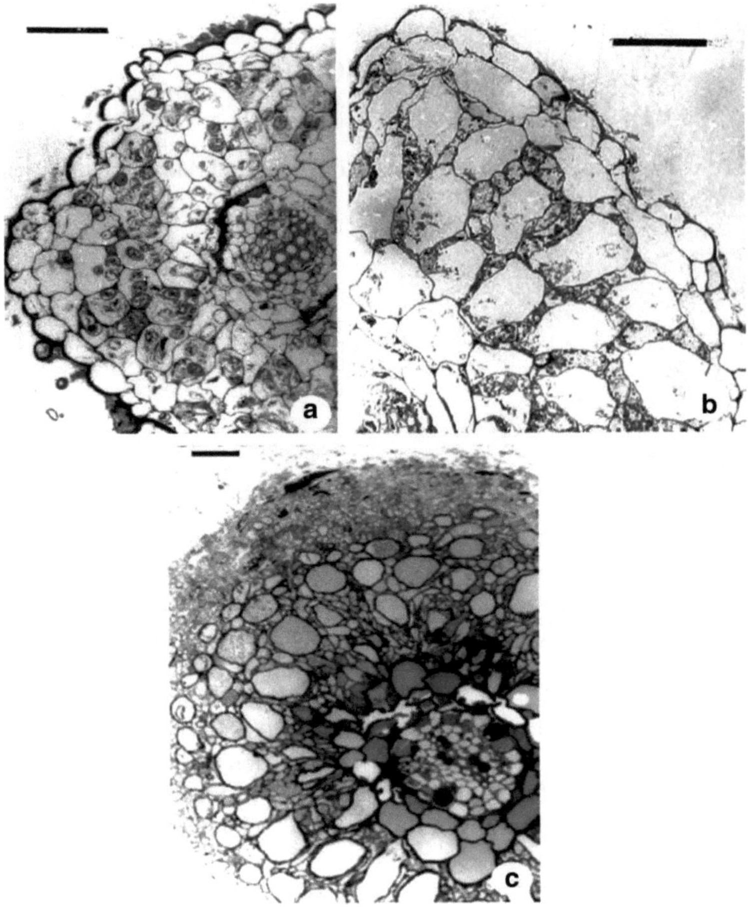

Fig. 5.1 Mycorrhizal roots of *Helianthemum almeriense* inoculated with *Terfezia claveryi*. (**a**) Collected from the field—endomycorrhiza with intracellular hyphae only. (**b**) Grown in greenhouse—ectomycorrhiza with Hartig net without mantle, (**c**) grown in vitro—ectomycorrhiza with a developed Hartig net and with mantle. From Gutiérrez et al. 2003, with kind permission of Springer Science and Business Media

six different Pezizales families (Lassoe and Hansen 2007). The most important of these from the standpoint of edible desert truffles is the Pezizaceae family, which includes, among others, the genera *Terfezia* and *Tirmania* (Awameh 1981; Alsheikh and Trappe 1983; Moreno et al. 2000, 2002; Díez et al. 2002; Kovács et al. 2011). Members of these genera colonize local shrubs or annuals, mainly species of *Helianthemum* (*Cistaceae*). The Pezizaceae also include the genera *Mattirrolomyces Terfezioides* (Lassoe and Hansen 2007), *Kalaharituber* and *Eremiomyces* (Ferdman et al. 2005), which form associations with acacia trees in the Kalahari Desert; there are no *Cistaceae* in the Kalahari (Riley 1963).

Three types of mycorrhiza are distinguished among desert truffles:

- Ectomycorrhizas, characterized by an intercellular network of hyphae—Hartig net—surrounding but not penetrating the cortex cells of plant roots; see Fig. 5.1b, c (Gutiérrez et al. 2003; Roth-Bejerano et al. 1990; Dexheimer et al. 1985).
- Endomycorrhizas, characterized by undifferentiated coil-shaped or globular intracellular hyphae penetrating the plant cells (Fig. 5.1a) (Gutiérrez et al. 2003; Awameh 1981; Kagan-Zur et al. 1999; Slama et al. 2010).

  Interestingly, endomycorrhizas predominate in Kuwait, the Kalahari, and the northern Sahara (Awameh 1981; Kagan-Zur et al. 1999). Moreover, Slama et al. (2010) report that in the northern Sahara the *Helianthemum sessiliflorum*–-*Terfezia boudieri* couple produces endomycorrhizas, while in Israel the same partners form ectomycorrhizas (Roth-Bejerano et al. 1990, Zaretsky et al. 2006).
- Ectendomycorrhizas, characterized by the presence of both intercellular Hartig net and intracellular hyphae penetrating the cortex cells (Navarro-Ródenas et al. 2012, 2013). It should be mentioned that in some instances more than one of the above mycorrhizal types may be observed along the root system of a single plant; Navarro-Ródenas et al. (2012) refer to this phenomenon as an "ectendomycorrhiza continuum."

5.2 Factors Affecting the Type of Mycorrhiza Formed

Dexheimer et al. (1985) found that inoculation of *Helianthemum salicifolium* with *Terfezia claveryi* led to the formation of an endomycorrhitic association characterized by intracellular hyphae surrounded by plasmalemma, whereas infection of the same host with *T. leptoderma* produced an ectomycorrhiza without mantle and with a typical Hartig net. No explanation was offered; however, these results appear to suggest that it is the fungus that determines whether the association will be endo- or ectomycorrhizal.

It has been demonstrated, on the one hand, that application of auxin alters root morphology and leads to the development of structures resembling mycorrhitic roots, (Barker and Tagu 2000); and, on the other hand, it is known that several mycorrhitic fungi secrete auxin (Barosso et al. 1978; Gogala 1991; Gay et al. 1994; Zaretsky et al. 2006; Turgeman Ph.D. thesis 2013). These findings inspired the notion that the development of mycorrhizas is founded on a dialog between the two partners—the auxin-secreting fungus and the plant—which secretes zeatin and hypaphorine (Barker and Tagu 2000; Martin et al. 2001). In line with this notion, the occurrence of different mycorrhizal types could be related to variations in the amount of auxin produced by the fungi (Gay et al. 1994; Gea et al. 1994; Zaretsky et al. 2006; Turgeman Ph.D. thesis 2013) (see Table 5.1).

Table 5.1 Endogenous content of IAA in two clones of *Cistus incanus* and two isolates of *Terfezia boudieri* and IAA excretion of the two isolates

	Clone/isolate	Free IAA ($\mu g\ g^{-1}$ fw)	Conjugated IAA (IAA-Asp) ($\mu g\ g^{-1}$ fw)
Endogenous IAA	Clone M-2	0.121 ± 0.011	0.072 ± 0.045
	Clone W-51	0.101 ± 0.009	0.076 ± 0.039
	Isolate 42a	0.064 ± 0.020	0.391 ± 0.140
	Isolate 27	0.256 ± 0.020	$0.035 \pm 1 \times 10^{-4}$
IAA excretion	Isolate 42a	6.700 ± 1.140	14.600 ± 6.279
	Isolate 27	38.460 ± 7.950	538.900 ± 249.209

Values are the mean of at least three replicates ± standard error. From Zaretsky et al. (2006), with kind permission of Springer Science and Business Media

5.2.1 The Effect of Auxin

Gay et al. (1994) reported that an auxin-overproducing mutant of the fungus *Hebeloma cylindrosporum* showed enhanced mycorrhization. Moreover, Gea et al. (1994) demonstrated that while the wild type formed regular ectomycorrhizas (Hartig net with mantle), the mutant produced a mantle and a thick Hartig net with intracellular hyphae penetrating the host cell wall. These were surrounded by host cell plasmalemma, in some cases with a narrow strip of host cell cytoplasm and with the host cell tonoplast. The intracellular fungal cells did not invade the entire cell.

Based on the acid growth theory (Rayle and Cleland 1992), Gea et al. (1994) discussed the possibility that fungal auxin excretes protons into the cell wall, lowering its pH and endowing it with greater elasticity, which in turn facilitates fungal cell penetration into the cortical cells. The intracellular *Hebeloma* mycelia, however, did not resemble coils or globuli, recalling the *Helianthemum salicifolium–Terfezia claveryi* mycorrhizas described by Dexheimer et al. (1985). The fact that an auxin-overproducing mutant of *Hebeloma* was able to penetrate cortical cells suggests that fungal auxin plays an important role in determining mycorrhizal type.

5.2.2 The Effect of Phosphate

Fortas and Chevalier (1992) found that, depending on the fertility of the substrate, *Terfezia arenaria*, *Terfezia claveryi*, and *Tirmania pinoyi* produced two different types of association with *Helianthemum guttatum*, namely, ectomycorrhizas without mantle in high phosphate soil (about 1 mM) and endomycorrhizas, also without mantle, at lower phosphate concentrations. By contrast, Navarro-Ródenas et al. (2012) report that the level of phosphate influenced the extent of inoculation of *Helianthemum almeriense* with *Terfezia claveryi* but had little influence on the form of the mycorrhiza. It has also been suggested that ions other than phosphate, such as iron or nitrate, may affect the type of mycorrhiza formed (see Kagan-Zur et al. 2008).

5.2.3 *Auxin–Phosphate Interaction*

Evidence exists that auxin and phosphate cooperate in root development. Low phosphate was found to be associated with increased lateral root number and density and decreased primary root length (López-Bucio et al. 2002, 2003, Pérez-Torres et al. 2008); subsequent treatment of P-deprived seedlings with auxins and auxin antagonists revealed that these changes were related to an elevated auxin sensitivity of the roots (López-Bucio et al. 2002, 2003, Pérez-Torres et al. 2008).

Zaretsky et al. (2006) examined the effect of phosphate and indole-3-acetic acid (IAA) on the sensitivity of two hairy root clones of *Cistus incanus*. The results (see Fig. 5.2) showed that:

(a) The different clones displayed different degrees of sensitivity to IAA, clone M2 being more sensitive to IAA than W51.
(b) Low phosphate increased sensitivity to external IAA in both isolates.

In mycorrhization experiments performed in vitro (Wenkart et al. 2001), transformed *Cistus* roots of the same clone were paired with two different isolates of *T. boudieri* displaying different endogenous IAA levels (as shown in Table 5.1) and exposed to five concentrations of 2,4-D (2,4-dichlorophenoxyacetic acid, a synthetic auxin analog).

The results (Fig. 5.3) revealed the following:

(a) In the presence of high P (96 mg l^{-1} KH_2PO_4)—under low 2,4-D—the low IAA-producing isolate (42a) did not form any mycorrhizas while the high IAA-producing isolate (27) produced ectomycorrhizas. Under high 2,4-D concentrations, both isolates produced endomycorrhizas.
(b) In the presence of low P (0.48 mg l^{-1} KH_2PO_4), the low IAA-producing isolate formed endomycorrhizas at low 2,4-D concentrations and ectomycorrhizas at high 2,4-D concentrations, while the high IAA-producing isolate formed ectomycorrhiza at all 2,4-D concentrations.

These results demonstrate the involvement of phosphate levels in ecto- or endomycorrhiza formation, as shown by Fortas and Chevalier (1992). As we saw in Sect. 5.2.2, auxin-induced increases in cell wall elasticity promote the formation of endomycorrhizas. However, when high levels of fungal auxin are coupled with high root sensitivity to auxin under low phosphate, ectomycorrhizas are produced, which points to a decline in the elasticity of the cell wall, inhibiting fungal penetration of the plant cell wall. This decrease in cell wall elasticity, in the presence of high auxin, may be explained by the bell shape of the dose–response curve for IAA-induced growth: gradual stimulation of growth as auxin increases is followed by inhibition at high IAA concentrations. This supraoptimal response to

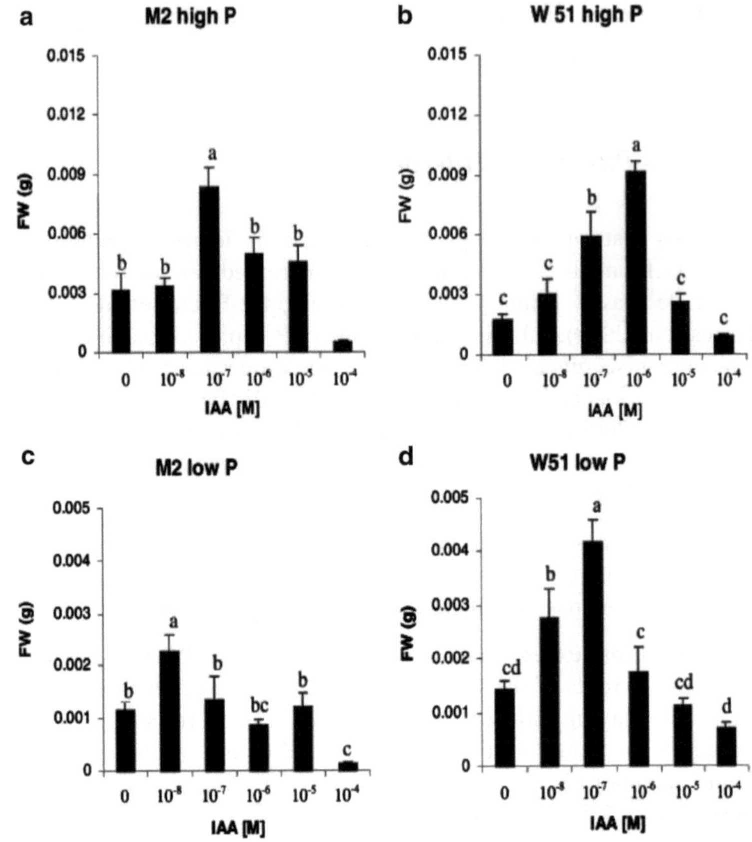

Fig. 5.2 Effect of exogenously applied IAA on growth of transformed *Cistus incanus* hairy roots (clones M2 and W51). Roots were grown for 12 days on medium with low (0.0035 mM) or high (0.70 mM) KH_2PO_4 concentrations (**a**, **b**—high P; **c**, **d**—low P). IAA was added to the media after autoclaving. Values are the mean for three plates containing three tips each ± standard error per treatment. Different letters within columns indicate a significant difference ($P = 0.05$). From Zaretsky et al. (2006), with kind permission of Springer Science and Business Media

IAA is linked to the generation of ethylene (Taiz and Zeiger 1998). Evidence exists that auxin stimulates ethylene production (Abeles 1965; Yu and Yanf 1979; Arteca et al. 1983) and that ethylene by itself, or together with auxin, inhibits root cell elongation (Pitts et al. 1998; Swarup et al. 2007). Hence the possibility cannot be ruled out that high levels of auxin actually reduce the elasticity of the cell wall, thereby inhibiting the formation of endomycorrhizas.

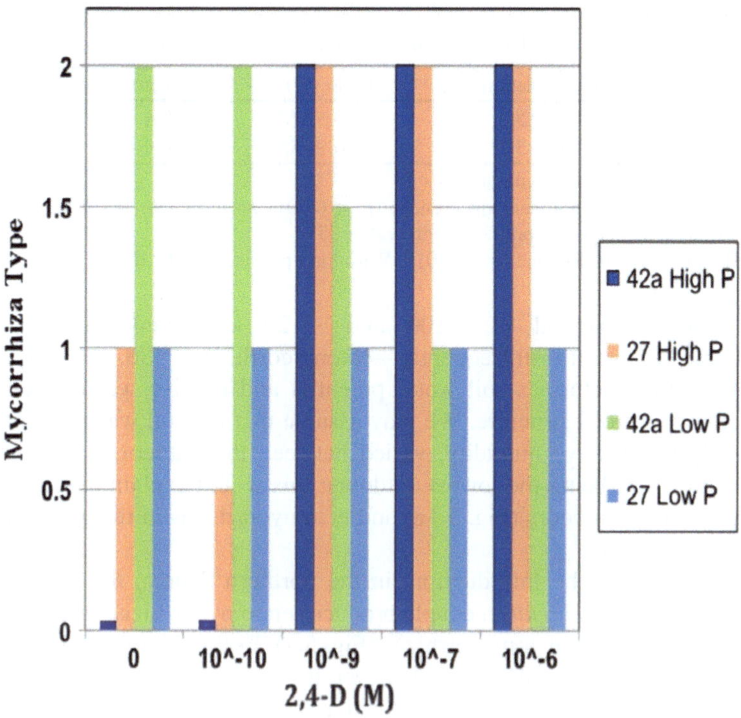

Fig. 5.3 Effect of exogenously applied 2,4-D on mycorrhization. All plant–fungus complexes were grown on M media modified with respect to KH_2PO_4 concentration. Roots were stained for assessment of mycorrhizas. The type of mycorrhiza obtained was ranked by assigning arbitrary numbers as follows: no mycorrhiza (*0*), poor Hartig net (*0.5*), ectomycorrhiza with Hartig net, without mantle (*1*), partly ecto- and partly endomycorrhiza (*1.5*), endomycorrhiza (*2*)

5.2.4 Microclimatic Conditions

As we saw in the Introduction, Gutiérrez et al. (2003) found that *Helianthemum almeriense–Terfezia claveryi* pairs produce endomycorrhizas without mantle in the field (Fig. 5.1a), ectomycorrhizas without mantle in the greenhouse (Fig. 5.1b), and ectomycorrhizas with developed Hartig net and a mantle in vitro (Fig. 5.1c). Phosphate concentrations were similar (28.7 ppm) outdoors and in the greenhouse, though somewhat higher in vitro (42.5 ppm). In view of the finding that low phosphate promotes formation of endomycorrhizas (Sect. 5.2.3), these differences in mycorrhizal architecture cannot be attributed to the influence of phosphate. Instead, it seems that the key factor here was either soil humidity or relative humidity. Navarro-Ródenas et al. (2013) found in greenhouse experiments that drought stimulates mycorrhization, as well as affecting the character of the mycorrhiza (Table 5.2). Well-watered plants with a soil matrix potential of 0–10 kPa generally produced ectendo- and ectomycorrhizas, while water-stressed plants with

Table 5.2 Percentages of total colonization and different types of mycorrhizal colonization in *H. almeriense* plants grown under greenhouse conditions with different irrigation treatments

Water treatment	Colonization (%)	Endo (%)	Endecto (%)	Ecto (%)
Drought	54.3 ± 5.5	53.3 ± 7.5	34.7 ± 7.2	11 ± 4.8
Watered	34.5 ± 3.7	15.0 ± 5.9	51.5 ± 6.7	33.5 ± 6.4

Data were subjected to chi-square analysis
Different letters mean significant differences ($P < 0.001$) according to Bonferroni correction. Values are the percentage ± confidence interval at 95 %
Reprinted from Navarro-Ródenas et al. 2013. With kind permission of MPMI

a soil matrix potential of −100 and −120 kPa predominantly formed endomycorrhizas and—infrequently—ectomycorrhizas. So, clearly, such humidity-related factors as soil water potential and relative water potential can influence mycorrhizal structure. We may assume that the soil water potential, and presumably the relative humidity, varied between the outdoor, greenhouse, and in vitro situations, being the lowest outdoors; this would explain the formation of endomycorrhizas outdoors (Fig. 5.1a) and ectomycorrhizas in the greenhouse and in vitro (Fig. 5.1b, c).

As mentioned in the Introduction, in the northern Sahara, the *Helianthemum sessiliflorum–Terfezia boudieri* couple produces endomycorrhizas whereas in Israel the same partners form ectomycorrhizas (Fig. 5.4). Both regions are classed as arid zones, but there may be differences between the two habitats of which we are not yet aware and which influence mycorrhizal architecture.

External conditions, among them drought, affect hormonal levels in plants. More specifically, drought depresses plant levels of auxin and cytokinins and increases those of abscisic acid (ABA) and ethylene (Itai 1999). According to the acid growth theory, a decline in auxin, brought about by dry conditions, could not explain the switch from ectomycorrhizas and ectendo- to endo-myorrhizas as shown by Navarro-Ródenas et al. (2013). However, there is evidence that dry conditions boost the concentration of ABA in shoots and even more so in roots, maintaining root growth and promoting the rate of cell expansion (Creelman et al. 1990). Moreover, Munns and Sharp (1993) and Wu et al. (1994) have shown that, in contrast to well-watered plants, at low water potentials root elongation requires increased levels of ABA.

Taken together, these findings point towards the possibility that under drought conditions it is ABA that preserves cortical cell wall extensibility, enabling the formation of endomycorrhizas. However, the possibility cannot be ruled out that auxins interact with other phytohormones or with other plant metabolites in a concerted manner, as suggested by Barker and Tagu (2000), and Swarup et al. (2007).

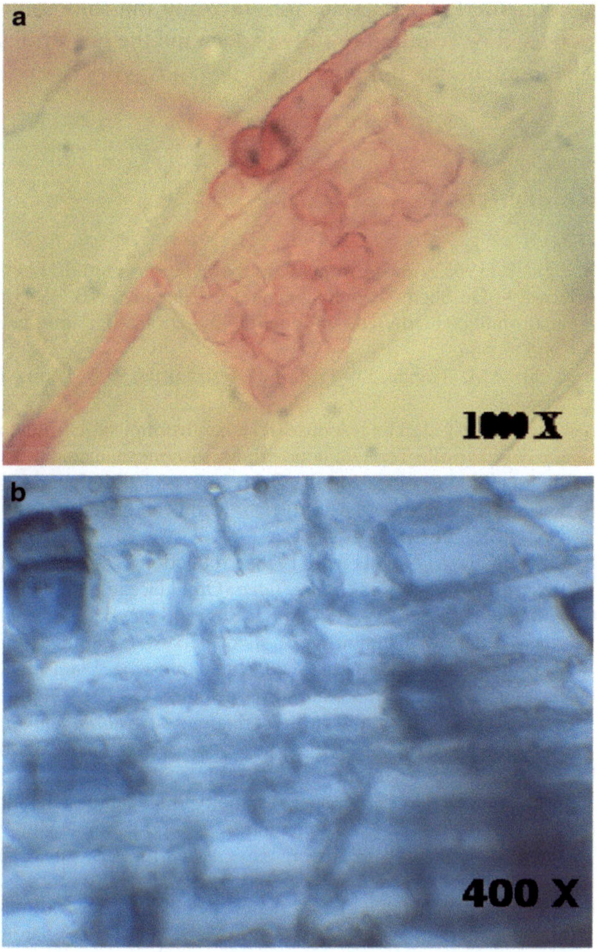

Fig. 5.4 *Helianthemum sessiliflorum* inoculated plant with *Terfezia boudieri*. (**a**) In Tunisia—Endomycorrhiza. From Slama et al. (2010), with Dr. Slama's kind permission and of African Journal of Microbiology Research (AJMAR). (**b**) Ectomycorrhiza from an experimental plot, Beer Sheva, Israel

5.3 Conclusions

This chapter summarizes current knowledge regarding the different types of mycorrhizal structure. The two key elements that determine mycorrhizal architecture appear to be concentration of auxin secreted by the fungi and root sensitivity to auxin. Obviously, any factor capable of affecting either of these two elements—such as phosphate level—will also play a role in determining the form of the mycorrhiza.

Another factor affecting mycorrhiza type is water stress. Drought is known to increase ABA levels in roots, and elevated levels of ABA in combination with drought have been shown to maintain root elongation. The physiological changes that occur in root cell walls under the influence of high ABA levels in drought are not known, but one possibility is that loosening of the cell wall takes place as a prerequisite for cell expansion, favoring the formation of endomycorrhizas. More work will be required to fully understand the conditions that modulate mycorrhizal development, but it seems that cell wall rigidity or elasticity is an important factor.

References

Abeles FB (1965) Auxin stimulation of ethylene evolution. Plant Physiol 41:585–588

Arteca RN, De-Sheng T, Schlagnhaufer C, Mandava NB (1983) The effect of brassinosteroid on auxin-induced ethylene production by etiolated mung bean segments. Pysiol Plantarum 59:539–544

Alsheikh AM, Trappe JM (1983) Desert truffles: the genus Tirmania. Trans Br Mycol Soc 81:83–90

Awameh MS (1981) The response of Helianthemum salicifolium and H. ledifolium to infection by the desert truffle Terfezia boudieri. Mushroom science. XI. Part II. In: Proceedings of the 11th International Congress on the Cultivation of Edible Fungi, Sydney, Australia, pp. 843–583

Barker SJ, Tagu D (2000) The roles of auxins and cytokinins in mycorrhizal symbioses. J Plant Growth Regul 19:144–154

Barosso J, Chaves N, Pais MS (1978) Production of indole-3-ethanol and indol-3-acetic acid by mycorrhizal fungus of *Ophrys lutea* (Orchidaceae). New Phytol 103:745–749

Creelman RA, Mason HS, Bensen RJ, Boyer JS, Mullet JE (1990) Water deficit and abscisic acid cause differential inhibition of shoot versus root growth in soybean seedlings. Analysis of growth, sugar accumulation, and gene expression. Plant Physiol 92:205–214

Dexheimer J, Gerard J, Leduc JP, Chevalier G (1985) Étude ultrastructurale comparée des associations symbiotiques mycorhiziennes *Helianthemum salicifolium–Terfezia claveryi Helianthemum salicifolium–Terfezia leptoderma*. Can J Bot 63:582–591

Díez J, Manjón JL, Moreno G (2002) Molecular phylogeny of the mycorrhizal desert truffles (Terfezia and Tirmania), host specificity and edaphic tolerance. Mycologia 94(2):247–259

Ferdman N, Aviram S, Roth-Bejerano N, Trappe J, Kagan-Zur V (2005) Phylogenetic Studies of *Terfezia pfeilii* and *Choiromyces echinulatus* (Pezizales) support new genera for southern African truffles: *Kalaharituber* and *Eremiomyces*. Mycol Res 109:237–245

Fortas Z, Chevalier G (1992) Effet des conditions de culture sur la mycorhization de l'Helianthemum guttatum par trois espèces de terfez des genres Terfezia et Tirmania d'Algérie. Can J Bot 70:2453–2460

Gay L, Normand L, Marmeisse B, Sotta B, Debaud JC (1994) Auxin overproducer mutants of Hebeloma cylindrosporum Romagnesi have increased mycorrhizal activity. New Phytol 128:645–657

Gea I, Normand L, Vian B, Gay G (1994) Structural aspects of ectomycorrhiza of *Pinus pinaster* (Ait.) Sol. formed by an IAA-overproducer mutant of *Hebeloma cylindrosporum* Romagnesi. New Phytol 128:651–670

Gogala N (1991) Regulation of mycorrhizal infection by hormonal factors produced by host and fungi. Experimentia 47:331–339

Gutiérrez A, Morte A, Honrubia M (2003) Morphological characterization of the mycorrhiza formed by *Helianthemum alm*eriense Pau with *Terfezia claveryi* Chatin and *Picoa lefebvrei* (Pat.) Maire. Mycorrhiza 13:299–307

Itai C (1999) Role of phytohormones in plant response to stresses. In: Lerner HR (ed) Plant responses to environmental stresses: from phytochrome to genome reorganization. Marcel Dekker, New York, NY

Kagan-Zur V, Kuang J, Tabak S, Taylor FW, Roth-Bejerano N (1999) Potential verification of a host plant for the desert truffle *Terfezia pfeilii* by molecular methods. Mycol Res 103:1270–1274

Kagan-Zur V (2001) Combating desertification with plants. In: Pasternak D, Schliessel A (eds) Terfezias a family of mycorrhizal edible mushrooms for arid zones. Kluwer Academic Publishers, New York, NY

Kagan-Zur V, Zaretsky M, Sitrit Y, Roth-Bejarano N (2008) Hypogeous pezizaceae: physiology and molecular genetics. In: Varma A (ed) Mycorrhiza: state of the art, genetics and molecular biology, eco-function, biotechnology, eco-physiology, structure and systematics, 3rd edn. Springer, Berlin

Kirk PM, Cannon PF, David JC, Stalpers JA (2001) Ainsworth and Bisby's dictionary of the fungi, 9th edn. CAB International, Wallingford

Kovács GM, Balázs TK, Calonge FD, Martín MP (2011) The diversity of *Terfezia* desert truffles: new species and a highly variable species complex with intrasporocarpic nrDNA ITS heterogeneity. Mycologia 103:841–853

Lassoe T, Hansen K (2007) Truffle trouble: What happened to the Tuberaeles? Mycol Res 111:1075–1099

López-Bucio J, Hernández-Abreu E, Sánchez-Calderón L, Nieto-Jacobo MF, Simpson J, Herrera-Estrella L (2002) Phosphate availability alters architecture and causes changes in hormone sensitivity in the *Arabidopsis* root system. Plant Physiol 129:244–256

López-Bucio J, Cruz-Ramírez A, Herrera-Estrella L (2003) The role of nutrient availability in regulating root architecture. Curr Opin Plant Biol 6:280–287

Martin F, Duplessis S, Ditengou F, Lagrange H, Voiblet C, Lapeyrie F (2001) Developmental cross talking in the ectomycorrhizal symbiosis: signals and communication genes. New Phytol 151:145–154

Moreno G, Díez J, Manjón JL (2000) *Picoa lefebvrei* and *Tirmania nivea*, two rare hypogeous fungi from Spain. Mycol Res 104:378–381

Moreno G, Díez J, Manjón JL (2002) *Terfezia boudieri*, first records from Europe of a rare vernal hypogeous mycorrhizal fungus. Persoonia (Leiden) 17:637–641

Munns R, Sharp RE (1993) Involvement of abscisic acid in controlling plant growth in sol of low water potential. Aust J Plant Physiol 20:425–437

Navarro-Ródenas N, Pérez-Gilabert M, Torrente P, Morte A (2012) The role of phosphorus in the ectendomycorrhiza continuum of desert truffle mycorrhizal plants. Mycorrhiza 22:565–575. doi:10.1007/s00572-012-0434-2

Navarro-Ródenas A, Bárzana G, Nicolas E, Carra A, Schubert A, Morte A (2013) Expression analysis of aquaporins from desert truffle mycorrhizal symbiosis reveals a fine-tuned regulation under drought. Mol Plant Microbe Interact, http://dx.doi.org/10.1094/MPMI-07-12-0178-R.

Pérez-Torres CA, López-Bucio J, Cruz-Ramírez A, Ibarra-Laclette E, Dharmasiri S, Estelle M, Herrera-Estrella L (2008) Phosphate availability alters lateral root development in Arabidopsis by modulating auxin sensitivity via a mechanism Involving the TIR1 auxin receptor. Plant Cell 20:3258–3272

Pitts RJ, Cernak A, Estelle M (1998) Auxin and ethylene promote root hair elongation in Arabidopsis. Plant J 16:553–560

Rayle DL, Cleland RE (1992) The acid growth theory of auxin-induced cell elongation is alive and well. Plant Physiol 99:1271–1274

Riley HP (1963) Families of flowering plants of Southern Africa. University of Kentucky Press, Lexington

Roth-Bejerano N, Livne D, Kagan-Zur V (1990) *Helianthemum-Terfezia* relations in different media. New Phytol 114:235–238

Slama A, Fortas Z, Boudabous A, Neffati M (2010) Cultivation of an edible desert truffle (*Terfezia boudieri* Chatin). Afr J Micobiol Res 4:2350–2356

Smith SE, Read DJ (2008) Mycorrhizal symbiosis, 2nd edn. Academic, London

Swarup R, Perry P, Hagenbeek D, Van Der Straeten D, Beemster GTS, Sandberg G, Bhalerao R, Ljung K, Bennetta MJ (2007) Ethylene upregulates biosynthesis in Arabidopsis seedlings to enhance inhibition of root cell elongation. Plant Cell 19:2186–2196

Zaretsky M, Kagan-Zur V, Mills D, Roth-Bejerano N (2006) Analysis of mycorrhizal associations formed by *Cistus incanus* transformed root clones with *Terfezia boudieri* isolates. Plant Cell Rep 25:62–70

Wenkart S, Roth-Bsejerano N, Mills D, Kagan-Zur V (2001) Mycorrhizal associations between Tuber melanosporum mycelia and transformed roots of Cistus incanus. Plant Cell Rep 20:369–373

Wu Y, Spollen WG, Sharp RE, Hetheringtone PR, Fry SC (1994) Root growth maintenance at low water potentials. Plant Physiol 106:607–615

Yu YB, Yanf SF (1979) Auxin-induced ethylene production and its inhibition by aminoethoxyvinyiglycine and cobalt ion. Plant Physiol 64:1074–1077

Taiz L, Zeiger E (1998) Plant physiology, 2nd edn. Sinauer Associates Inc., Sunderland, MA

Turgeman T (2013) Pre-symbiotic signal exchange between the host plant *Helianthemum sessiliflorum* and the desert truffle *Terfezia boudieri*. Submitted to the Senate of Ben-Gurion University of the Negev, Israel

Chapter 6
Pre-symbiotic Interactions Between the Desert Truffle *Terfezia boudieri* and Its Host Plant *Helianthemum sessiliflorum*

Yaron Sitrit, Nurit Roth-Bejerano, Varda Kagan-Zur, and Tidhar Turgeman

6.1 Introduction

The desert truffle *Terfezia boudieri* (Pezizaceae) inhabits the arid and semiarid regions of the Middle East stretching from the northern parts of Africa near the Mediterranean Sea through the Arabian deserts up to the Persian Gulf (see Chaps. 9–12).

In the Negev desert, *T. boudieri* forms ectomycorrhizal associations with its main host plant, *Helianthemum sessiliflorum* (Cistaceae), a small perennial shrub inhabiting the dunes and sandy soils of these areas (see Chaps. 5 and 11).

Survival in arid zones requires adaptations to extreme and hostile environmental conditions by both partners, the desert truffle and its host plant (Turgeman et al. 2011). Most of the year the host plant is exposed to severe water stress, high ambient temperatures, and excessive radiation; the strategy it adopts in order to survive the long, harsh summer is to shed leaves and contract the root system through die-back; under severe water stress the branches also die back. In the short wet seasons of winter and early spring the canopy recovers and new roots are formed. This is the only time that both host plant and fungus have available resources to complete their life cycles. In spring the plant flowers and bears fruits, and in parallel the fungus forms fruit bodies. Mature seeds as well as spores germinate in the next winter. Germinating spores face two problems: first, the wet season is very short and symbiosis must be accomplished before the onset of the dry season; second, the exploring hyphae develop in soils poor in organic matter, and to secure a carbon supply before internal reserves are exhausted, it is necessary for the mycorrhizal association with the host to be established without delay. Yet the

Y. Sitrit (✉) • T. Turgeman
The Jacob Blaustein Institute for Desert Research, Ben-Gurion University of the Negev, Beer-Sheva 84105, Israel
e-mail: sitrit@bgu.ac.il

N. Roth-Bejerano • V. Kagan-Zur
Department of Life Sciences, Ben-Gurion University of the Negev, Beer-Sheva 84105, Israel

V. Kagan-Zur et al. (eds.), *Desert Truffles*, Soil Biology 38,
DOI 10.1007/978-3-642-40096-4_6,

environmental conditions are not conducive to the establishment of new mycorrhizal relations between germinating host seeds and fungal spores. In the course of evolution both partners have responded by developing special mechanisms that facilitate establishment of symbiosis.

This review describes the pre-symbiotic signals exchanged between partners, focusing on secretion of auxin—the plant growth regulator indole-3-acetic acid (IAA)—by the fungus and of chemoattractant(s) by the host roots.

6.2 The Role of IAA at the Pre-mycorrhizal Phase

While the ability of ectomycorrhizal fungi to synthesize the auxin IAA has been securely established (Ek et al. 1983), we still lack a complete picture of the functions of the plant hormone in ectomycorrhizal symbiosis, and more especially at the pre-symbiosis phase. More than 40 years have passed since it was first proposed that auxin alters the architecture of mycorrhizal roots (Slankis 1973), but new evidence indicates this is only one facet of its impact on symbioses. To date there is no accepted comprehensive model describing the role and functions of auxin at the pre-symbiosis phase. However, we have discovered several instances where IAA secretion by the fungus enhanced chances of mycorrhizal formation.

6.2.1 Coordination of Growth Rates Between Partners by Inhibition of Taproot Growth

Truffles in general, and desert truffles in particular, regulate plant root morphogenesis via secretion of IAA and emission of ethylene, as was shown for *Tuber borchii*, *Tuber melanosporum* (Splivallo et al. 2009), and *T. boudieri* (Zaretsky et al. 2006). These changes are already evident before physical contact is formed between the symbionts (Splivallo et al. 2009). The most prominent effect of auxin on root morphology is reduction of taproot elongation (Gogala 1991; Barker and Tagu 2000). As the latter authors put it, the function of auxin inhibition of taproot growth is: “I [the root] will halt, you [the hyphae] go there” (Barker and Tagu 2000). In other words, degeneration of the host’s root system will force the plant to rely on the exploring hyphae for its supply of minerals and water, increasing the plant’s dependence on the symbiosis (Hetrick 1991); in this way, the fungus will “secure” its carbon supply from the plant.

Another function of taproot growth inhibition becomes evident when the growth rates of hyphae and host roots are compared: roots grow about ten times faster than hyphae (10–20 and 1–5 mm/week, respectively; see Fig. 6.1). Under these conditions of mismatched growth rates, the fast growing root may readily cross the mycelia-populated layers, and since the root cap is the sole fungal invasion site,

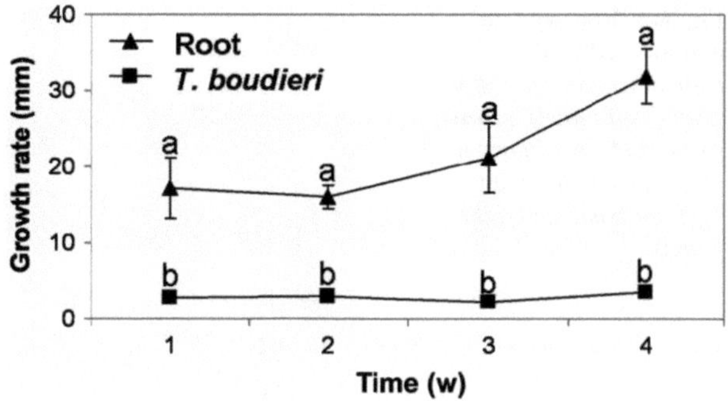

Fig. 6.1 Growth rate of *Terfezia boudieri* hyphae versus growth rate of roots of host plant *Helianthemum sessiliflorum* (Cistaceae). The plant and fungus were grown in Petri dishes containing medium M

the fast growing root may evade inoculation, leaving the slow-growing mycelia behind. Coordination of the growth rates of the partners via fungal-auxin secretion increases the chances of a successful encounter taking place and favors the establishment of symbiosis.

6.2.2 Proliferation of Penetration Sites by Induction of Lateral Root Formation

Secretion of auxin induces lateral root formation (Gogala 1991; Barker et al. 1998). Since, as mentioned above, attachment of hyphae occurs only around the cap region of roots, any increase in the number of roots will also increase the likelihood of inoculation (Horan et al. 1988). Accordingly, strains of the ectomycorrhizal fungus *Paxillus involutus* producing higher levels of IAA (and inducing more lateral root formation in *Pinus sylvestris*) were better colonizers than strains that secreted lower levels of the plant hormone (Rudawska and Kieliszewska-Rokicka 1997). Stimulation of lateral root formation takes place through the specific induction of numerous genes belonging to different classes and with different functions. Included among them are genes involved in auxin transport and distribution, as was shown for the ectomycorrhizal fungus *Laccaria bicolor* when co-cultured with poplar and the nonhost plant Arabidopsis (Felten et al. 2009). It was found that the auxin influx carriers *PtaAUX3* and *PtaAUX6* were up-regulated, as were several efflux carriers belonging to the *PIN* genes family.

Co-culturing of *T. boudieri* with its host *H. sessiliflorum* induced lateral root formation and inhibited taproot elongation (Fig. 6.2). The lateral roots of *H. sessiliflorum* formed in this dual culture displayed the typical phenotype of slow-growing, swollen, short lateral roots (Fig. 6.2). Transcript induction of auxin carriers from both gene families—*AUX* and PIN—was detected in *H. sessiliflorum* before physical contact was made between the partners (Sitrit unpublished).

Fig. 6.2 Root system of non-mycorrhized *Helianthemum sessiliflorum* plants (*left panel*) versus roots of *H. sessiliflorum* plants inoculated with *Terfezia boudieri* (*right panel*)

6.2.3 Root Hairs

If auxin secretion is designed to inhibit taproot growth in order to enhance the plant's dependence on the mycorrhizing fungus, we may suppose that inhibition of root hairs is intended to further "force" host reliance on symbiosis. Mineral and water acquisition occur through root hairs, which have a larger absorptive surface area than lateral roots (Peterson and Farquhar 1996). Inhibition of root hair formation by fungal-secreted compounds may be expected to facilitate the establishment of mycorrhizal relations owing to the need to compensate for the impairment of the water and mineral absorption capabilities of the roots. Roots of plants that form mycorrhizal associations typically produce root hairs in the non-mycorrhized state; loss of root hairs is one the dramatic changes in root morphology associated with mycorrhization (Linderman 1988; Gutierrez et al. 2003). Non-mycorrhizal plant species have been shown to produce abundant root hairs compared to mycorrhizal species (Hetrick 1991; Miller et al. 1999). When the nonhost plant *Arabidopsis thaliana* is grown in the presence of *T. boudieri*, the IAA secreted by the truffle induces root hair formation (Fig. 6.3) quite unlike the smooth root morphology exhibited by *H. sessiliflorum* when grown in a similar dual culture. The roots of *H. sessiliflorum* may form root hairs when grown without the fungus, as has already been noted for other ectomycorrhizal plants, such as *Eucalyptus globulus* (the latter's partner is *Pisolithus tinctorius*) (Ditengou et al. 2000). *P. tinctorius* excretes the indole alkaloid hypaphorine, which counteracts IAA induction of root hair elongation (Beguiristain and Lapeyrie 1997; Ditengou et al. 2000; Reboutier et al. 2002). Inhibition of root hair development by hypaphorine is another strategy used by the fungus to promote the dependence of the host plant on mycorrhizal

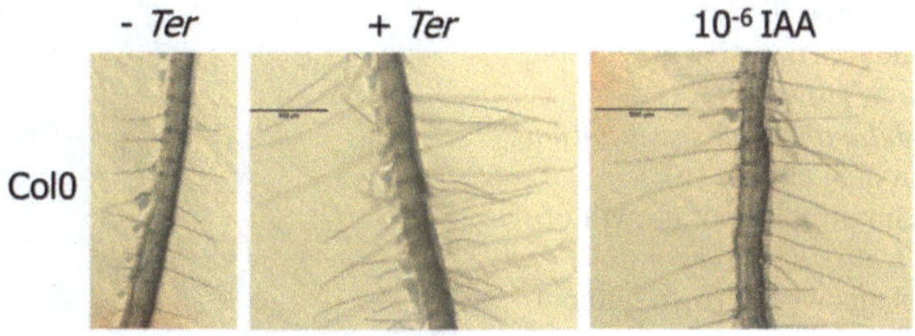

Fig. 6.3 Induction of root hair formation and elongation by *Terfezia boudieri* (+*Ter*) and IAA (10^{-6} M) in roots of the nonhost plant *Arabidopsis thaliana* Col-0 ecotype

symbiosis. Whether hypaphorine or other compounds with similar activity are secreted by desert truffles has yet to be determined.

6.2.4 IAA-Induced Root Reorientation by Negative Gravitropism

In a study of *Arabidopsis* mutants impaired in auxin transport Rigas et al. (2012) were able to show that root gravitropism and root hair development are coupled auxin-regulated processes. Recently we discovered that root gravitropism was disturbed already at the pre-mycorrhizal phase when *Terfezia* and *H. sessiliflorum* were grown in dual culture. Root gravitropism is a complex process that comprises perception of root growth deviation from the gravity axis, signal transmission, and organ curvature (Strohm et al. 2012). Root bending involves auxin transport, gradient generation, and transient accumulation of auxin at the lower side of the root to promote downward curvature (Young et al. 1990). When *H. sessiliflorum* was grown in dual culture with *Terfezia*, downward root growth was impaired and taproots grew horizontally, probably due to exposure of the root to a front of high levels of auxin secreted by the fungus (Fig. 6.4). More than 65 % of the roots in the dual culture grew at 40°–60° deviation from the gravity axis, while 74 % of the control roots grew at only 0°–20° deviation. Our previous work (Zaretsky et al. 2006) had demonstrated that *Terfezia* secretes high levels of IAA, leading us to hypothesize that fungal-originated IAA was involved in the root reorientation. The concentration of fungal-secreted auxin in the growth medium (30 ml agar plate) measured after 2 weeks of mycelial growth was about 10^{-5} M, which is high enough to reduce root growth and induce root redirection. Furthermore, the effect of *Terfezia* on root orientation could be mimicked by exogenous application of the synthetic auxin naphthalene acetic acid at a similar concentration (Fig. 6.5).

Since auxin perception mutants were not available for *H. sessiliflorum*, we pursued our investigation using *Arabidopsis* mutants impaired in auxin perception. When nonmutant *Arabidopsis* plants were grown in dual culture with *Terfezia*, acute bending of roots towards the horizontal was observed, recalling the effect we had observed in *H. sessiliflorum. Arabidopsis* triple mutants of the TIR1/AFB family of auxin receptors did not show any abnormal root orientation when grown in the presence of *Terfezia*, further supporting our hypothesis that fungal-originated

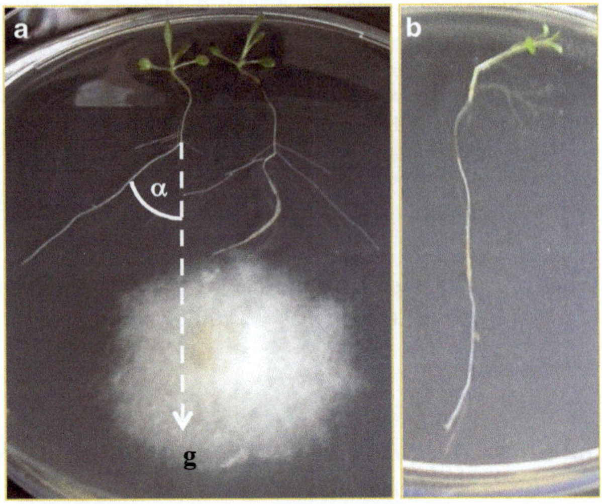

Fig. 6.4 *Terfezia boudieri* induces negative gravitropic root growth of *Helianthemum sessiliflorum* in dual culture (**a**). Gravitropic root growth of *H. sessiliflorum* control plants when grown without *T. boudieri* (**b**)

Fig. 6.5 The synthetic auxin 1-naphthaleneacetic acid mimics the negative gravitropic root growth induced by IAA in *Helianthemum sessiliflorum*

IAA is indeed involved in reorientation of the host root. Expression analysis of the auxin influx transporter *Aux1* and efflux carriers *Pin1* and *Pin3* revealed that all three were induced in *Arabidopsis* plants grown in the presence of *Terfezia* prior to actual physical contact between them. Specifically, the induction level of *Aux1* was 30-fold higher than the control, implying that the auxin in the growth medium can be transported into the roots to establish the gradient necessary for root bending.

6.2.5 Ecological Implications and Mode of IAA Action in Mycorrhizal Relations

Under desert conditions in which both partners have only a short time in which to grow and establish mycorrhizal relations, the problem of coordinated development assumes considerable importance. In the course of co-evolution with its plant partners, *Terfezia* acquired capabilities that enable it to secure successful encounters with the host's roots. One such mechanism is the ability to secrete auxin. Secretion of high levels of auxin in the vicinity of the developing roots directly affects their architecture and orientation. When host seeds and fungal spores germinate, fungal-secreted IAA induces lateral root proliferation, multiplying the invasion sites available to the fungus. High levels of auxin also slow down the pace of root growth, giving the fungus time to develop towards the root. Auxin induced bending in the direction of horizontal growth means that the roots remain confined to the same plane as the developing mycelia. Otherwise the roots would rapidly cross the soil layer inhabited by the fungal spores, leaving the developing mycelia behind.

At the pre-symbiotic stage, then, the function of auxin is to induce the roots to remain in the fungal layer. Indeed, when strains of *Terfezia* differing in auxin secretion capacity were compared in relation to their ability to inoculate the host, it was found that high auxin producers had a better inoculation capability (Sitrit et al. unpublished), as previously shown for auxin overproducer mutants of *Hebeloma cylindrosporum* and *Paxillus involutus* (Rudawska and Kieliszewska-Rokicka 1997; Gay et al. 1994). Although the latter experiments were carried under laboratory conditions, which presumably favor encounter, effective auxin gradients may be formed under natural desert conditions as well, supporting successful mycorrhizal establishment.

6.3 Chemotropism: A Key Component in the Establishment of Mycorrhizas

In the course of co-evolution, mycorrhizal fungi and their host plants have developed ways of facilitating successful encounters. The effects of auxin secretion by the fungus were discussed in the preceding section. Another mechanism, chemotropism, confers oriented growth, enabling the fungus to save time and energy that would otherwise have been invested in nontargeted hyphal exploration for a host. Oriented growth towards the host plant requires signal exchange between the partners, leading to host plant recognition by the exploring hyphae. The signals secreted by the fungus at the pre-symbiotic phase include molecules that affect host gene expression and root architecture. Among them are Myc factors (similar to Nod factors) that induce root branching and the expression of some mycorrhizal-responsive genes in the host plant (Bonfante and Requena 2011; Frettinger et al. 2006; Maillet et al. 2011; Weidmann et al. 2004). Perception by the host of the molecules secreted by the fungus is mediated by specific proteins and receptors (Antolín-Llovera et al. 2012; Gherbi et al. 2008). However, such receptors for hosts' signals are yet to be identified in hyphae tips (Akiyama et al. 2010; Badri et al. 2009; Bouwmeester et al. 2007).

Secretion of molecules from the host plant into the rhizosphere is well documented for mycorrhizas, although little is known about the chemical characteristics of these signal molecules (Giovannetti et al. 1993; Horan and Chilvers 1990; Sbrana and Giovannetti 2005; Suriyapperuma and Koske 1995). Fungal tip reorientation is probably implemented through modulation of the conserved machinery involved in polarized growth (Bonfante and Requena 2011; Brand and Gow 2009). Several molecules have been postulated to act as chemoattractants, not only in mycorrhizal relations but also in other fungal-plant communications. These signals belong to diverse groups of compounds, including flavonoids, diterpenes, phenolics, and sesquiterpene lactones (strigolactones), as well as volatile organic compounds (Akiyama et al. 2005; Besserer et al. 2006; Bouwmeester et al. 2007; Koske 1982; Lagrange et al. 2001; Sbrana and Giovannetti 2005; Shaw et al. 2006; Siqueira et al. 1991; Steinkellner et al. 2007). However, little progress has been made in the last decades towards chemical characterization of these molecules, probably due to their low concentration and the complexity of isolation and analysis. Although several authors have hypothesized that rutin, strigolactones and flavonoid compounds may serve as chemoattractants, no direct evidence for such activity has been presented (Akiyama and Hayashi 2006; Lagrange et al. 2001; Siqueira et al. 1991).

Most of the knowledge summarized in this section comes from studies of arbuscular mycorrhizal fungi, in which the compounds were shown to induce spore germination, hyphal growth, hyphal branching, and increased inoculation (Akiyama et al. 2005; Lagrange et al. 2001; Scervino et al. 2005). However, much less is known about chemotropism in other mycorrhizal fungi and even less about chemotropism in mycorrhizal fungi developing under desert conditions.

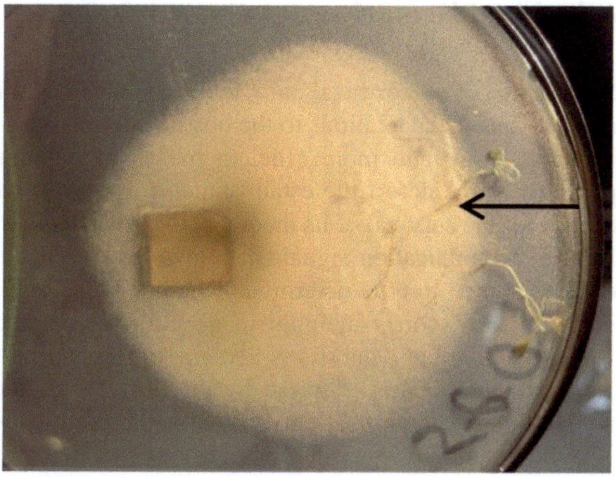

Fig. 6.6 Chemotropic growth of *Terfezia boudieri* towards the host plant *Helianthemum sessiliflorum*. A cube of agar containing the fungus was placed at the center of the plate and seedlings (marked by *arrow*) were placed at one the side of the plate

6.3.1 Mycorrhizal Chemotropic Growth Under Desert Conditions

Desert soils such as those of the Negev are characterized by low levels of organic matter 0.44–1.18 % (Saul-Tcherkas et al. 2012) and low water content, which restricts saprophytic growth of free living mycelia. Under these environmental conditions, where the ability of developing hyphae to search for and identify a host is limited by the necessity to complete the process within the narrow interval between spore germination and the end of the winter-spring season (Bonfante and Genre 2010), prompt recognition of the host and rapid reoriented hyphal-tip growth towards the plant are crucial. Thus in extreme environments an efficient signal exchange system is vital for the survival of both partners. We have developed a biological assay for chemotropism that we use for analysis of the active molecule(s) (Fig. 6.6). *T. boudieri* shows chemotropic growth when cultured in the presence of the host plant *H. sessiliflorum* (Fig. 6.6). When our biological assay was run successively with five different members of the Cistaceae, the results all showed a positive response indicating chemotropic growth (Sitrit et al. unpublished), which suggests that members of the Cistaceae family contain a common signaling molecule. This chemoattracting molecule is specific to the Cistaceae-*Terfezia* combination, since a dual culture of *P. tinctorius* with *H. sessiliflorum* did not show induction of chemotropic growth. The strength of the chemotropic response depended on the sucrose concentration in the growing media. Increasing media sucrose from 0.25 to 1 % weakened the chemotropic response, implying that when a carbon source is readily accessible, the formation of mycorrhizal relations is less essential for the fungus (Sitrit et al. unpublished). Isolation of potential plant chemoattractants was undertaken, and a compound, as yet unidentified, with a molecular weight of about 535, was found to possess chemotropic activity (Sitrit et al. unpublished).

6.4 Final Remarks

Subsistence of mycorrhizal pairs under the harsh conditions that prevail in deserts requires special adaptations to the hostile environment—adaptations providing, for instance, faster and more efficient means of communication between potential partners. In the desert the establishment of a symbiotic association is even more crucial for the survival of its members than it is in other, milder environments. That is why communication signals have evolved in the course of co-evolution. However, we have yet to determine whether these signals are species specific or common to mycorrhizal fungi in general, as was shown to be the case for the Nod factors. Many questions remain unresolved; for example, how do fungi perceive the signal(s) excreted by the host? What are the receptors involved in recognition in the two partners? What is the chemical composition of the signals?

With recent advances in techniques of chemical analysis and the advent of the powerful "omics" methods, one should expect to see a fundamental breakthrough in this basic research in the near future.

References

Akiyama K, Hayashi H (2006) Strigolactones: chemical signals for fungal symbionts and parasitic weeds in plant roots. Ann Bot 97:925–931

Akiyama K, Matsuzaki K, Hayashi H (2005) Plant sesquiterpenes induce hyphal branching in arbuscular mycorrhizal fungi. Nature 435:824–827

Akiyama K, Ogasawara S, Ito S, Hayashi H (2010) Structural requirements of strigolactones for hyphal branching in AM fungi. Plant Cell Physiol 51:1104–1117

Antolín-Llovera M, Ried MK, Binder A, Parniske M (2012) Receptor kinase signaling pathways in plant–microbe interactions. Annu Rev Phytopathol 50:451–473

Badri DV, Weir TL, van der Lelie D, Vivanco JM (2009) Rhizosphere chemical dialogues: plant-microbe interactions. Curr Opin Biotechnol 20:642–650

Barker SJ, Tagu D (2000) The roles of auxins and cytokinins in mycorrhizal symbioses. J Plant Growth Regul 19:144–154

Barker SJ, Tagu D, Delp G (1998) Regulation of root and fungal morphogenesis in mycorrhizal symbioses. Plant Physiol 116:1201–1207

Beguiristain T, Lapeyrie F (1997) Host plant stimulates hypaphorine accumulation in *Pisolithus tinctorius* hyphae during ectomycorrhizal infection while excreted fungal hypaphorine controls root hair development. New Phytol 136:525–532

Besserer A, Puech-Pagès V, Kiefer P, Gomez-Roldan V, Jauneau A, Roy S, Portais JC, Roux C, Bécard G, Séjalon-Delmas N (2006) Strigolactones stimulate arbuscular mycorrhizal fungi by activating mitochondria. PLoS Biol 4:e226

Bonfante P, Genre A (2010) Mechanisms underlying beneficial plant–fungus interactions in mycorrhizal symbiosis. Nat Commun 27:1–11

Bonfante P, Requena N (2011) Dating in the dark: how roots respond to fungal signals to establish arbuscular mycorrhizal symbiosis. Curr Opin Plant Biol 14:451–457

Bouwmeester HJ, Roux C, Lopez-Raez JA, Bécard G (2007) Rhizosphere communication of plants, parasitic plants and AM fungi. Trends Plant Sci 12:224–230

Brand A, Gow NA (2009) Mechanisms of hypha orientation of fungi. Curr Opin Microbiol 12:350–357

Ditengou FA, Beguiristain T, Lapeyrie F (2000) Root hair elongation is inhibited by hypaphorine, the indole alkaloid from the ectomycorrhizal fungus *Pisolithus tinctorius*, and restored by indole-3-acetic acid. Planta 211:722–728

Ek M, Ljungquist PO, Stenstrom E (1983) Indole-3-acetic acid production by mycorrhizal fungi determined by gas chromatography–mass spectrometry. New Phytol 94:401–407

Felten J, Kohler A, Morin E, Bhalerao RP, Palme K, Martin F, Ditengou FA, Legue V (2009) The ectomycorrhizal fungus *Laccaria bicolor* stimulates lateral root formation in poplar and arabidopsis through auxin transport and signaling. Plant Physiol 151:1991–2005

Frettinger P, Herrmann S, Lapeyrie F, Oelmüller R, Buscot F (2006) Differential expression of two class III chitinases in two types of roots of *Quercus robur* during pre-mycorrhizal interactions with *Piloderma croceum*. Mycorrhiza 16:219–223

Gay G, Normand L, Marmeisse R, Sotta B, Debaud JC (1994) Auxin overproducer mutants of *Hebeloma cylindrosporum* Romagnesi have increased mycorrhizal activity. New Phytol 128:645–657

Gherbi H, Markmann K, Svistoonoff S, Estevan J, Autran D, Giczey G, Auguy F, Péret B, Laplaze L, Franche C, Parniske M, Bogusz D (2008) SymRK defines a common genetic basis for plant root endosymbioses with arbuscular mycorrhiza fungi, rhizobia, and Frankia bacteria. Proc Natl Acad Sci U S A 105:4928–4932

Giovannetti M, Sbrana C, Avio L, Citernesi AS, Logi C (1993) Differential hyphal morphogenesis in arbuscular mycorrhizal fungi during pre-infection stages. New Phytol 125:587–594

Gogala N (1991) Regulation of mycorrhizal infection by hormonal factors produced by hosts and fungi. Experientia 47:331–340

Gutierrez A, Morte A, Honrubia M (2003) Morphological characterization of the mycorrhiza formed by *Helianthemum almeriense* Pau with *Terfezia claveryi* Chatin and *Picoa lefebvrei* (Pat.) Maire. Mycorrhiza 13:299–307

Hetrick BAD (1991) Mycorrhizas and root architecture. Experientia 47:355–362

Horan DP, Chilvers GA (1990) Chemotropism—the key to ectomycorrhizal formation? New Phytol 116:297–301

Horan DP, Chilvers GA, Lapeyrie FF (1988) Time sequence of the infection process in eucalypt ectomycorrhizas. New Phytol 109:451–458

Koske RE (1982) Evidence for a volatile attractant from plant roots affecting germ tubes of a VA mycorrhizal fungus. Trans Br Mycol Soc 79:305–310

Lagrange H, Christian J-A, Lapeyrie F (2001) Rutin, the phenolglycoside from eucalyptus root exudates, stimulates *Pisolithus* hyphal growth at picomolar concentrations. New Phytol 149:349–355

Linderman RG (1988) Mycorrhizal interactions with the rhizosphere microflora: the mycorrhizosphere effect. Phytopathology 78:366–371

Maillet F, Poinsot V, André O, Puech-Pagès V, Haouy A, Gueunier M, Cromer L, Giraudet D, Formey D, Niebel A, Martinez EA, Driguez H, Bécard G, Dénarié J (2011) Fungal lipochitooligosaccharide symbiotic signals in arbuscular mycorrhiza. Nature 469:58–63

Miller RM, Smith CI, Jastrow JD, Bever JD (1999) Mycorrhizal status of the genus *Carex* (*Cyperaceae*). Am J Bot 86:547–553

Peterson RL, Farquhar ML (1996) Root hairs: specialized tubular cells extending root surfaces. Bot Rev 62:1–40

Reboutier D, Bianchi M, Brault M, Roux C, Dauphin A, Rona JP, Legué V, Lapeyrie F, Bouteau F (2002) The indolic compound hypaphorine produced by ectomycorrhizal fungus interferes with auxin action and evokes early responses in nonhost *Arabidopsis thaliana*. Mol Plant Microbe Interact 15:932–938

Rigas S, Ditengou FA, Ljung K, Daras G, Tietz O, Palme K, Hatzopoulos P (2012) Root gravitropism and root hair development constitute coupled developmental responses regulated by auxin homeostasis in the *Arabidopsis* root apex. New Phytol 197:1130–1141

Rudawska ML, Kieliszewska-Rokicka B (1997) Mycorrhizal formation by *Paxillus involutus* strains in relation to their IAA-synthesizing activity. New Phytol 137:509–517

Saul-Tcherkas V, Unc A, Steinberger Y (2012) Soil microbial diversity in the vicinity of desert shrubs. Microb Ecol. doi:10.1007/s00248-012-0141-8

Sbrana C, Giovannetti M (2005) Chemotropism in the arbuscular mycorrhizal fungus *Glomus mosseae*. Mycorrhiza 15:539–545

Scervino JM, Ponce MA, Erra-Bassells R, Vierheilig H, Ocampo JA, Godeas A (2005) Flavonoids exhibit fungal species and genus specific effects on the presymbiotic growth of Gigaspora and Glomus. Mycol Res 109:789–794

Shaw LJ, Morris P, Hooker JE (2006) Perception and modification of plant flavonoid signals by rhizosphere microorganisms. Environ Microbiol 8:1867–1880

Siqueira JO, Safir GR, Nair MG (1991) Stimulation of vesicular-arbuscular mycorrhiza formation and growth of white clover by flavonoid compounds. New Phytol 118:87–93

Slankis V (1973) Hormonal relationships in mycorrhizal development. In: Marks GC, Kozlowski TT (eds) Ectomycorrhizae: their ecology and physiology. Academic, New York, pp 231–298

Splivallo R, Fischer U, Gobel C, Feussner I, Karlovsky P (2009) Truffles regulate plant root morphogenesis via the production of auxin and ethylene. Plant Physiol 150:2018–2029

Steinkellner S, Lendzemo V, Langer I, Schweiger P, Khaosaad T, Toussaint JP, Vierheilig H (2007) Flavonoids and strigolactones in root exudates as signals in symbiotic and pathogenic plant-fungus interactions. Molecules 12:1290–1306

Strohm AK, Baldwin KL, Masson PH (2012) Multiple roles for membrane-associated protein trafficking and signaling in gravitropism. Front Plant Sci 3:274

Suriyapperuma SP, Koske RE (1995) Attraction of germ tubes and germination of spores of the arbuscular mycorrhizal fungus *Gigaspora gigantean* in the presence of roots of maize exposed to different concentrations of phosphorus. Mycologia 87:772–778

Turgeman T, Ben Asher J, Roth-Bejerano N, Kagan-Zur V, Kapulnik Y, Sitrit Y (2011) Mycorrhizal association between the desert truffle *Terfezia boudieri* and *Helianthemum sessiliflorum* alters plant physiology and fitness to arid conditions. Mycorrhiza 21:623–630

Weidmann S, Sanchez L, Descombin J, Chatagnier O, Gianinazzi S, Gianinazzi-Pearson V (2004) Fungal elicitation of signal transduction-related plant genes precedes mycorrhiza establishment and requires the dmi3 gene in *Medicago truncatula*. Mol Plant Microbe Interact 17:1385–1393

Young LM, Evans ML, Hertel R (1990) Correlations between gravitropic curvature and auxin movement across gravistimulated roots of *Zea mays*. Plant Physiol 92:792–796

Zaretsky M, Kagan-Zur V, Mills D, Roth-Bejerano N (2006) Analysis of mycorrhizal associations formed by *Cistus incanus* transformed root clones with *Terfezia boudieri* isolates. Plant Cell Rep 25:62–70

Chapter 7
Benefits Conferred on Plants

Varda Kagan-Zur, Tidhar Turgeman, Nurit Roth-Bejerano, Asunción Morte, and Yaron Sitrit

7.1 Introduction

Under stress conditions, the level of mycorrhizal associations formed is higher, as compared to that of well-nourished or well-watered plants. This was demonstrated for both AM endomycorrhiza (e.g., Douds et al. 1993) and ectomycorrhiza (e.g., Davies et al. 1996). Partnerships with desert truffles follow the same pattern: in a study performed on field-grown *Helianthemum almeriense* plants mycorrhized by *Terfezia claveryi*, Morte et al. (2010) found that water stress enhanced mycorrhizal establishment (70 % in nonirrigated plants versus 48 % in irrigated plants). Navarro-Rodenas et al. (2013) reported a similar pattern in water-stressed greenhouse plants of the same mycorrhizal pair (54.5 % mycorrhization in stressed plants versus 34.5 % in well-watered ones). Regarding mineral stress, Kagan-Zur et al. (1994) demonstrated that mycorrhizal colonization of *Helianthemum sessiliflorum* roots by the desert truffle *Terfezia boudieri* (thought to be *leonis* at the time) was strongly enhanced by low iron availability (70 % under iron deficiency versus 35 % in well-nourished plants); these experiments were performed using well-watered *Helianthemum* seedlings. Navarro-Ródenas et al. (2012a) observed a similar phenomenon in phosphate stressed *T. claveryi*-mycorrhized *H. almeriense* plants (30 % mycorrhization in P stressed plants versus 19 % in non-stressed plants).

V. Kagan-Zur (✉) • N. Roth-Bejerano
Department of Life Sciences, Ben-Gurion University of the Negev, Beer-Sheva 84105, Israel
e-mail: zur@bgu.ac.il

T. Turgeman • Y. Sitrit
The Jacob Blaustein Institute for Desert Research, Ben-Gurion University of the Negev, Beer-Sheva 84105, Israel

A. Morte
Departmento Biologia Vegetal (Botanica), Facultad de Biologia, Universitad di Murcia, Campus de Espinardo, 30100 Murcia, Spain

V. Kagan-Zur et al. (eds.), *Desert Truffles*, Soil Biology 38,
DOI 10.1007/978-3-642-40096-4_7, © Springer-Verlag Berlin Heidelberg 2014

The higher level of mycorrhizal colonization of roots under stress conditions reflects the ability of mycorrhizal fungi to assist plants in overcoming different types of deprivation. The beneficial effect of mycorrhizal fungi on their plant host's development and adaptation to various harsh conditions has been thoroughly studied over several decades, leading to the publication of numerous papers summarizing the results. A number of books have been published on the subject, starting with Hacskaylo's in 1971 and continuing into the current millennium (see list compiled by Auge 2012). It has long been known that mycorrhized plants have, under water stress or restricted nutrient supply, a higher biomass as compared with non-mycorrhized plants, produce higher overall yields, and exhibit higher nutrient contents, higher photosynthesis rates, and better water use efficiency, as well as greater resistance to soil pathogens (e.g., Bethlenfalvay and Linderman 1992; Fan et al. 2008; Harley and Smith 1983; Jung et al. 2012; Smith and Read 2008). In these respects desert truffle-mycorrhized hosts are no different from other mycorrhizal plants. This chapter sums up knowledge gathered by studying desert truffles in an effort to understand the benefits conferred by the latter on their plant hosts.

7.2 Plant Water Status

7.2.1 Transpiration and Stomatal Conductance

Mycorrhizal fungi alter the water relations of their host plants. Most, if not all, of the relevant parameters that have been studied (stomatal conductance, transpiration, and water potential) exhibit higher values in arbuscular (AM)-mycorrhized plants, at least under drought conditions (e.g., Auge 2001). Fini et al. (2011) found that both ectomycorrhizal and AM plants manifest higher transpiration under certain growth conditions.

Plants mycorrhized by desert truffles show higher rates of survival under drought conditions (Morte et al. 2000, 2010). Morte et al. (2000), studying *H. almeriense* mycorrhized by *T. claveryi*, demonstrated conclusively that this pair had higher transpiration and stomatal conductance rates under both drought and well-watered conditions (Fig. 7.1) than non-mycorrhized plants. Furthermore, the same authors (Morte et al. 2010) showed, in a follow-up paper, that stomatal conductance was more sensitive to drought than shoot water potential. Turgeman et al. (2011), studying *H. sessiliflorum* mycorrhized by *T. boudieri*, corroborated the finding regarding higher transpiration. They also reported that, under well-watered conditions, stomatal closure occurred 1 h later in mycorrhized than in non-mycorrhized plants. In addition, in an unpublished work, the same group demonstrated (Fig. 7.2) that water usage was higher among mycorrhized plants.

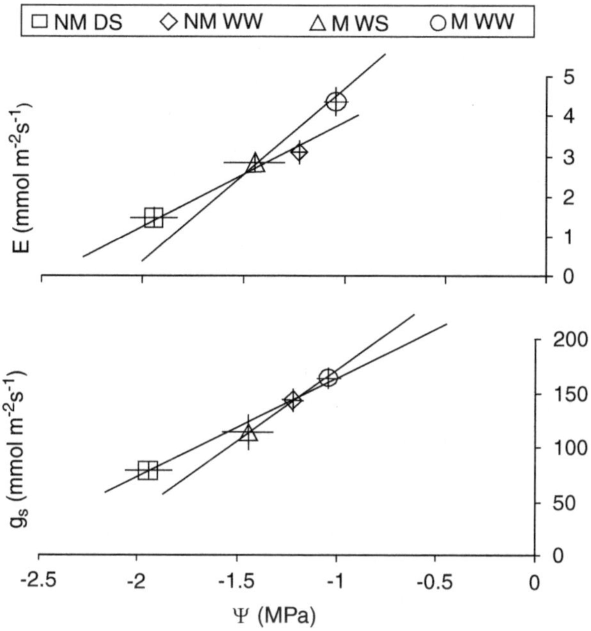

Fig. 7.1 Transpiration rate (E) and stomatal conductance (g_s) plotted versus leaf water potential (ψ) in *Helianthemum almeriense* plants inoculated (M) and not inoculated (NM) with *Terfezia claveryi*. Plants were either subjected to a drought stress (DS) or well watered (WW). (Modified from Morte et al. 2000, with kind permission of Springer Science and Business Media)

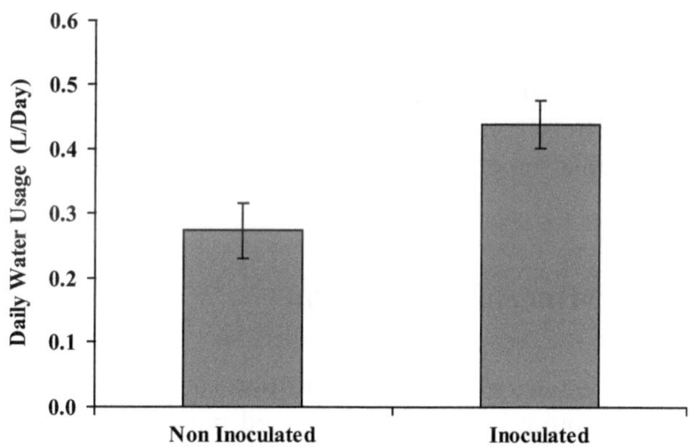

Fig. 7.2 Daily water usage of *Helianthemum sessiliflorum* plants inoculated and non-inoculated by *Terfezia boudieri*. Plants were grown in 10 L pots under greenhouse conditions. The ambient temperatures during the experiment were 36–42 °C in daytime and 20–23 °C at night. Sitrit et al., unpublished

7.2.2 Role of Aquaporins

While seeking to elucidate the mechanisms underlying the phenomena described above, Mushin and Zwiazek (2002) demonstrated that the presence of ectomycorrhiza increases hydraulic conductivity in *Ulmus americana* seedlings. This finding, which implies changes in apoplastic water transport, was corroborated by Marjanovic et al. (2005) for mycorrhized poplar seedlings. The latter authors also examined aquaporins in these seedlings and detected an increase in the water transport capacity of the mycorrhized plants' plasma membrane, suggesting that symplastic water transport also contributes to the higher water transport capacity of mycorrhized plants. Thus both apoplastic and symplastic transport capabilities are affected by mycorrhizal associations. Barazana et al. (2012) showed recently that the same applies to AM mycorrhized plants under both well-watered and drought conditions. Aquaporin genes of mycorrhizal fungi have recently been isolated and studied in both AM (Aroca et al. 2009) and ectomycorrhizas (Dietz et al. 2011). Navarro-Ródenas et al. (2012b), in a study of an aquaporin gene of the desert truffle fungus *T. claveryi*, were able to demonstrate that the fungus possessed elevated water and CO_2 conductivities, boosting the in vitro drought tolerance of the mycelium. The same authors (2013) later studied five aquaporins isolated from *H. almeriense*, the *T. claveryi* host plant, and reported that each had its own characteristic effect on water and/or CO_2 transport when expressed in a heterologous system (yeast plasma membrane). They demonstrated that in mycorrhized *H. almeriense* plants expression of one gene (HaPIP1.1) was down regulated under drought stress conditions.

Several negative correlations between the *H. almeriense* aquaporins and either photosynthesis or stomatal conductance, mostly under drought conditions, were identified by the above authors (Table 7.1). They interpreted these negative correlations to mean that the expression of aquaporins is controlled either by photosynthesis or by stomatal conductance, i.e., aquaporin expression is induced when there is a drop in either parameter. In the mycorrhized *H. almeriense* plants, the *T. claveryi* aquaporin gene was negatively correlated with plant physiological parameters, pointing to some form of communication between the symbionts. This negative correlation, along with the more extensive root colonization observed under stress conditions (see Sect. 7.1), could be among the factors that promote the adaptation of mycorrhized plants to arid (and semiarid) conditions.

7.3 Mineral Acquisition

Fungal modes of mineral acquisition can differ from those of plants. For example, fungi acquire iron through secreted molecules—siderophores (ferric iron specific chelators)—that mobilize extracellular iron to the fungus (Haas 2003; Haselwandter 2008), while plants absorb iron through acidification of the rhizosphere and reduction of Fe^{3+} (Romheld 1987). Also, fungi can spread densely in the soil, thereby exploiting the available soil minerals more efficiently than plants. Transfer of minerals from fungus to plant has been proven to exist and is well documented for both endo- and ectomycorrhizas (e.g., Hahn and Mendgen 2001; Chalot et al. 2002). Thus, in general, mycorrhizal fungi are capable of improving the mineral status of their host plant. Though comparatively few studies have been made of desert truffle-assisted mineral acquisition by host plants, some information exists: Morte et al. (2000) reported higher levels of N, P, and K in shoots of mycorrhized *H. almeriense* plants versus non-mycorrhized plants under drought conditions (Table 7.2). Similar results for P and K, but not for N, were obtained by

Table 7.1 Correlations between net photosynthesis (*A*) or stomatal conductance (g_s) and aquaporin genes expression of mycorrhizal (M) and non-mycorrhizal (NM) *H. almeriense* plants under drought (DS) or well-watered (WW) conditions

	Leaf					Root					
	HaTIP1.1	HaPIP1.1	HaPIP1.2	HaPIP2.1	HaPIP2.2	HaTTP1.1	HaPIP1.1	HaPIP1.2	HaPIP2.1	HaPIP2.2	TcAQP1
NM-DS											
A	−0.983*	−0.692	−0.108	0.888*	−0.617	−0.935*	−0.385	0.478	0.250	0.199	–
g_s	−0.968*	−0.946*	−0.233	0.573	−0.760	−0.953*	−0.716	0.555	0.387	0.103	–
M-DS											
A	−0.024	0.327	0.119	−0.128	0.100	−0.792	0.566	−0.891*	0.548	0.846	−0.889*
g_s	−0.164	0.643	0.024	−0.270	−0.116	−0.922*	0.732	−0.952*	0.175	0.776	−0.755
NM-WW											
A	−0.471	−0.327	0.355	0.157	0.527	−0.420	−0.571	−0.901*	−0.226	−0.882*	—
g_s	−0.688	0.473	0.466	0.218	0.601	−0.781	0.116	−0.666	−0.422	−0.826	—
M-WW											
A	0.259	0.403	0.491	0.024	0.536	0.247	0.563	−0.247	−0.503	−0.250	−0.863
g_s	0.517	0.677	0.655	0.988	0.681	0.098	0.497	−0.381	−0.550	−0.358	−0.866

Values are the Pearson's coefficient

Reprinted from Navaro Rodenas et al. (2013) with kind permission of MPMI

*Indicates significance at $P < 0.05$

Table 7.2 Effect of mycorrhization with *T. claveryi* and drought acclimation on N, P, and K contents (shoot) of *H. almeriense* plants

Plants	Treatment	%N	%P	%K
Well-watered	Control	1.03a	0.13a	0.67a
	T. claveryi	1.09a	0.14a	0.83b
Drought-stressed	Control	0.99a	0.15a	0.71a
	T. claveryi	1.32b	0.18b	0.92b

Values followed by the same letter are not significantly different according to Student's *t*-test
Adapted from Morte et al. (2000), with kind permission of Springer Science and Business Media

Slama et al. (2012) in a study of *H. sessiliflorum* mycorrhized by *T. boudieri* grown on two soil types, under a normal irrigation regime.

7.4 Effects on Photosynthesis

7.4.1 Diurnal Effects

Mycorrhizal associations are known to promote net photosynthesis (Fan et al. 2008; Reid et al. 1983). While the strong carbon demand of the mycorrhized roots may be the factor driving this increase, Kaschuk et al. (2009) showed that in certain cases the net additional photosynthate gain is greater than the quota allocated to the fungus. The ability of mycorrhized plants to achieve higher photosynthesis levels may be explained in part by their better water status (see Sect. 7.2) and higher mineral contents (see Sect 7.3); however, the mechanisms involved remained unclear.

Turgeman et al. (2011) studied diurnal changes in photosynthesis and transpiration in well-watered *H. sessiliflorum* plants that had been mycorrhized by *T. boudieri* (M) versus non-mycorrhized plants (NM). The M plants exhibited perceptible photosynthetic rates in the early morning (7:30 a.m.), and by 2 h after sunrise the photosynthetic rates of the M and NM plants were distinctly different, the differentials between the two groups being maintained throughout the day. The M plants exhibited two distinct peaks, one at 10 a.m. and the other at 5 p.m.; the 10 a.m. peak was typically followed by a decline—between 12.00 and 2:30 p.m.—but later in the afternoon their photosynthetic activity revived and reached a level similar to that measured in the morning. In some experiments the NM plants too exhibited two peaks; however, peak levels were consistently higher in the M plants (Fig. 7.3).

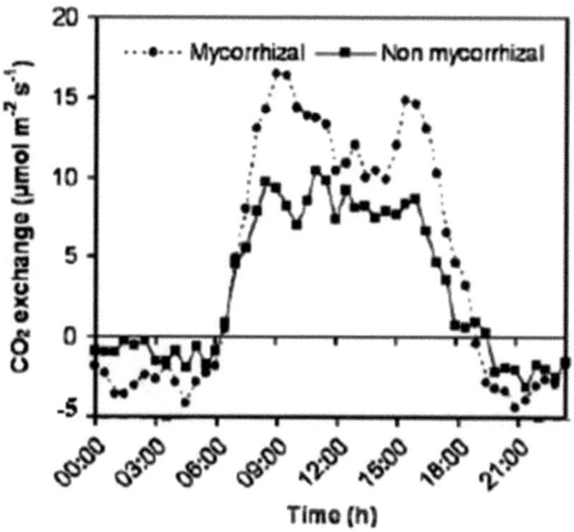

Fig. 7.3 Mycorrhized and non-mycorrhized *Helianthemum sessiliflorum* plants' daily CO_2 exchange. (From Turgeman et al. 2011, with kind permission of Springer Science and Business Media)

7.4.2 Effects of Carbon Sink-Source Manipulation on Photosynthesis

Spring is the fruiting season for host plants and desert truffles alike. With the initial aim of evaluating how much photosynthate is translocated from the canopy to the roots in this period, we girdled adult branches of mycorrhized (M) and non-mycorrhized (NM) plants in such a way that phloem photosynthate flow was disrupted but xylem water flow from roots to canopy was not, thereby maintaining the transport of water (and minerals) to girdled branches while halting photosynthate translocation to the roots and the mycorrhizing fungus. We expected that girdling would reduce photosynthesis, the reasoning being that, as photosynthates accumulated, branch reserves would fill up rapidly, triggering feedback inhibition (e.g., Goldschmidt and Huber 1992; Herold 1980). We had also expected the girdled M branches to exhibit a greater reduction in photosynthesis, reflecting the larger amount of photosynthates remaining in the branch. However, results over the 9 days of measurement revealed a rather small reduction of photosynthetic activity in the M plants and a larger reduction in the NM plants (Fig. 7.4). Two possible explanations, neither of which has been investigated to date, are:

1. Branches of mycorrhized plants, having shifted a large portion of their photosynthates to the roots and fungus, are depleted of starch and are now filling up reserves.
2. Because of their more favorable water and mineral status, the M girdled branches may be capable of diverting the extra photosynthate to vigorous growth of the branch during the limited study period.

Whatever the case, at the initial stage the NM plants were more affected by girdling-induced inhibition of photosynthesis than the M plants.

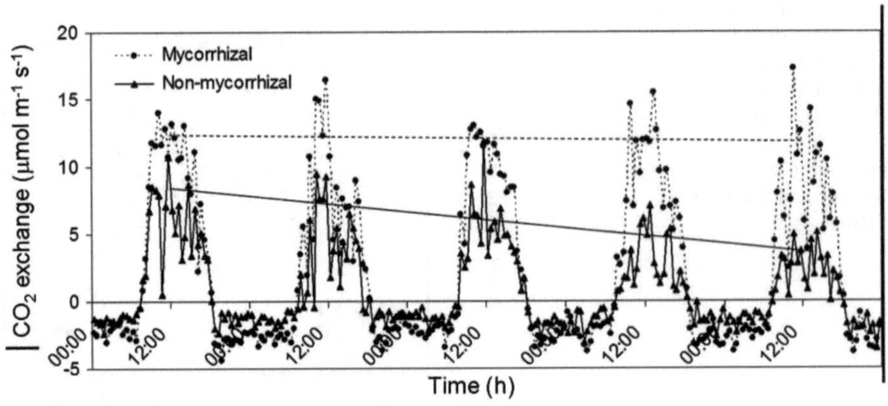

Fig. 7.4 Effect of Girdling on CO_2 assimilation by *Helianthemum sessiliflorum* plants either mycorrhized or not mycorrhized with *Terfezia boudieri*. Main branches of fully developed plants were girdled to a point which disrupted the phloem tissue but not the xylem. Photosynthetic activity was measured using a PTM-48 device

7.4.3 Mycorrhizal Effects on Chlorophyll b *Level*

The highest photosynthesis activity (Fig. 7.3) of M plants occurred under low photosynthetic photon flux level, when photon flux should be a limiting factor for efficient photosynthesis. This might be attributed to the M plants' having an elevated total chlorophyll content, and indeed Morte et al. (2000), studying *H. almeriense* inoculated with *T. claveryi*, found a higher level of total chlorophyll in mycorrhized plants under water stress (but not under well-watered conditions). Turgeman et al. (2011) found, by means of a detailed analysis of chlorophyll composition, that although both Chl *a* and Chl *b* concentrations were significantly higher in M plants (by comparison with NM plants), the increase in Chl *b* was larger (2.4-fold increase versus 1.52-fold for Chl *a*; Fig. 7.5). An elevated Chl *b* content improves light absorbance at lower energy wavelengths and as such supports light harvesting under the lower irradiance conditions typical of mornings and afternoons (Chow et al. 1990; Melis and Harvey 1981).

7.4.4 Photosynthesis Activation Energy

Taken together, the elevated Chl *b* levels of the mycorrhized plants, their higher photosynthesis rates under low photosynthetic photon flux, and their significantly higher photosynthetic rates under suboptimal temperatures (data not shown) imply a lower activation energy for onset of photosynthesis. Indeed, Turgeman et al. (2011) calculated the activation energy using the Arrhenius equation and

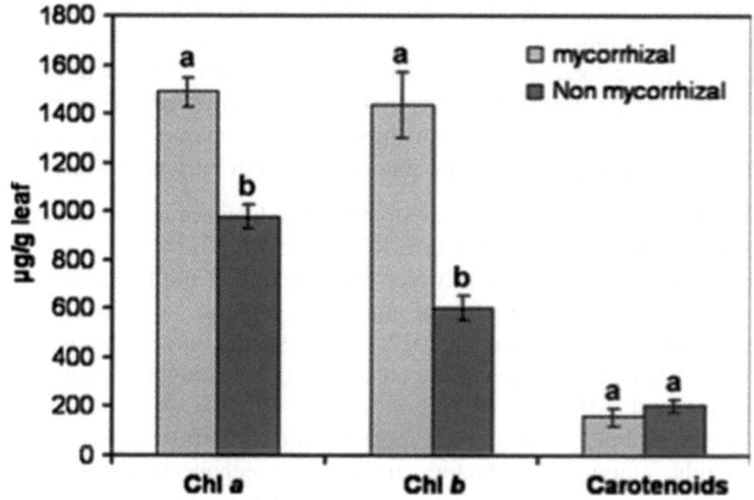

Fig. 7.5 Chl *a*, Chl *b*, and carotenoid contents of M and NM plant leaves sampled from plants 8 months post-germination; values are means ± SE ($n = 10$). Different *lowercase letters* denote significant differences ($P < 0.05$) (From Turgeman et al. 2011 with kind permission of Springer Science and Business Media)

found it to be 48.62 kJ mol^{-1} for M plants versus 61.56 kJ mol^{-1} for NM plants, or about 21 % lower.

Thus the higher photosynthesis rates of mycorrhized plants are due to:

(a) An elevated total chlorophyll, which facilitates assimilation by the photosynthetic machinery
(b) An elevated proportion of Chl *b*, which promotes photosynthesis under less favorable light and temperature conditions
(c) Longer hours of stomatal opening and higher stomatal water conductance (see Sect. 7.2)
(d) Higher removal of CO_2, through assimilation or higher stomatal conductance of CO_2

All these elements are reflected in the lower photosynthesis activation energy found for mycorrhized plants. However, the fungal signals responsible for the elevated chlorophyll content and the change in chlorophyll composition in M plants have yet to be elucidated.

7.5 Conclusions

Various aspects of the benefits conferred by mycorrhizal fungi on their plant hosts have been the object of research. Of these, beneficial effects on agricultural yields (e.g., Bethlenfalvay and Linderman 1992) are not relevant to desert truffles, whose natural plant hosts are of no agricultural value. Another benefit, namely, higher resistance to soil pathogens (e.g., Pozo and Azcon-Aguilar 2007; Jung et al. 2012), has never been investigated with reference to desert truffles and their host plants.

In this chapter we examined three areas in which mycorrhization benefits the plant: plant water status, mineral acquisition, and photosynthesis. The research we described reveals that these areas are interconnected in intricate ways:

- The level of photosynthesis and/or stomatal conductance affects the expression of a fungal aquaporin gene (Navarro-Ródenas et al. 2012b, 2013).
- In addition to playing a role in water transfer, one of the five plant aquaporin genes studied, as well as a fungal aquaporin gene, are also involved in CO_2 transfer (Navarro-Ródenas et al. 2012a, b, 2013), and thus in photosynthesis. In mycorrhized plants these genes are enhanced under drought (Morte et al. 2000; Turgeman et al. 2011).
- The elevated photosynthesis rates recorded for mycorrhized plants (which are attributable, among other factors, to a higher enhancement of *Chl b* gene expression as compared with *Chl a*) are accompanied by a higher water use efficiency and longer duration of stomatal opening (Turgeman et al. 2011).

The actual controls involved in the processes that trigger the cascade of changes observed upon transition from the non-mycorrhized to the mycorrhized state are still far from clear.

References

Aroca R, Bago A, Sutka M, Paz JA, Cano C, Amodeo G, Ruiz-Lozano JM (2009) Expression analysis of the first arbuscular mycorrhizal fungi aquaporin described reveals concerted gene expression between salt-stressed and nonstressed mycelium. Mol Plant Microbe Interact 22:1169–1178. doi:org/10.1094/MPMI-22-9-1169

Auge RM (2001) Water reactions, drought and vesicular-arbuscular mycorrhizal symbiosis. Mycorrhiza 11:3–42

Auge RM (2012) Mycorrhizal books. http://mycorrhiza.ag.utk.edu/mbook.htm

Barazana G, Aroca R, Pas JA, Chaumont F, Martinez-Ballestra MC, Carvajal M, Ruiz-Lozanol JM (2012) Arbuscular mycorrhizal symbiosis increases relative apoplastic water flow in roots of the host plant under both well-watered and drought stress conditions. Ann Bot. doi:10.1093/aob/mcs007

Bethlenfalvay GJ, Linderman RG (1992) Mycorrhizae in sustainable agriculture, vol 54, ASA Special Publication. Agronomy Society of America, Madison, WI, p 124. ISBN 0-89118-112-1

Chalot M, Javelle A, Blaudez D, Lambilliote R, Cooke R, Sentenac H, Wipf D, Botton B (2002) An update on nutrient transport processes in ectomycorrhizas. Plant Soil 244:165–175

Chow WS, Melis A, Anderson JM (1990) Adjustments of photosystem stoichiometry in chloroplasts improve the quantum efficiency of photosynthesis. Proc Natl Acad Sci U S A 87:7502

Davies FT Jr, Svenson SE, Cole JC, Phavaphutanon L, Duray SA, Olalde-Portugal O, Meier CE, Bo SH (1996) Non-nutritional stress acclimation of mycorrhizal woody plants exposed to drought. Tree Physiol 16:985–993

Dietz S, von Below J, Beitz E, Nehls U (2011) The aquaporin gene family of the ectomycorrhizal fungus *Laccaria bicolor*: lessons for symbiotic functions. New Phytol 190:927–940

Douds DD, Janke RR, Peters SE (1993) VAM fungus spore populations and colonization of roots of maize and soybean under conventional and low-input sustainable agriculture. Agric Ecosyst Environ 43:325–335

Fan Y, Luan Y, An L, Yu K (2008) Arbuscular mycorrhizae formed by *Penicillium pinophilum* improve the growth, nutrient uptake and photosynthesis of strawberry with two inoculum-types. Biotechnol Lett 30:1489–1494

Fini A, Frangi P, Amoroso G, Piatti R, Faoro M, Bellasio C, Ferrini F (2011) Effect of controlled inoculation with specific mycorrhizal fungi from the urban environment on growth and physiology of containerized shade tree species growing under different water regimes. Mycorrhiza 21:703–719

Goldschmidt AE, Huber SC (1992) Regulation of photosynthesis by end-product accumulation in leaves of plants storing starch, sucrose, and hexose sugars. Plant Physiol 99:1443–1448, http://dx.doi.org/10.1104/pp.99.4.1443

Haas H (2003) Molecular genetics of fungal siderophore biosynthesis and uptake: the role of siderophores in iron uptake and storage. Appl Microbiol Biotechnol 62:316–330

Hacskaylo E (1971) Mycorrhizae. U.S. Government Printing Office, Washington, DC, p 255. ISBN 0-409-068

Hahn M, Mendgen K (2001) Signal and nutrient exchange at biotrophic plant–fungus interfaces. Curr Opin Plant Biol 4:322–327

Harley JL, Smith SE (1983) Mycorrhizal symbiosis. Academic, New York

Haselwandter K (2008) Structure and function of siderophores produced by mycorrhizal fungi. Mineral Magazine 77:61–64

Herold A (1980) Regulation of photosynthesis by sink activity—the missing link. New Phytol 86:131–144. doi:10.1111/j.1469-8137.1980.tb03184.x

Jung SC, Martinez-Medina A, Lopez-Raez JA, Pozo MJ (2012) Mycorrhiza-induced resistance and priming of plant defenses. J Chem Ecol 38:651–654

Kagan-Zur V, Raveh E, Lischinsky S, Roth-Bejerano N (1994) *Helianthemum–Terfezia* association is enhanced by low iron in the growth medium. New Phytol 127:567–570

Kaschuk G, Kuyper TW, Leffelaar PA, Hungria M, Giller KE (2009) Are the rates of photosynthesis stimulated by the carbon sink strength of rhizobial and arbuscular mycorrhizal symbioses? Soil Biol Biochem 41:1233–1244

Marjanovic A, Uehlein N, Kaldenhoff R, Zwiazek JJ, Weiss M, Hampp R, Nehls U (2005) Aquaporins in poplar: what a difference a symbiont makes! Planta 222:258–268

Melis A, Harvey G (1981) Regulation of photosystem stoichiometry chlorophyll *a* and chlorophyll *b* content and relation to chloroplast ultrastructure. Biochim Biophys Acta (BBA)—Bioenergetics 637:138–145

Morte A, Lovisolo C, Schubert A (2000) Effect of drought stress on growth and water relations of the mycorrhizal association Helianthemum almeriense-Terfezia claveryi. Mycorrhiza 10:115–119

Morte A, Navarro-Rodenas A, Nicolas E (2010) Physiological parameters of desert truffle Mycorrhizal *Helianthemum almeriense* plants cultivated in orchards under water deficit conditions. Symbiosis 52:133–139

Mushin TM, Zwiazek JJ (2002) Ectomycorrhizas increase apoplastic water transport and root hydraulic conductivity in Ulmus Americana seedlings. New Phytol 153:153–158

Navarro-Ródenas N, Pérez-Gilabert M, Torrente P, Morte A (2012a) The role of phosphorus in the ectendomycorrhiza continuum of desert truffle mycorrhizal plants. Mycorrhiza 22:565–575

Navarro-Ródenas A, Ruíz-Lozano JM, Kaldenhoff R, Morte A (2012b) The aquaporin *TcAQP1* of the desert truffle *Terfezia claveryi* is a membrane pore for water and CO_2 transport. Mol Plant Microbe Interact 25:259–266

Navarro-Rodenas A, Barzana G, Nicolas E, Carra A, Schubert A, Morte A (2013) Expression analysis of aquaporins from desert truffle mycorrhized symbiosis reveals a fine-tuned regulation under drought. Mol Plant Microbe Interact, http://dx.doi.org/10.1094/MPMI-07-12-0178-R

Pozo MJ, Azcon-Aguilar C (2007) Unraveling mycorrhiza-induced resistance. Curr Opin Plant Biol 10:393–398

Reid CPP, Kidd FA, Ekwebelam SA (1983) Nitrogen nutrition, photosynthesis and carbon allocation in ectomycorrhizal pine. Plant Soil 71:415–431

Romheld V (1987) Different strategies for iron acquisition in higher plants. Physiol Plant 70:231–234

Slama A, Gorai M, Fortas Z, Boudabous A, Neffati M (2012) Growth, root colonization and nutrient status of *Helianthemum sessiliflorum* Desf. inoculated with a desert truffle *Terfezia boudieri* Chatin. Saudi J Biol Sci 19:25–29

Smith SE, Read DJ (2008) Mycorrhizal symbiosis, 3rd edn. Academic, London. ISBN 978-0-12-370526-6

Turgeman T, Ben-Asher Y, Roth-Bejerano N, Kagan-Zur V, Kapulnik Y, Sitrit Y (2011) Mycorrhizal association between the desert truffle *Terfezia boudieri* and *Helianthemum sessiliflorum* alters plant physiology and fitness to arid conditions. Mycorrhiza 21:623–630. doi:10.1007/s00572-011-0369-z

Part III
Distribution

Chapter 8
Ecology and Distribution of Desert Truffles in Western North America

James M. Trappe, Gábor M. Kovács, and Nancy S. Weber

8.1 Introduction

The four major North American desert regions occur in the far west of the continent (Fig. 8.1). The Great Basin Desert (411,000 km^2) extends from southeastern Oregon and southern Idaho through Nevada, much of Utah, southwestern Wyoming, far western Colorado, the northwestern corner of New Mexico, and far northern Arizona. The Mojave (65,000 km^2) occupies southeastern California, southern Nevada, and part of western Arizona. The Sonoran (313,000 km^2) reaches from southeastern California and southwestern Arizona south into Baja California and northwestern Mexico. The Chihuahuan (455,000 km^2) extends from south central New Mexico and western Texas into the central Mexican plateau. Altogether, these desert systems cover about 1,244,000 km^2 (Britannica online 2012), somewhat less than the Australian Outback deserts or a bit more than the Kalahari (see Chaps. 13 and 14). Adding the area of semiarid land associated with these deserts would probably increase the total of very dry habitats to more than 2,000,000 km^2 in North America.

Two truffle genera, *Carbomyces* and *Stouffera*, together comprising four species, have been reported from these deserts (Trappe and Weber 2001; Kovács

J.M. Trappe (✉)
Department of Forest Ecosystems and Society, Oregon State University, Corvallis, OR 97331-5752, USA

USDA Forest Service, Pacific Northwest Research Station, Forestry Sciences Laboratory, Corvallis, OR 97331, USA
e-mail: trappej@onid.orst.edu

G.M. Kovács
Department of Plant Anatomy, Institute of Biology, Eötvös Loránd University, Pázmány Péter sétány 1/C, 1117 Budapest, Hungary

N.S. Weber
Department of Forest Ecosystems and Society, Oregon State University, Corvallis, OR 97331-5752, USA

V. Kagan-Zur et al. (eds.), *Desert Truffles*, Soil Biology 38,
DOI 10.1007/978-3-642-40096-4_8, © Springer-Verlag Berlin Heidelberg 2014

Fig. 8.1 Map of North American deserts.

et al. 2011). *Mattirolomyces* with one species is tentatively included, having been collected in or near a desert habitat. We exclude from this chapter taxa occurring in riparian zones along rivers traversing deserts, on high mountains surrounded by the deserts but receiving enough precipitation to support xeric woodlands or forests (e.g., Fogel and Pacioni 1989; Kropp et al. 2012), and desert secotioid basidiomycetes not strictly hypogeous (Moreno et al. 2007).

8.2 North American Deserts

The North American deserts are characterized by high diversity. They extend more than 2,200 km from the northern edge of the Great Basin Desert at about 42° N. lat. in Oregon south nearly to the Tropic of Cancer at about 22° N. lat. at the southern margins of the Sonoran and Chihuahuan Deserts in Mexico (Fig. 8.1). West to east they lie in the rain shadows of the Cascade, Sierra Nevada, and Sierra Madre Occidental mountains, high north–south aligned ranges that intercept rainstorms blowing in from the Pacific Ocean. The broad Rocky Mountains and their subsidiary ranges rise to high elevations on the east boundaries of these deserts, capturing precipitation from continental storms to nourish forests but also producing localized rain shadows. In northwestern Mexico, the trade winds drop little precipitation, so that even the Baja California peninsula is mostly Sonoran Desert despite being surrounded by water on three sides (Phillips and Comus 2000).

These deserts have been inhabited by indigenous Americans for thousands of years. Major archeological sites abound, especially in Utah (Great Basin Desert), Arizona (Sonoran Desert), and New Mexico (Chihuahuan Desert): many of these cultures developed advanced agricultural practices that extensively influenced plant communities as far back as 3,000 years (Phillips and Comus 2000; Columbian Electronic Encyclopedia 2005).

8.2.1 Landscapes

All of the North American deserts encompass mountain ranges, large and small. This diversity of topography engenders diversity of soils, climate, and vegetation.

The Great Basin Desert (Fig. 8.1) is the largest and northernmost of the four North American deserts and averages the highest elevations, which range from about 1,200 to 2,000 m or more in elevation between mountain ranges that may reach as high as nearly 4,000 m (USDA Forest Service 1994). Most of the Great Basin Desert creeks and rivers flow into low areas and disappear, creating lakes in wet years that dry into playas in dry years and saline lakes such as Utah's Great Salt Lake (Grayson 2011). Exceptions to drainage into the basin are in the north and northeast of the Great Basin Desert, where streams flow into the Snake River drainage to find their way to the Pacific Ocean, and in southeastern Utah and northeastern Arizona, where streams empty into the ocean-bound Colorado River. These exceptions are not part of the Great Basin geographically but are considered extensions of the Great Basin Desert. Magnificent canyons crisscross the Great Basin Desert, including the Grand Canyon of the Colorado and Canyonlands National Park.

At its south margin the Great Basin Desert merges into the Mojave Desert (Fig. 8.1), notable for California's Death Valley, the lowest place in North America at about 86 m below sea level. The Mojave is bounded to the west by the Sierra

Nevada range, including the highest peak of the lower 48 states, Mt. Whitney, elevation 4,420 m; to the north by the ranges that enclose Death Valley and the Great Basin; to the east by the Colorado Plateau; and to the south by the Sonoran Desert (Webb et al. 2009).

The Sonoran Desert (Fig. 8.1) extends south from the Mojave into southwestern Arizona and Mexico's Baja California peninsula and northwest mainland. It ranges from its hundreds of kilometers of ocean and gulf beaches to colorful canyons and mesas and boasts a wealth of cactus and shrub species. Spectacular canyons and the lower Grand Canyon of the Colorado traverse the Sonoran Desert. It has been said that the Sonoran region encompasses most of the world's biomes, from dry tropical forest through thornscrub, desert, grassland, chaparral, temperate deciduous forest, coniferous forest, and tundra (Phillips and Comus 2000). Of these, the Sonoran desert and its ecotones with grassland and thornscrub are the potential habitats for desert truffles.

The Chihuahuan Desert (Fig. 8.1) is separated geographically from the other three North American desert regions. It extends from the Jornada Basin in south central New Mexico, east to far western Texas and the Rio Grande Basin and south between the Sierra Madre Occidental and Sierra Madre Oriental through the Central Mexican Plateau to slightly south of the Tropic of Cancer. The Chihuahuan Desert contains north–south mountain ranges with broad, intervening desert basins (Haystad et al. 2006). Much of the Rio Grande Basin is below elevation 910 m, but most of the rest of the Chihuahuan Desert ranges from 1,200 to 1,830 m (Lee et al. 2011).

The metropolitan areas within all four deserts profoundly change the habitats over their large surrounding areas. These include Boise, Idaho, Salt Lake City, Utah and Reno, Nevada within the Great Basin Desert, while Las Vegas, Nevada, strikingly exemplifies post-World War II urban sprawl in the Mojave Desert. The Sonoran Desert includes Phoenix, Tucson, and Yuma Arizona in the USA and Hermosillo, Sonora in Mexico. The Chihuahuan desert includes Albuquerque, New Mexico, and El Paso, Texas in the USA as well as Chihuahua and Durango in Mexico.

The geology of the huge extent of North American deserts is characterized by complex and often localized events, from seafloor depositions to tectonic, volcanic, and glacial, much of it ongoing. We cannot do justice here: suffice it to say that all of these deserts have developed extensive areas of sand potentially suitable for the desert truffles and their hosts.

8.2.2 Climate

The North American deserts differ strongly in temperatures and rainfall due to latitude, elevation, and mountains that constrain maritime influences on weather. Moreover, the north–south alignment of the major mountain ranges leaves relatively little impediment to southward flow of arctic cold fronts. For purposes of this chapter, only the desert communities are considered. For example, the higher mountains studding the Great Basin Desert get enough precipitation to support

Table 8.1 Average annual and monthly maximum and minimum precipitation for selected weather stations in North American Deserts (Western Regional Climate Center 2013; Colegio de postgraduados 2013)

Weather station	Mean annual (mm)	Mean monthly max. (mm)	Month of mean annual max.	Mean monthly min. (mm)	Month of mean annual min.
GB—Burns, Oregon	264	41	Dec	6	Aug
GB—Reno, Nevada	185	26	Jan	4	Jul
MO—Las Vegas, Nevada	106	19	Feb	3	May
MO—Death Valley, California	60	13	Feb	1	Jun
SO—Phoenix, Arizona	200	27	Jul	1	Jun
SO—Yuma, Arizona	76	42	Jan	1	May
CH—Albuquerque, New Mexico	240	40	Jul	10	May
CH—El Paso, Texas	246	51	Jul	6	Apr
CH—Chihuahua, Mexico	327	87	Aug	3	Feb

GB Great Basin, *MO* Mojave, *SO* Sonoran, *CH* Chihuahuan

vegetation communities ranging from woodlands to forests to subalpine and alpine communities, but only the deserts will be discussed. Precipitation and temperatures are taken from the Western Regional Climate Center (2013) and the Colegio de Postgraduados (2013) as summarized online in Wikipedia (search "weather records" and name of city in Google).

As storms from the Pacific hit the Cascade and Sierra Nevada ranges, most of their moisture precipitates onto the western slopes, leaving the rain shadow to the east that produces the Great Basin Desert. The Burns, Oregon weather station at the northern edge of the Great Basin records an average yearly precipitation of about 264 mm (Table 8.1), mostly falling in November through March. The average maximum monthly temperature of 31 °C usually occurs in July (Table 8.2), as did the record high of 42°C. The average monthly low temperature is −9 °C, in January, which also produced the record low of −30°C. Further to the south, Reno, Nevada averaged only 190 mm annual precipitation and is somewhat warmer than Burns, Oregon in both summer and winter (Tables 8.1 and 8.2). The Great Basin Desert is the coldest of the North American deserts, with the growing season confined to summer months (Lee et al. 2011). Only one desert truffle collection is recorded there, and it was at its relatively warm southeast margin adjoining the Sonoran Desert.

Located at relatively low elevations in the rain shadow of the highest peaks of the Sierra Nevada, the Mojave Desert is notably drier and warmer than most of the

Table 8.2 Average maximum and minimum monthly temperatures and extreme records for selected weather stations in North American Deserts (Western Regional Climate Center 2013; Colegio de postgraduados 2013)

Weather station	Month max. (°C)	Month of mean max.	Record max. (°C)	Month min. (°C)	Month of mean min.	Record min. (°C)
GB—Burns, Oregon	31	Jul	42	−9	Jan	−30
GB—Reno, Nevada	34	Jul	42	−5	Dec	−28
MO—Las Vegas, Nevada	40	Jul	47	4	Dec	−13
MO—Death Valley, California	47	Jul	57	−1	Dec	−6
SO—Phoenix, Arizona	42	Jul	48	7	Dec	−2
SO—Yuma, Arizona	42	Jul	51	7	Jan	−4
CH—Albuquerque, New Mexico	33	Jul	42	−4	Dec	−27
CH—El Paso, Texas	36	Jun	48	0	Dec	−22
CH—Chihuahua, Mexico	33	Jun	40	7	Jan	−10

GB Great Basin, *MO* Mojave, *SO* Sonoran, *CH* Chihuahuan

Great Basin Desert to its North (Tables 8.1 and 8.2). At Las Vegas, the Mojave precipitation averages only 106 mm annual precipitation, the rainiest month on average being January with 17.5 mm. July produced the average maximum temperature at 40°C and the record high of 47°C. The hottest months have been June through September. The coldest months have been December and January at an average of 4°C, the record low being −13°C in January. The Mojave also contains Death Valley, the driest and hottest place in North America. Its average annual precipitation is 60 mm, and in most years some months receive none; 1919 and 1953 were totally rain-free, and no rain fell for 40 consecutive months in 1931–1934. Its record high temperature was 56.7°C in July 1913, the hottest temperature ever recorded worldwide (World Meteorological Organization 2013). July also produces the highest average temperature of 47°C. The coldest month in Death Valley is December, averaging −1°C.

The Sonoran comes close to rivaling the Mojave for heat and aridity. The Phoenix, Arizona weather station records an average annual precipitation of 200 mm; July averages the wettest month with 27 mm (Table 8.1); most of the rain falls in July through September when wet air surges in from the Pacific Ocean. The average monthly maximum temperature is 42 °C in July, the record high of 48°C also occurred in July (Table 8.2). The average monthly winter minimum is 7°C in December. Yuma, Arizona has similar ranges of precipitation and maximum temperatures as Phoenix, but the average minimum (16°C) is warmer, the record low is −4°C, somewhat colder than Phoenix. Cabo San Lucas in Baja California

Sur, Mexico, is near the tip of the Baja California Peninsula (data not included in Tables 8.1 or 8.2); its climate is moderated by the maritime environment with somewhat more rain, lower average maximum temperatures, and higher average minima than Phoenix or Yuma.

The Chihuahuan Desert extends more to the south than the other deserts but is generally at higher elevations than the Mojave and Sonoran Deserts. In their temperature patterns, the three Chihuahuan weather stations (Table 8.2) more closely resemble those of the Great Basin Desert far to the north than those of the intervening Mojave and Sonoran Deserts. The Chihuahuan stations report higher precipitation on average than the other three deserts, and their rainy season is in summer as is that of the Sonoran Desert, thus differing from the winter rains of the Great Basin and Mojave Deserts (Table 8.1).

8.3 North American Desert Truffles

8.3.1 History of Discovery

The history of desert truffle discoveries in North America is relatively prosaic: no heroic exploring expeditions, no encounters with native tribesmen to reveal their use of truffles. This is not to say that indigenous people in North American deserts did not use truffles, it is only to acknowledge that no such use has been recorded. Moreover, some early collectors may have had adventures not recorded for posterity.

E. Forges found a truffle along the banks of the Red River in northwestern Louisiana in 1886. We know nothing about Forges, but the truffle collection was given to Rev. A. B. Langlois, of whom it was said "Louisiana is at this time the fortunate possessor of a most industrious and acute botanist in the person of Rev. A. B. Langlois, of St. Martinville" (Lamson-Scribner 1893). Langlois was a prolific collector of plants and fungi. His herbarium, now housed at the University of North Carolina, numbers about 20,000 collections (McCormick 2012). He, in turn, sent some of Forges' truffles to New Jersey mycologist Job Ellis, who in collaboration with B. M. Everhart had started the sets of exsiccata known as *North American Fungi* sent to dozens of herbaria around the world (Kaye 1986). Forges' collection was large enough to be split for inclusion in *North American Fungi Ser. 2* as Nr. 1782, *Terfezia leonis* (Tul. & C. Tul.) Tul. & C. Tul. The Californian mycologist H. W. Harkness had Nr. 1782 and concluded it was not *T. leonis* but rather a new species, which he described as *Terfezia spinosa* Harkn. (Harkness 1899). Harkness' split of the Forges collection thus becomes the holotype of *T. spinosa*, now in the Harkness Collection in the US National Fungus Collections. The other splits scattered around the world are isotypes. The species rested with this name for more than a century, until Kovács et al. (2011) subjected it to phylogenetic analysis to demonstrate it belongs in the genus *Mattirolomyces*.

Louisiana is not a desert, so what is the relevance of this story? As we will show in the following section, *Mattirolomyces spinosus* occurs in or near North American deserts as well.

William H. Long of the US Bureau of Plant Industry was the first mycologist to specialize in fungi of the arid southwestern USA from his retirement in 1937 into the early 1940s. During those years he accounted for many collections of the desert truffle genus *Carbomyces* and the type species of *Carbomyces* and *Stouffera*, the latter named for his companion collector David J. Stouffer (Kovács et al. 2011). Helen M. Gilkey, for 50 years North America's renowned expert on truffle taxonomy, examined Long's ascomycete collections and determined them to represent a new genus. She described *Carbomyces* and two species, *Carbomyces emergens* and *Carbomyces longii* (1954). Later, Nancy Weber discovered that a third species had been misinterpreted as a basidiomycete, so it was named *Carbomyces gilbertsonii* in honor of the much respected Arizona mycologist, Robert Gilbertson (Trappe and Weber 2001). Meanwhile, North American desert truffles were collected sporadically and opportunistically until Mexican mycologist Marcos Lizárraga and his associates collected a large number of *Carbomyces* spp. in northern Mexico in 2008–2011 (Moreno et al. 2012). Occasional recent collections by others such as Robert M. Chapman in New Mexico have added to our knowledge of distributions, and phylogenetic analyses have clarified taxonomic relationships and revealed new taxa (Kovács et al. 2011).

8.3.2 Taxonomy, Endemism, and Distribution

The desert truffle taxa discussed in this chapter are described in detail and illustrated by Trappe and Weber (2001), Kovács et al. (2011), and Moreno et al. (2012). Of the three genera of desert truffles known from North America, two are endemics (*Carbomyces*, *Stouffera*), whereas the third, *Mattirolomyces*, is widely distributed in both northern and southern hemispheres from mesic forests to semiarid and arid habitats (see Chaps. 13 and 14).

Carbomyces (Carbomycetaceae) contains three species, *C. emergens* Gilkey (Fig. 8.2), *C. gilbertsonii* N. S. Weber and Trappe, and *C. longii* Gilkey (Trappe and Weber 2001). The genus is readily differentiated from all other desert truffles worldwide by its brown asci that disintegrate by maturity. It occurs in the USA from New Mexico through Arizona to California in the Chihuahuan, Sonoran and Mojave deserts, respectively, and in the Mexican state of Chihuahua in the Chihuahuan Desert. The other two species are known in the USA only from the type collections. Both have recently been found in the Chihuahuan Desert in Mexico (Moreno et al. 2012).

Stouffera, with its single species *Stouffera longii*, is known only from the type locality in the southeast corner of the Great Basin Desert in northwestern New Mexico. Its spores are distinctive by being reticulate but with the spore surface within the reticular walls having minute rounded bumps (Kovács et al. 2011).

Fig. 8.2 Ascomata of *Carbomyces emergens* from the Chihuahuan Desert. Scale bar = 25 mm. *Top*: cross-section of fresh ascoma showing white tramal veins separating dark pockets of asci and spores (image courtesy of John Zak and *Mycologia*). *Middle*: surface view of dry ascoma found lying loose on ground; note breaks in peridium (image courtesy of Robert M. Chapman). *Bottom*: cross-section of dry ascoma showing the powder of spores and collapsed asci (image courtesy of Robert M. Chapman)

Two species of *Mattirolomyces* have been found in North America, but only *M. spinosus* has been found in arid or semiarid environments (Kovács et al. 2011). As noted above in the history section, the type of *M. spinosus* was found along the banks of a river in Louisiana, not an arid habitat. However, the genus *Mattirolomyces* occurs in desert habitats in Australia and southern Africa (Trappe et al. 2010). One North American collection of *M. spinosus* was from Arizona, but included no data on time or specific location. For the present we do not know if it occurred in desert, semi-desert, dry woodland, or forest. Because Arizona is mostly occupied by deserts (Great Basin, Mojave, Sonoran), we elected to include *M. spinosus* here until its total distribution becomes better known.

Edibility of the North American desert truffles is unknown. Presumably they are not toxic in keeping with the edibility of the west Asian, northern African, southern African Kalahari, and Australian desert truffles. However, the North American species have mostly been found in a dry, powdery state that seems unpalatable.

8.3.3 Ecology: Key to Distribution

C. emergens (Fig. 8.2) is the only North American truffle documented well enough to offer a glimpse of its habitat. Notes accompanying collections by W. H. Long in the Chihuahuan and Mojave Deserts report its habitat variously as "hypogeous. . . in sandy soil on ridge. . .in soil on sagebrush area. . .in mesquite sandhill area" (mesquite is *Prosopis glandulosa* Torr., a member of the Fabaceae). Zak and Whitford (1986), who found *C. emergens* fruiting in the Chihuahuan Desert in the same general area where it had been collected 40 years earlier by Long, describe the habitat: "The site consisted of coppiced dunes vegetated with *Atriplex canescens* (Pursh) Nutt., *P. glandulosa* Torr., several spring flowering annuals, *Lepidium lasiocarpum* Nutt., and *Lesquerella gordonii* (Gray) Wats. The interdune spaces were generally devoid of vegetation. The ascocarps were discovered 2–5 cm below the soil surface in and around recent rodent digs located in the interdune areas."

That *C. emergens* was evidently dug by rodents suggests mycophagy as a means of spore dispersal. Given the time of year and habitat in which they found the rodent diggings, Zak and Whitford (1986) reckoned that spotted ground squirrels (*Xerospermophilus spilosoma*) might be the mycophagists (Fig. 8.3). This truffle also has an alternative method of spore dispersal. The species epithet *emergens* indicates that it can emerge through the soil surface. Once it emerges, it may dry in situ and be transported by wind or water. This is supported by notes on some of Long's collections: "Loose in sandy wash north of airport garbage dump. . .in wash where water washed (the ascocarps). . .loose in sand wash. . .on top of ground but loose." Zak and Whitford (1986) found ascocarps lying loose on the soil surface 15 km N of the original collection site. Dr. Robert Gilbertson provided a specimen of *C. emergens* for examination by Trappe and Weber (2001); he found it caught in a brush pile in his backyard at the edge of the desert (R. Gilbertson personal communication).

Clearly *C. emergens* indeed emerges and is moved around to be caught in arroyos, clumps of vegetation, brush piles, etc. The success of this mechanism for spore dispersal lies in its anatomy. The fresh ascomata studied by Zak and Whitford (1986) had a solid gleba, which was also evident in young specimens dried for herbarium accession. The tissues of these specimens rehydrated readily in the laboratory and consisted of large, thin-walled cells and asci (Trappe and Weber 2001). The ascomata lift themselves to the surface as they expand and the spores mature. As they dry in situ, they detach from the sand, and their inflated, thin-walled

Fig. 8.3 Spotted ground squirrel, a likely eater of desert truffles in the Chihuahuan Desert. Adapted from http://commons.wikimedia.org/wiki/File:Spotted_ground_squirrel.jpg. April 2013

glebal hyphae collapse. When wind or water move them about, the peridia break open or are abraded away by sand while the glebae become reduced to a powder of spores and fragmented asci and tramal cells (Fig. 8.2). It is easy to envision this spore-bearing powder escaping through the broken peridia to be dispersed as the dry, wind-blown ascomata bounce along the ground.

The mycorrhizal hosts of North American desert truffles are unknown. The notes accompanying *Carbomyces* collections mention several common woody perennials that often occur together. All are regarded as forming arbuscular mycorrhizae or being nonmycorrhizal (Wang and Qiu 2006), although *Prosopis* has been found to be ectomycorrhizal as well in one case (Frioni et al. 1999). Desert truffles in the Pezizaceae, however, are recorded as forming unusual types of mycorrhizae on annual and perennial plants (see Chap. 5). DNA needs to be analyzed to resolve the question.

Trappe and Weber (2001) and Kovács et al. (2011) together list 21 collections of all species of North American desert truffles in the USA; Moreno et al. (2012) added about 30 from Mexico. As noted in Sect. 8.2, the Great Basin and Chihuahuan Deserts average the highest in elevation and coolest of the four North American deserts, the Mojave and Sonoran being generally warmer and drier. Other factors such as soils are little known in terms of truffle distribution, except that sands are indicated for those collections for which the substrate was recorded. Associated vegetation is similarly little recorded. Of the roughly 50 total collections, one each is from the relatively cool Great Basin Desert and the relatively warm Mojave, two are from the relatively warm Sonoran, and 46 from the relatively cool Chihuahuan.

Judging from data available to date, the most productive areas for North American desert truffles, specifically *C. emergens*, are within the Chihuahua desert, ranging from central New Mexico, USA south, to adjacent northern Chihuahua, Mexico, and that species may fruit any month of the year (Trappe and Weber 2001; Moreno et al. 2012). The Chihuahua desert has relatively high precipitation and cool temperatures compared to the other deserts (Tables 8.1 and 8.2). The relationship of specific fruiting events to weather phenomena needs more detailed examination to determine how seasonal weather patterns affect fruiting. The human factor also needs to be considered: both the New Mexican and Chihuahua high-production areas have had nearby academic and research institutions with active mycological research programs when and where most of the collections have been found. Other of the North American desert regions may prove to be productive as well, when the weather is right and collecting is vigorously pursued.

8.4 The Outlook for North American Desert Truffles

Prior to World War II, overgrazing by livestock and mining were the primary threats to truffles of the North American deserts. They changed composition of plant communities, promoted invasion of exotic weeds, compacted soil, and exacerbated fire hazard and erosion. Then the damming of rivers provided irrigation water that enabled expansion of agriculture. Since World War II urban sprawl, recreation, water overuse, energy development, road construction, and air pollution have gained prominence in environmental degradation of these deserts (Phillips and Comus 2000; Haystad et al. 2006; Webb et al. 2009; Finch 2012; Ford et al. 2012). Because so little is known about the habitat requirements of the North American desert truffles, it is impossible to specify how much these disturbances would affect desert truffle production other than to note that large areas of the ecosystems in which these truffles might have grown have been drastically affected or taken out of truffle production.

As is true of most deserts, climate varies markedly over the course of the year and between years. Moreover, the changes will differ between deserts and habitats within those deserts. Climate change in the North American deserts has received considerable recent attention (Ford et al. 2012). The present climate models predict overall warming and drying for the North American deserts through 2090. Longer and more severe droughts will increase potential for "mega fires," susceptibility to insect pests and diseases, invasion of exotic weeds, and conflicts over use of diminishing freshwater resources, to mention several. Some of these events may move habitats suitable for truffle fungi northward into the Great Basin cold desert from the warmer Mojave and Sonoran deserts. But desert truffle populations have not been systematically monitored so far and are not likely to be monitored in the future. With no solid baseline data in hand, we may never learn how climate change affects truffle populations in North American deserts except in a broad, hypothetical way.

Acknowledgements Gábor M. Kovács is grateful for support of his participation in this chapter by the Hungarian Research Fund (OTKA K72776) and János Bolyai Scholarship (Hungarian Academy of Sciences). The image in Fig. 8.2 (top) by J. C. Zak and W. G. Whitford (1986) is reprinted with permission from Mycologia, ©The Mycological Society of America.

References

Brittanica Online (2012) Deserts of the world. http://www.britannica.com/EBchecked/topic/1854209/Deserts-of-the-World. Accessed 21 Jan 2013

Columbia Electronic Encyclopedia (2005) American Indian archaeological sites. http://www.infoplease.com/spot/aihmsites1.html. Accessed 23 Feb 2013

Finch DM (ed) (2012) Climate change in grasslands, shrublands, and deserts of the interior American West: a review and needs assessment. USDA Forest Service, Rocky Mountain Research Station Gen Tech Rep RMRS-GTR 285, Fort Collins, Colorado

Fogel R, Pacioni G (1989) Materials for a hypogeous mycoflora of the Great Basin and adjacent cordilleras of the western United States. Mem N Y Bot Gard 49:119–128

Ford PL, Chambers JK, Coe SJ, Pendleton BC (2012) Disturbance and climate change in the interior West. In: Finch DM (ed) Climate change in grasslands, shrublands, and deserts of the interior American West: a review and needs assessment. USDA Forest Service, Rocky Mountain Research Station Gen Tech Rep RMRS-GTR 285, Fort Collins, Colorado. pp 80–96

USDA Forest Service (1994) Ecological subregions of the United States. http://www.fs.fed.us/land/pubs/ecoregions/toc.html. Accessed 1 Feb 2013

Frioni L, Minasian H, Volfovicz R (1999) Arbuscular mycorrhizae and ectomycorrhizae in native tree legumes in Uruguay. For Ecol Manage 115:41–47

Gilkey HM (1954) Order Tuberales. N Am Flora Ser 2, 1:1–36

Grayson DH (2011) The Great Basin: a natural prehistory. University of California Press, Berkeley

Harkness HW (1899) California hypogeous fungi. Proc Calif Acad Sci Ser 3, 1:241–292

Haystad KM, Huenneke LF, Schlesinger WH (eds) (2006) Structure and function of a Chihuahuan desert ecosystem. Oxford University Press, New York

Kaye GC (1986) Job Bicknell Ellis 1829–1905. Mycotaxon 26:29–45

Kovács GM, Trappe JM, Alsheikh AM, Hansen K, Healy RA, Vági P (2011) *Terfezia* disappears from the American truffle mycota as two new genera and *Mattirolomyces* species emerge. Mycologia 103:831–840

Kropp BR, Albee-Scott S, Castellano MA, Trappe JM (2012) *Cryptolepiota,* a new sequestrate genus in the Agaricaceae with evidence for adaptive radiation in western North America. Mycologia 104:164–174

Lamson-Scribner F (1893) Southern botanists. Bull Torr Bot Club 30:326–328

Lee C, Arroyo R, Lee J, Dimmitt M (2011) Deserts of North America. Encyclopedia of earth, http://www.eoearth.org/article/Deserts_of_North_America. Accessed 7 Dec 2012

McCormick CA (2012) Collectors of the UNC Herbarium. http://www.herbarium.unc.edu/Collectors/Langlois_AB.htm. Accessed 21 Jan 2013

Moreno G, Esqueda M, Pérez Silva E, Herrera T, Altés A (2007) Some interesting gasteroid and secotioid fungi from Sonora, Mexico. Persoonia 19:265–280

Moreno G, Lizárraga M, Esqueda M, Galán R, Alvarado P (2012) New records of little-known species of *Carbomyes* (*Carbomycetaceae, Ascomycota*). Mycotaxon 120:89–98

Phillips SJ, Comus PW (eds) (2000) A natural history of the Sonoran Desert. Arizona Desert Museum Press, Tucson

Colegio des Postgraduados (2013) Normales climatológicas para Chihuahua, Chihuahua. http://www.cm.colpos.mx/meteoro/progde/norm/norm24/33010.htm. Accessed 23 Mar 2013

Trappe JM, Weber NS (2001) North American desert truffles: the genus *Carbomyces* (Ascomycota, Carbomycetaceae). Harvard Pap Bot 6:209–214

Trappe JM, Kovács GM, Claridge AW (2010) Comparative taxonomy of desert truffles of the Australian outback and the African Kalahari. Mycol Prog 9:131–143

Wang V, Qiu Y-L (2006) Phylogenetic distribution and evolution of mycorrhizas in land plants. Mycorrhiza 16:299–363

Webb RH, Fenstermaker LF, Heaton JS, Hughson DL, McDonald EV, Miller DM (eds) (2009) The Mojave desert—ecosystem processes and sustainability. University of Nevada Press, Reno

Western Regional Climate Center (2013) Summaries for the western United States. http://www.wrcc.dri.edu/summary/lcdus08.html. Accessed 23 Mar 2013
World Meteorological Organization (2013) World weather/climate extremes archive. http://wmo.asu.edu/world-highest-temperature. Accessed 23 Mar 2013
Zak JC, Whitford WG (1986) The occurrence of a hypogeous ascomycete in the northern Chihuahuan desert. Mycologia 78:840–841

Chapter 9
The European Desert Truffles

Gérard Chevalier

9.1 Introduction

The name terfez evokes hypogenous mushrooms that grow in the desert and are collected by the Bedouins in the Middle East, hence 'desert truffles'. It is less known that there are also some terfez in Europe and that these are consumed and marketed in some countries. The European terfez are, from the north to the south, *Mattirolomyces terfezioides*, (Matt.) Fischer, which is not a true 'desert' truffle as it grows in temperate climate zone (see Sect. 9.2); *T. leptoderma* Tul. & C. Tul and *T. olbiensis* Tul.; *T. arenaria* (Morris) Trappe; *T. boudieri* Chatin and *T. claveryi* Chatin; as well as *Tirmania nivea* (Des:. Fr) Trappe and *Picoa juniperi* Vittad. There are other terfez species that were discovered in the last few years in Spain. Two have been recognised as new species: *T. alsheikii* Kovacs, Martin and Calonge (Valladolid, Burgos), *T. canariensis* Bordallo and Rodriguez (Fuerte Ventura, La Gomera, Lanzarote, Las Palmas de Gran Canarias). Four have been accepted and are to be published in the scientific magazine Mycotaxon: *T. albida* Rodriguez, Mohedano and Bordallo (Albacete); *T. eliocrocae* Bordallo, Morte and Honrubia (Burgos); *Terfezia pini* Bordallo, Rodriguez and Mohedano (Burgos); *T. peudoleptoderma* Bordallo, Rodriguez and Mohedano (Burgos, Caceres) (Honrubia pers. comm.) (see also Bordallo and Rodriguez, this volume).

The local usual names 'sand truffles' or 'desert truffles' are numerous: in Algeria, terfàs, terfez, torfez, torfàs and kama (in Kabylia) (Chatin 1892); in Morocco, red terfass of Tafilalet, white terfass of Tafilalet, pink terfass of Mamora, black terfass of Zaïr and male terfass (Khabar et al. 2001); and in Tunisia, ahmer terfess (red truffle), abyadh terfess (white truffle) and zouber (Khabar et al. 2001). In Near and the Middle East: kamés, hama, thama, tama (Chatin 1892), khamé, zoubaidi (Alsheikh 1994). The terfez and khamé from southern Europe, North

G. Chevalier (✉)
INRA researcher retired, Résidence Cristelle, 33 rue des Vergers, 63800 Cournon d'Auvergne, France
e-mail: charlemagne33@orange.fr; gerard.chevalier@clermont.inra.fr

V. Kagan-Zur et al. (eds.), *Desert Truffles*, Soil Biology 38,
DOI 10.1007/978-3-642-40096-4_9, © Springer-Verlag Berlin Heidelberg 2014

Fig. 9.1 Fruiting body of *Mattirolomyces terfezioides* (Italy, Reggio-Emilia, Castelnovomonte) (photo L. Riousset)

Africa and the Middle East correspond to the Hydnum, Misy and Mison that the Greeks and the Romans obtained from Lesbos and Carthage (Chatin 1892).

9.2 *Mattirolomyces terfezioides* (Mat.) E. Fischer 1938 (syn. *Choiromyces terfezioides* Matt. 1887; *Terfezia terfezioides* (Mat.) Trappe 1971) (Fig. 9.1)

9.2.1 *Geography*

Central and southern Europe: France, Italy, not reported in Spain and Portugal, Hungary, north of Serbia (Lawrynowicz et al. 1997) and Greece (Cephalonia island); not reported in Near and the Middle East; Asia: Pakistan, India and China (Alsheikh 1994).

France: rare and collected only once in Provence (Vaucluse's department, Le Thor, near Avignon) (Riousset pers. comm.). It may also have been collected in the Lot's region (Quercy, southwest of France).

Hungary: common: Budapest (Pestimre, Pestlörinc, Prope, Rakoshegy, Ujpest) and surroundings (Csevharaszt, Gyal, Kiskunlachaza, Sülysap); around Gödöllö (Orbottyan, Mogyorod), Kiskunhalas (Kecel, Kunfeherto), Szekszard (Kunfeherto), Tolna, near Kajdacs, Kiskajdacs, close to the Danube (Alsheikh 1994; Babos 1981; Hollos 1933; Kiraly and Bratek 1992; Kovacs et al. 2002; Novak and Zeller 1959; Szemere 1965).

Italy: Piedmont (Asti, Cavoretto); Venetia (Vigonono), Emilia-Romagna (Ferrara, Rovigo) (Alsheikh 1994), Sardinia (province of Nuoro) (Brotzu 1994; Mattirolo 1900; Montecchi and Sarasini 2000). The holotype of Mattirolo comes from Testone (near Moncalieri, Piedmont).

Between about 47°41′ (Gödöllö, Hungary) and 40°19′N (Nuoro, Italy); 4°49′ (Avignon, France) and 19°29′ E (Kunfeherto, Hungary).

9.2.2 Habitat

Continental habitat, not linked to coastal zones with cistus and *Helianthemum*.

France: a specimen weighing 500 g collected in December 1974 under *Diospyros kaki* in the Vaucluse's region, in calcareous soil (quaternary Rhône's alluviums).

Italy: in cultivated fields under *Prunus avium* (*Cerasus avium* var. d*uracina*) and *Solanum, Helianthus* and *Tubera* spp. (Alsheikh 1994) in ruderal environments and cultivated soils; a species of the continental inland, also under various trees as *Ficus* and *Prunus*, in cultivated fields, along dikes along rivers (Montecchi and Sarasini 2000), quaternary alluviums of the Po's plain.

Hungary and *North Serbia*: under *Robinia pseudoacacia* or *Celtis, Robinia* stands and/or *Acacia* trees in forests or cemeteries in sandy soil (Alsheikh 1994). The majority of records come from the Carpathian Basin. The habitats can be found in the floristic area of Eupannonicum. Sandy soils are reported in every case and mainly presence of *Robinia pseudoacacia* from native habitats. As a result of ecological investigations, the association of *Bromo sterilis–Robinetum* was mainly detected. Phytoindication reflects some sub-Mediterranean climate effects, slightly basic soils, semi-humid and intermediate moisture conditions, and disturbed, secondary and artificial *Robinietum cultum* habitats (Fig. 9.2). All habitats can be found on the sandy soils deposited by the river Danube. Neutral-slightly basic soils with low $CaCO_3$, slightly humiferous, very low P205 concentration, middle level of K20, rich in N-sources (Bratek et al. 1994, 1996, 2004).

9.3 The Terfezia

9.3.1 Terfezia leptoderma Tul. 1851 (syn. *Choiromyces leptoderma* Tul. 1845; *T. fanfanii* Matt. 1900) (Fig. 9.3)

9.3.1.1 Geography

Central and southern Europe: Portugal, Spain, France, Italy and Hungary (Alsheikh 1994; Montecchi and Sarasini 2000). Morocco (Khabar et al. 2001).

France: Aquitaine, harvested near Bordeaux (middle October 1843) and in the vicinity of Bazas (Gironde's department), under *Cistus guttatus* (Tulasne and Tulasne 1851). Provence: Bouches-du-Rhône's department, near Avignon (Barbentane, at the foot of the Montagnette) or near Arles (Fontvieille, at the at the foot of the Alpilles) (Dexheimer et al. 1985; Janex-Favre et al.1988); Var's region: Hyères islands (Donadini 1979).

Fig. 9.2 Site of production of *Mattirolomyces terfezioides* in Hungary (photo Z. Bratek)

Fig. 9.3 Fruiting body of *Terfezia leptoderma* (Sardinia, Zurratile) (photo L. Riousset)

Auvergne: Puy-de-Dôme's region, in Clermont-Ferrand, altitude 250 m, 45°46′N, 03°08′E, in a garden under an ornamental *Cistus*, November 1999 (Chevalier unpublished).

Hungary: in the vicinity of Pamuk, Somogyvar, comm. Somogy (Szemere 1965).

Italy: Sardinia, Oristano's province (Zurratile, Palmas Arborea, Tora Grande, Costa verde) (Montecchi and Sarasini 2000; Riousset pers. comm.).

Spain: near Madrid, Estrémadure (Diez and Manjon 2001); Caceres (Villafranca de los Barros, Trujillo), Murcia (Alsheikh 1994; Honrubia pers. comm.).

Portugal: Barca d'Alva (near Coimbra) (Alsheikh 1994).

9.3.1.2 Habitat

Forests environments or associated with *Helianthemum guttatum*, on sites also containing *T. arenaria* and *Tuber asa*; in pine forests under *Quercus* sp. (Alsheikh 1994).

France: in sandy soil, under tufts of grass, pine forests edges; scarce, slightly hidden (Tulasne and Tulasne 1851).

In Fontvieille, the site is a former olive grove colonised with sparse grass/lawn where the tree stratum composition is made up of Aleppo pines (*Pinus halepensis*), evergreen oaks (*Quercus ilex*) and olive trees (*Olea europaea*). The phytosociologic terminology classifies the woodland type in the *Quercetalia ilicis*, *Quercion ilicis* alliance, *Quercetum ilicis galloprovinciale* association (Janex-Favre et al. 1988) (Fig. 9.4). Together with evergreen oaks and Aleppo pines, several plant species are potential hosts to *T. leptoderma*: *Q. coccifera*, *Cistus* spp. (*Cistus albidus*), *Helianthemum* spp. (*H. apenninum* and *H. marifolium*). The soils found are brown soils, calcareous-clayey, slightly sandy, and on bedrock of the upper Cretaceous. The site is also productive for other hypogenous fungi (*Balsamia, Picoa, Genea, Genabea, Hydnocystis Glomus* and numerous *Hymenogaster* (not all of which are desert truffles) and many species of forest truffles (*Tuber aestivum, borchii, brumale, dryophilum, excavatum* var. *sulphureum, melanosporum, nitidum, oligospermum, requieni, rufum, panniferum*). From March to April–May (but maturity only in May).

Italy: in sand, under shrubs or at the edge of coastal pine forests; fruiting bodies are generally deep (10–20 cm below the surface) in pine forests soil; spring species (April–May); under *Cistus salviifolius* and *Helianthemum* spp. (Montecchi and Sarasini 2000).

In Sardinia, near Oristano, the habitat is different from the one of Fontvieille: acidic soil, lawn with herbaceous stratum without shrubs and trees. The terfez seems to be linked to *Helianthemum guttatum* ssp. *plantagineum* always mixed with *Tuber asa* and *Terfezia arenaria*; the beginning of the season is March (Janex-Favre et al. 1988).

Spain: mixed with *T. claveryi* in open Mediterranean silvo-pastoral systems (called 'dehasa'). Such a dehesa-like vegetation exists in the Extremadura region of western Spain. Most of these woodland/pastures type areas are extensively managed, coupes of open woodland with farm livestock or deer grazing. These savannoid woodlands are Mediterranean open forests of holm (=evergreen) oaks (*Quercus ilex*) and/or cork trees (*Quercus suber*) and *Helianthemum guttatum* (Diez and Manjon 2001).

Beside *H. guttatum*, in the plant communities where the desert truffles are collected, *Rumex bucephalophorus* subsp. *gallicus* indicates that the soil is acidic. White flowers belonging to other plant species, such as *Spergula arvensis* or *Bellis annua*, which give the white colour of the grazing land, also include *Poa bulbosa*, *Cytisus multiflorus*, *Molineriella levis*, *Erica vesicaria*, *Malcomia triloba*, *Mibora minima*, *Aphanes microcarpa*, *Crassula tillaea*, *Moenchia erecta*, *Vulpia myuros* and *Hypochaeris* (Diez et al. 2001).

During spring, local inhabitant harvest desert truffles near the annual plant *Helianthemum guttatum* (Moreno et al. 1986, 1991). They use a pointed stick with two iron prongs, one with a scoop shape end the other with a pointed tip. The hunter probes the soil repetitively with the pointed tip until he detects a truffle, and then he uses the scoop tip to uproot it.

Due to commercial market value, *T. leptoderma* and *T. claveryi* do not only mean food for local inhabitants but represent also a valuable additional income. Therefore, they are an important natural resource in these regions. Furthermore,

Fig. 9.4 Site of production of *Terfezia leptoderma* (France, Provence, Fontvieille) (Photo L. Riousset)

these mycorrhizal fungi play an important role in the maintenance of vegetation that prevents erosion and desertification.

In the Canary Islands *T. leptoderma* is associated with *Erica arborea* and *Smilax* spp. or with *Pinus canariensis* (Alsheikh 1994)

Hungary: Szemere (1965) harvested some specimens in July 1951 under *Quercus cerris*, near Pamuk, in the proximity of *Tuber aestivum*, *T. puberulum* and *Elaphomyces muricatus*.

9.3.2 *Terfezia olbiensis* Tul. 1851 (=*Choiromyces olbiensis* Tul. 1844) (Fig. 9.5)

Malençon (1973), Moreno (1980) and Diez et al. (2002) consider *T. olbiensis* to be an immature form and a synonym of *T. leptoderma*. On the contrary, Alsheikh (1994) maintains that *T. olbiensis* is a distinct species. Gutierrez et al. (2004) clearly separate *T. olbiensis* from *T. leptoderma* using morphological, ecological and molecular characteristics.

9.3.2.1 Geography

Central and Mediterranean Europe (East Portugal to Spain, France, and Italy); north to the former Czechoslovakia and south to the Canary Islands; North Africa: Morocco to Tunisia and the Near East, a single mention from Israel (Alsheikh 1994).

Found more to east than *T. leptoderma*.

France: Provence, Barbentane; Hyère Islands (Porquerolles) (Riousset pers. comm.; Tulasne and Tulasne 1851).

Italy: Sienna; Sardinia (Orune, Cagliari, San Leonardo, several sites on the coast); Sicily (Catania's province, Torre Armerina) (Alsheikh 1994; Montecchi and Sarasini 2000; Morara et al. 2009; Riousset pers. comm.). The species

Fig. 9.5 Fruiting body of *Terfezia olbiensis* (Sardinia, San Leonardo) (photo L. Riousset)

T. olbiensis was named by Tulasne according to a sample collected in 1844 in Olbia (north east coast of Sardinia).

Spain: Alcinty, Arjona and Elgalapagar (Madrid); Caceres; Catalonia; Salamanca; Truijillo.

Canary Islands: Hierro, Tenerife (Alsheikh 1994).

Portugal: Alentejo, Barca d' Alva, Coimbre, Val de Bosal (Alsheikh 1994).

9.3.2.2 Habitat

In pine forests and shrub areas, December to June, but mainly March to May (Alsheikh 1994).

France: in Barbentane, under Aleppo pine, evergreen oak, cistus; in plains, open places, with cistus, *Helianthemum* spp., rare evergreen oaks and Aleppo pines. Acidic soils originated from shale. December to May (December 1974, May 1976) (Alsheikh 1994 Riousset pers. comm.). According to Tulasne and Tulasne (1851), from December to February, not rare, partially epigeous, solitary, under a thickness of oak leaves or pine needles.

Italy: in Sardinia, in *Quercus* and *Pinus* forests, from late autumn to spring, generally in spring (Montecchi and Sarasini 2000). In the province of San Leonardo, at altitude, in a plantation of *Pinus radiata* (Janex-Favre et al. 1988).

9.3.3 ***Terfezia arenaria*** **(Moris) Trappe 1971 (syn.** ***Tuber arenarium*** **Moris 1827;** ***Tuber algeriense*** **Montagne 1851;** ***Choiromyces leonis*** **Tul. 1845;** ***Terfezia leonis*** **Tul. 1851;** ***T. goffartii*** **Chatin 1895;** ***T. heterospora*** **Chat. 1896;** ***T. mellerionis*** **Chat. 1895 (Fig.** 9.6**)**

Terfez, torfès, terfàs of the Arabs, camha of the Kabyles, turma of the Spaniards, kamé of Smyrne, turiena de arena of the Italians.

Fig. 9.6 Fruiting body of *Terfezia arenaria* (Sardinia, San-Giusta) (photo L. Riousset)

9.3.3.1 Geography

Around the Mediterranean basin: Southern Europe (France, Spain, Portugal, Italy, former Yugoslavia, Greece, Cyprus), Near East (Turkey), Rumania, southern regions of the old Soviet Union, North Africa (Morocco, Algeria, Tunisia), rare in the Middle East (Iran), Asia (Pakistan) and in the Arabian Peninsula (Saudi Arabia) (Alsheikh 1994).

France: southwest, in the vicinity of Bordeaux, Tartas (Landes' department), sandy soils; Meilhan (Lot-et-Garonne's department, in Garonne river's alluviums); south of France (Moyen 1989) Corsica?

Italy: promontory of Circeo (Rome's province, Campania, near the village of Santa Euphemia where it is commonly named 'tartufo bianco') (Tulasne and Tulasne 1851). Piedmont (near Vercelli), Anzio, Nettuno, Forli (Martorano) (Alsheikh 1994).

Sardinia: near Oristano (Arborea, Marrùbiu), Cagliari, Caltagirone, Ragusa (near Vittoria), Sangiusta, Sassari, Sorso, Terralba, Terracina (Alsheikh 1994; Langiu 1979; Riousset pers. comm.; Tulasne and Tulasne 1851). *T. arenaria* is always mixed with *Tuber asa* and *T. leptoderma*; however it is more abundant than the two last species. The *T. arenaria* holotype (as *Tuber arenarium*) comes from Sardinia (Moris 1827).

Spain: Andalucia, Castilla Leon, Castilla la Mancha, Estremadure, Caceres, Catalonia, Huelva (Coto de Donana), Madrid (Piutado), Santa Cruz de Betamar, Toledo,Trujillo,Turcia, Villafranca de los Barros (Alsheikh 1994; Diez and Manjon 2001; Honrubia pers. comm.).

Portugal: Alemtejo, Portuguese Estremadure, Algarve, Coimbra, Evora Alentejo, Samora Correia (Mattirolo 1905, 1906, Alsheikh 1994).

Rumania: Casimcea (Casidavecchia) (Alsheikh 1994).

Cyprus (Alsheikh 1994).

Greece: Lesbos ('truffle' that the Romans obtained from Lesbos) (Chatin 1892), Prevesas (near Epirus) (Alsheikh 1994).

Former Yougoslavia: Istria (Isola di Colius) (Alsheikh 1994).

Former Soviet Union): Halia? (Alsheikh 1994).

Turkey: Izmir (Smyrne) (Chatin 1892); Istanbul (Alsheikh 1994).
Pakistan: Iran (Alsheikh 1994).
T. arenaria is the most common terfez in Southern Europe.

9.3.3.2 Habitat

Mediterranean and in southern regions of the old Soviet Union; in deserts, coastal zones and forests of *Quercus robur* in southern Europe; in Morocco in association with *Helianthemum guttatum*, *Helianthemum* spp., other species of *Cistaceae* as well as other annual and perennial mycorrhizal hosts. December to August, but mostly March to May (Alsheikh 1994).

Italy: Sardinia, in grassy and uncultivated flat lands, rich in *Helianthemum guttatum* ssp. *plantagineum*, April–May (Montecchi and Sarasini 2000). Mixed with *T. leptoderma* but is more abundant. Grows in lawns in herbaceous stratum where shrubs and trees are absent. Acidic soil (Fig. 9.7).

T. arenaria is much sought after from as early as March by the local inhabitants. The fruiting bodies are located using a long stick fitted with an iron spike at its end for about 20 cm, used as a detector ('arrudera'). This method was already mentioned by Tulasne and Tulasne (1851).

Sicily: near Caltagirone (Morara et al. 2009).

Spain: in Estremadure in a forest of evergreen oaks and cork oaks (*Q. suber*) in stony soil (Montecchi and Sarasini 2000). In areas of dominating cistus called 'tumera' (Chatin 1892). With *T. leptoderma* in open sylviculture-pastoral systems ('dehesa') under evergreen oaks and/or cork oaks and *Helianthemum* (Diez and Manjon 2001).

9.3.4 *Terfezia boudieri* Chatin 1891 (=*T. boudieri* var. *arabica* Chat. 1892; *T. deflersii* Pat. 1894) (Fig. 9.8)

Kamé of Damas (Chatin 1892).

Defined by Chatin in 1891 from specimens originating from the south of Algeria (Biskra, Barika, Batna, Bou-Saada, El Golea) and from Syria (Damas).

9.3.4.1 Geography

From Southern Europe (France?, Italy, Spain, (Moreno et al. 2002) Cyprus) to the Near East (Israël, Syria, Turkey) towards the Middle East (Iraq) and the Arabian Peninsula (Kuwait, Bahrain Islands, Saudi Arabia); North Africa (Morocco,

Fig. 9.7 Site of production of *Terfezia arenaria* (Sardinia, province of Oristano)

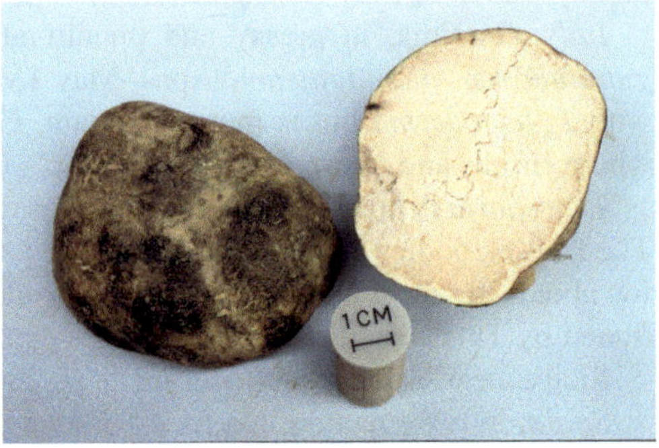

Fig. 9.8 Fruiting body of *Terfezia boudieri* (photo L. Riousset)

Algeria, Tunisia, Lybia, Egypt) (Agaoglu et al. 1992; Alsheikh 1994; Bawadikji 2004; Khabar et al. 2001).

Further south than *T. arenaria*.

France: reported by Montecchi and Sarasini (2000, p. 218) without registering the location.

Italy: Sardinia (Oristano's province) (Montecchi and Sarasini 2000).

Spain: Canary Islands (Lanzarote) (Alsheikh 1994).

Cyprus: (Alsheikh 1994).

9.3.4.2 Habitat

In natural areas probably mycorrhizal with *Cistus albidus*, *C. monspeliensis*, *C. salviifolius*, *Fumana procumbens*, *Halimium halimium*, *Helianthemum apenninum*, *H. eremophilum*, *H. hirtum* var. *deserti*, *H. kahiricum*, *H. ledifolium*, *H. lipii*, *H. salicifolium*, *H. guttatum (Cistaceae)*, *Plantago albicans* (*Plantagineae*), *Artemisia monosperma* (*Compositae*), sometimes associated with *Schismus barbatus* (*Gramineae*) (Alsheikh 1994). In deserts, from January to May, but mostly February to April (Alsheikh 1994), in Morocco with *Helianthemum* spp., *H. lipii* and *H. apertum* (Khabar et al. 2001).

Fig. 9.9 Fruiting body of *Terfezia claveryi* (photo L. Riousset)

9.3.5 Terfezia claveryi Chatin 1891 (= *T. hafizi* Chat. 1892) (Fig. 9.9)

Kamé of Damas, white Kamé of Bagdad (Chatin 1892), red Terfez of Tafilalet (Malençon 1973).

9.3.5.1 Geography

Southern Europe (France?, Spain, Italy, Cyprus); Africa: from Morocco, Algeria, Tunisia to Libya, Egypt; Near East (Israël, Syria, Jordan, Liban, Turkey); the Middle East (Iraq, Iran); Asia (Armenia, Transcaucasia, Azerbaïdjan, Turkmenistan) and Arabian Peninsula (Bahrain Islands, Kuwait, Saudi Arabia) (Alsheikh 1994; Bawadikji 2004; Chatin 1892; Khabar et al. 2001; Malençon 1973).

France: reported by Montecchi and Sarasini (2000) without the location (p. 222).

Italy: Sardinia (near Oristano, Las Arenas) (Riousset pers. comm.)

Spain: Aragon, Alicante, Almeria, Castillon la Mancha, Catalunia, Granada, Jaén, Murcia (Cano et al. 1991; Honrubia et al. 1992; Honrubia pers. comm.). Canary Islands: Lanzarote (Alsheikh 1994).

9.3.5.2 Habitat

In deserts. January to May but mostly February to April (Alsheikh 1994). Southern semidesert (Montecchi and Sarasini 2000). Semidesert zones of the sub-Mediterranean basin and of the Middle East. Sandy soils. Under *Helianthemum (aegyptiacum, guttatum, salicifolium, ledifolium, lipii, sessiliflorum)* (Awameh and Alsheikh 1979, see also Chap. 11 by Kagan-Zur and Akyuz). Spring: March–April.

Mycorrhizal with *Helianthemum almeriense* (Morte et al. 1994) *Helianthemum ledifolium*, *H. salicifolium*, *H. lipii*, *H. eremophilum*, *Cistus* spp., *Atractylis serratuloides* (?), *Thymelaea hirsuta* (?), *Plantago albicans* (?), *Artemisia herba-alba* (?) *A. monosperma* (?), *Acacia hebeclada* (?) (Alsheikh 1994; Maire 1907); in Morocco with *Helianthemum lipii* and *H. apertum* (Khabar et al. 2001); in Algeria with *H. guttatum* (Fortas 1990).

9.4 The Tirmania

Tirmania nivea (Desf.) Trappe 1971) (=*Tuber niveum* Desf. 1823; *Terfezia ovalisperma* Pat. 1890; *Tirmania ovalisperma* (Pat.) Pat. 1892; *Tirmania africana* Chat. 1891; *T. cambonii* Chat. 1892; *Terfezia Africana* (Chatin) Maire 1916) (Fig. 9.10).

Big white terfass of southern Algeria (Chatin 1892), white terfass of Tafilalet (Malençon 1973), zoubaïdi (Khabar et al. 2001).

9.4.1 Geography

From Mediterranean basin (Spain, Italy; North Africa: Morocco, Algeria, Tunisia, Libya, Egypt and Israel) to the Arabic peninsula (Alsheikh 1994; Awameh and Alsheikh 1979; Alsheikh and Trappe 1983).

France: not reported

Spain: very rare; desert lands in the south (Diez et al. 2002).

9.4.2 Habitat

Linked to the semidesert plains; in relation with the presence of several species of *Helianthemum* among which many annual species especially *H. aegyptiacum*, *H. salicifolium* and *H. ledifolium* in Kuwait; *H. guttatum* in Algeria; and *H. hirtum* in Morocco. In calcareous soils. February to April (June).

9.5 *Picoa juniperi* Vitt. 1831

(etym.: from Latin *juniperus* = juniper)

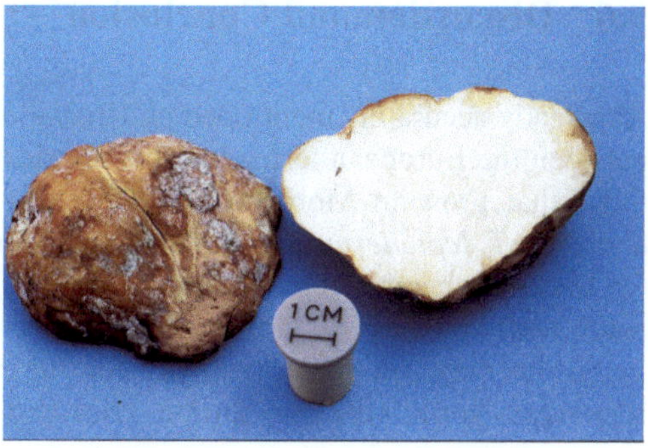

Fig. 9.10 Fruiting body of *Tirmania nivea* (market of Batna) (photo L. Riousset)

9.5.1 Geography

France: Alpilles (Riousset pers. comm.); Auvergne (Clermont-Ferrand) (Chevalier Match 2001).

Italy: Sardinia (Montecchi and Sarasini 2000).

Spain: Southern Spain: University Campus in Espinardo (Murcia); near Cullar, on the roadside from Murcia to Granada (Honrubia et al. 1992).

9.5.2 Habitat

France: scarce; in Alpilles, same ecology as *Terfezia leptoderma* (Riousset pers. comm.) harvested only one time in March 2001 in the experimental truffiere of the INRA of Clermont-Ferrand under *Helianthemum apenninum.*

Italy: Mediterranean environments with *Cistus, Quercus and Helianthemum* in spring (Montecchi and Sarasini 2000). In the forests of the hills and the mountains of the area situated beyond the Po river, especially around the junipers. Harvested in late autumn and early winter (Vittadini 1831).

Spain: under *Helianthemum almeriense* in marl calcareous soils or in marl-gypsum soil. Together with *T. claveryi*, the most frequent hypogeous fungus in semiarid lands in Southern Spain. Mostly found under species of *Helianthemum* with which it establishes mycorrhizal symbiosis as does *T. claveryi*. Known in this area as 'la turma negra' ('black turma') or 'el chivato de la turma' ('turma's informant'), because it produces its sporocarps 2–3 weeks earlier than *T. claveryi.*

Calonge (1982) and Calonge et al. (1985) previously reported this species in Spain. They considered *P. junipery* as an infrequent species, collected under *Stipa tenacissima* and in an uncultured site.

9.6 Discussion and Conclusion

As already accurately mentioned by Khabar et al. (2001), there are strong analogies between the European terfez in the countries bordering the Mediterranean sea and those that grow in Morocco. The terfez at the north of Morocco like *Terfezia arenaria*, *T. leptoderma* and *olbiensis* are found also in several European countries. On the other hand species like *T. claveryi*, *T. boudieri* and *Tirmania nivea* of the south and southeast of Morocco are known only to Spain and Italy. The Iberian Peninsula may provide a pathway for fungal migration from arid and semiarid regions of North Africa to Europe (Diez and Manjon 2001).

9.6.1 Distribution

The distribution of the terfez is conditioned by various factors: climate, soil and vegetation (Riousset et al. 2004).

The climate plays an essential part in the characteristic distribution of truffle species between Europe, North Africa, Near East, Middle East and Arabic peninsula. If the *Tuber* dominate in the temperate humid and subhumid zones of Europe, the terfez, (genera *Terfezia* (*claveryi*, *boudieri*) and *Tirmania* (*nivea*, *pinoyi*), are rather xerothermic species, characteristically occupy the arid and Saharan zones of the countries south of the Mediterranean Sea.

The European most northern terfez is *Mattirolomyces terfezioides* as it is present and common in all the Hungarian plains and in northern Serbia. In Hungary, it grows as far as the Budapest area (47–48°N). It is unknown in Spain and Portugal. It is also the terfez with the widest geographic distribution (Pakistan, India, China). It is an exception among the terfez as its climatic requirements are closer to those of the *Tuber* species.

T. arenaria is the terfez of the Western Mediterranean basin. It has the widest geographic distribution in Europe. With Hungary, it is also common in Portugal, Spain and Italy (Sardinia, Sicilia). It is found in Greece (Lesbos island), on Cyprus, in former Yugoslavia and Rumania. In North Africa, it is common in Morocco, in the Mamora forest (north of Rabat). It is rare in France where its northern limit is near Bordeaux (45°N). Its favourite zone is situated between 35 and 40°N. It is a truffle of semiarid climate (Diez et al. 2002).

T. leptoderma is also a terfez of the Western Mediterranean basin (eastern limit around 20°E). Its known northern limit is Pamuk, in Hungary (46°N). It is common in Portugal, Spain and Italy (Sardinia). It is scarce in France and in Hungary. It is also common in Morocco in the Mamora forest. It is a species of semiarid zones (Diez et al. 2002).

T. olbiensis is also common in Portugal, Spain and Italy (Sardinia). It is rare in France and former Czechoslovakia (about 50°N). Its geographic distribution is almost similar to that of *T. leptoderma* (with which it has often been confused), but it has been reported more to the east (former Czechoslovakia, up to Israel) (see Sect. 8.3.2.1). It is also a terfez of semiarid zones.

T. boudieri and *T. claveryi*, which are two desert truffles species, are less common in Europe. However, they have been collected in many Spanish provinces and in Sardinia (Oristano's province), as well as on Cyprus. They are common in North Africa (southeast of Morocco, Algeria and Tunisia as well as Israel (Alsheikh 1994; Kagan-Zur and Roth-Bejerano 2009). They are terfez of semiarid and arid zones (Diez et al. 2002).

Finally *Tirmania nivea*, a desert truffle (arid climate), has been collected in the desert of southern Spain (Diez et al. 2002). It is a very rare species in Europe. The other species of *Tirmania, T. pinoyi*, has not yet been reported as found in Europe.

Except *Mattirolomyces terfezioides*, the European desert truffles of semiarid climate (*T. arenaria, T. oligosperma and T. olbiensis*) do not generally go higher than 44–45°N (Bordeaux, Clermont-Ferrand in France). The presence of *T. leptoderma* in Hungary and *T. olbiensis* in former Czechoslovakia is exceptional. The terfez of the arid zones (*T. boudieri* and *claveryi*) are found as far as 40°N. One *Tirmania* (*nivea*) reaches a latitude superior to 35°N in Spain and in Italy. *Terfezia leptoderma* reaches approximately 37°N.

The furthest southern and eastern European regions where terfez grow are the Canary Islands (28°N, 15°W) (Alsheikh 1994).

Ranking by increased resistance to drought and high temperatures, the position of the species can be the following: *Mattirolomyces terfezioides*, *T. leptoderma* and *T. olbiensis*, *T. arenaria*, *T. boudieri* and *T. claveryi*, *Tirmania nivea and T. pinoyi*. *T. boudieri* and *T. claveryi* and above all *Tirmania nivea* and *T. pinoyi* are the great «classics» of the deserts of North Africa and the Middle East. The Iberian Peninsula builds bridges between Africa and Europe. That is why almost the same species of terfez grow in Morocco and Spain.

In total, seven species grow in Italy: three are common (*Terfezia arenaria*, *T. leptoderma*, *T. olbiensis*), three are less common (*M. terfezioides*, *T. boudieri*, *T. claveryi*) and one is scarce (*Tirmania nivea*). Six traditional species (+seven new species) grow in Spain; three are common: *T. arenaria*, *T. leptoderma*, *T. olbiensis;* two less common: *T. boudieri*, *T. claveryi*; and one is rare: *Tirmania nivea*. France has four (six?) species: *M. terfezioides*, *T. arenaria*, *T. leptoderma*, *T. olbiensis*, (*T. boudieri*?, *T. claveryi*?). All are rare.

9.6.2 Typical Soils

Another important factor for the distribution of terfez is the soil.

From a physical point of view, the soils where the terfez grow are light, sandy, rather formed from a fine and soft loam than the famous 'desert sands' defined by Chatin (1892), sandy or sandy-loamy trend (Khabar et al. 2001).

From a chemical point of view they are not all calcareous contrary to Chatin's assertions.

In Morocco, *Terfezia arenaria* and *T. leptoderma* prefer acidic soils. On the contrary in France *T. leptoderma* grows in calcareous soils where *Tuber melanosporum* and *T. aestivum* also grow. In the Alpilles, *Terfezia olbiensis* grows in calcareous soils originated from Mesozoïc bedrocks, whereas on the Porquerolles Island, situated at 150 km as the crow flies, the soil originated from Paleozoïc bedrock is not calcareous.

According to Diez et al. (2002), *Terfezia boudieri* and *T. claveryi* grow in Spain in marl-gypsum soils. In contrast, *T. arenaria* grows in siliceous sands. *T. leptoderma* was found in association with *Helianthemum guttatum* in acidic soils and with *Cistus ladanifer* in slate-derived soils. *T. leptoderma* might be a species with wide edaphic tolerance and host range (*Helianthemum* spp., *Cistus* spp., *Q. ilex*, *P. halepensis*) or a mix of species; fruiting bodies growing on *Cistus* shrubs, pine or *Quercus sclerophyllous* woodlands could belong to distinct species with different hosts or/and edaphic specialisation. It is the same in France between the fruiting bodies collected in the Alpilles in calcareous soils under *Quercus ilex* and *Pinus halepensis* and those collected in the Hyères Islands under pines in acidic soils.

In the case of *Tirmania* spp., *T. nivea* was found in basic soils and associated with basophilous plant *H. salicifolium*. In contrast, *T. pinoyi* specimens were collected under the acidophilous plant *H. guttatum*.

The distribution pattern of the desert truffle species seems to correlate so strongly with host, that the two factors (host specialisation and soil pH) might have played a key role (Diez et al. 2002). Furthermore, most species of *Helianthemum* forming mycorrhizae with desert truffles in the Mediterranean region show different edaphic tolerance. Whereas *H. salicifolium* and H. *ledifolium* occur in basic soils, other species of *Helianthemum* occur only in acid soils, e.g. *H. guttatum*. Soil features therefore have an important impact on the distribution of the host plants. Distribution of desert truffle species according to soil features might therefore just reflect the edaphic tolerance of their hosts. Because the different species of desert truffles (*Terfezia* and *Tirmania*) are adapted to different types of soils, the differences in the edaphic tolerance may account for species diversity (Diez et al. 2002, see also Chap. 4 by Bonifacio and Morte).

9.6.3 Host Plants

The vegetation of the symbiotic plants is the third main factor for the geographic distribution of the truffles and the terfez. Chatin (1892) is too categorical when he asserts that the plants favourable to the *Tuber* are trees whereas those favourable to the terfez are 'grasses' or 'sub-ligneous'. However he recognises that in the north of the zone of harvest of the terfez and in the mountains of Algeria the terfez grow under oaks, pines and cedars. The ability of the *Helianthemum* and the cistus to produce terfez has been demonstrated for more than 40 years, when the first mycorrhizal syntheses were carried out by Awameh and Alsheikh at the K.I.S.R.

(Kuwait Institute for Scientific Research) (Awameh et al. 1979) and confirmed by Dexheimer et al. (1985), Leduc et al. (1986), Fortas (1990), Fortas and Chevalier (1990), Fortas and Chevalier (1992), Gutierrez et al. (2003), Morte et al. (1994), Roth-Bejerano et al. (1990) and Slama et al. (2004). However in France *Terfezia leptoderma* was found under evergreen oaks and Aleppo pines (Janex-Favre et al. 1988). Because the shrub stratum was constituted of *Cistus albidus* and the herbaceous one from *Helianthemum apenninum*, *marifolium* and *salicifolium*, all symbionts of terfez, it was difficult to identify the plants that produced the fruiting bodies. Moreover in some producing sites the *Helianthemum* were absent. For this reason we have supposed that other *Cistaceae* were able to produce terfez. In fact our works showed in 1984 that *Terfezia leptoderma*, a 'French' terfez, is able to form ectomycorrhizae (although incomplete) with ectomycorrhizal perennial *Cistaceae (Cistus* spp., *Fumana procumbens*) (Chevalier et al. 1984) like the *Tuber* (Chevalier et al. 1975). The results have led us to think that *T. leptoderma* was also able to form mycorrhizal associations with the tree species that produce *Tuber* (*Pinus* spp., *Quercus* spp.). The evidence was given by Chafi et al. (2004), then by El-Houaria Zitouni-Haouar et al. (2012), who obtained, in Algeria, in controlled conditions (greenhouse) the ectomycorrhizal association between *Terfezia leptoderma* and *Pinus halepensis*. In this way some southern European and North African terfez would be able to mycorrhize pines and oaks, while at the other end of the zone of production of the terfez other *Terfezia* (*boudieri)* and *Tirmania* (*nivea, pinoyi*) would mycorrhize only the *Helianthemum,* only plants able to develop in the extreme conditions of the desert. From the point of view of the functioning of the symbiotic association, the terfez coming from the coastal areas of the Mediterranean basin with incomplete ectomycorrhizae would make the transition between the *Tuber* with typical ectomycorrhizae and the *Terfezia* and *Tirmania* with coiled ('terfezioïd') ectomycorrhizae (Chevalier et al. 1984).

9.6.4 Climate

The prime factor for the development of the terfez is the climate that acts in two ways: on the development and the fructification of the fungus but also on the distribution of the symbiotic species. If *Mattirolomyces terfezioides* does not grow further south than France and Sardinia, it is because this mushroom develops only in temperate climate. If the *Tirmania* have been collected only under *Helianthemum*, it is because cistus, oaks and pines are not able to live in the desert. In the semiarid climates of some zones of Spain and Sardinia, the climatic conditions allow the development not only of Therophytes, Chamaephytes and Phanerophytes of the family of *Cistaceae* but also of Phanerophytes of other families (oaks, pines).

The main limiting factor of the development of the terfez in Europe is not the soil because Europe presents a wide variety of light soils, basic and acidic, nor the absence of plant species able to form an association with the terfez (annual and perennial *Cistaceae*, oaks, pines) but the climate that must be warm enough. With the global warming the terfez will have a radiant future in Europe.

9.6.5 Nutritional Value

Although the main areas of consuming terfez are North Africa, the Middle East and the Arab peninsula the terfez are also harvested in Hungary, Italy and Spain. From a gastronomic point of view, due to their bland taste, the terfez are rather a food than a condiment like the *Tuber*. They are rich in proteins and poor in carbohydrates and lipids (see Chap. 17 by Martínez-Tomé et al.).

T. leptoderma is a good mushroom although its aroma is weak. Because sand particles become easily embedded in the fruiting body, it is preferable to peel it to clean it before consumption. When it is cooked it is reminiscent of *Boletus edulis* (Janex-Favre et al 1988). *T. arenaria* has a 'sweet and agreeable' aroma and a taste (Chatin 1892). I found *T. arenaria* from the Mamora forest to have a mushroom aroma when it is cooked and a taste of cooked artichoke heart (unpublished).

In Hungary, there is a great demand for *M. terfezioides* due to its success in gastronomy and the fact that is the only sweet fungus. It is excellent for sweets, cream, sorbet, and soups. Possibly in the near future, the demand will exceed the offer, and it may lead to the damage of its natural habitats (Bratek et al. 2004).

Acknowledgments I am grateful to M.H. Awameh and A.M. Alsheikh (Kuwait Institute for Scientific Research) for initiating me at the discovery of the terfez in the desert of Kuwait at the end of the 1970s. I also thank very much L. and G. Riousset for letting me discover the French terfez in Provence in the early 1980s and for providing me with plenty of fruiting bodies for my studies. Many thanks too to M. Honrubia for giving me precious indications on the seven new *Terfezia* species in Spain and to Marie-Anne French for helping me with the translation of the French text to English.

References

Agaoglu YS, Artik N, Ercan R, Guler M, Baspinar R (1992) Chemical composition of desert truffle (*Terfezia boudieri*) grown in Turkey. Micol Veg Medit 7(2):295–301

Alsheikh AM (1994) Taxonomy and mycorrhizal ecology of the desert truffles in the genus *Terfezia*. Ph.D. thesis, Oregon State University, Oregon, USA, 239 p

Alsheikh AM, Trappe JM (1983) Desert truffles: the genus *Tirmania*. Trans Br Mycol Soc 81:83–90

Awameh MS, Alsheikh AM (1979) Laboratory and field study on four kinds of truffle (kamah), *Terfezia* and *Tirmania* species for cultivation. Mushr Sci 10(2):507–517

Awameh MS, Alsheikh AM, Al-Ghawas S (1979) Mycorrhizal synthesis between *Helianthemum ledifolium*, *H. salicifolium* and four species of the genera *Terfezia* and *Tirmania* using ascospores and mycelial cultures obtained from ascospore germination. In: Proceedings of the fourth North American Conference on Mycorrhizae, 24–28 June 1979, Colorado State University, Fort Collins. pp. 22–23

Babos M (1981) Distribution of *Choiromyces venosus* and *Terfezia terfezioides* in Hungary Mikologiai. Közlem\u00e9nyck 1–2:47–56 (in Hungarian)

Bawadikji AH (2004) La truffe du désert en Syrie: aspects écologiques et économiques. In: Actes 1er Symposium sur les champignons hypogés du bassin méditerranéen, 6–8 avril 2004, Rabat, Maroc. pp. 43–48

Bratek Z, Gogan A, Berez B, Dimeny J (2004) Habitat preferences of *Terfezia terfezioides* in Hungary. In: Actes 1er Symposium sur les Champignons Hypogés du bassin méditerranéen, 6–8 avril 2004, Rabat, Maroc. pp. 256–264

Bratek Z, Jakucs E, Kiraly I (1994) Establishment of an *in vitro* mycorrhizal connection between black locust (*Robinia pseudoacacia*) and *Terfezia terfezioides*. In: Proceedings Intern. symposium on biotechnology of ectomycorrhizae. Molecular approaches. 9–11 Nov 1994, Urbino, Italy. Abstract: 66

Bratek Z, Jakucs E, Boka K, Szedlay G (1996) Mycorrhizae between black locust (*Robinia pseudoacacia*) and *Terfezia terfezioides*. Mycorrhiza 6:271–274

Brotzu R (1994) Specie fungine interessanti del territorio della provincia di Nuoro. MicologiaItaliana 2:131–138

Calonge FD (1982) Adiciones al catalogo de hoàngos hipogeos de Espana. Garcia de orta, Sér Est Agronom Lisboa 9:143–146

Calonge FD, Rocabruna A, Tabares M (1985) Nuevos datos sobre los hongos hipogeos de Espana. Bol Soc Micol Castellana 9:45–54

Cano A, Honrubia M, Molina-Ninirola (1991) Mycorrhizae in semiarid ecosystems: synthesis of mycorrhizae between *Terfezia claveryi* Chat., *Picoa juniperi* Vitt. and *Helianthemum almeriense* (Cistaceae). Abstract 3d European Symposium on Mycorrhizas. Mycorrhizasin ecosystems-Structure and function, 19–23 Aug 1991 Sheffield, UK

Chafi MEH, Fortas Z, Bensoltane A (2004) Bioclimatic survey of the terfez zones of the South West of Algeria and an essay of inoculation of *Pinus halepensis* Mill. with *Tirmania pinoyi*. Egypt J Appl Sci 19(3):88–100

Chatin AD (1892) La truffe: Botanique de la truffe et des plantes truffières. JB Baillière, Paris, France, 330 p

Chevalier G, Mousain D, Couteaudier Y (1975) Associations ectomycorhiziennes entre Tuberacées et Cistacées. Ann Phytopathol 7:355–356

Chevalier G, Riousset L, Dexheimer J, Dupré C (1984) Synthèse mycorhizienne entre *Terfezia leptoderma* Tul. et diverses Cistacées. Agronomie 4:210–211

Dexheimer J, Gérard J, Leduc JP, Chevalier G (1985) Etude ultrastructurale comparée des associations symbiotiques mycorhiziennes *Helianthemum salicifolium-Terfezia claveryi* et *H. salicifolium-Terfezia leptoderma*. Can J Bot 63:582–591

Diez J, Manjon JL (2001) Biology of desert truffles in Mediterranean dehesa type forests. In: Actes 5ème Congrès International Science et culture de la truffe, 4–6 mars 1999, Aix-en-Provence, France. Fédération Française des Trufficulteurs, Paris. pp. 140–141

Diez J, Manjon JL, Moreno G (2001). Taxonomy and ecology of desert truffles: the genera *Picoa* and *Tirmania* in the Iberian peninsula. In: Actes 5ème Congrès International Science et culture de la truffe, 4–6 mars 1999, Aix-en-Provence, France. Fédération Française des Trufficulteurs, Paris. pp. 64–65

Diez J, Manjon JL, Martin F (2002) Molecular phylogeny of the mycorrhizal desert truffles (*Terfezia* and *Tirmania*), host specificity and edaphic tolerance. Mycologia 94(2):247–259

Donadini (1979) Les champignons hypogés des îles d'Hyères (Ascomycètes et Basidiomycètes). Trav Sci Parc national Port-Cros 5. pp. 9–18

Fortas Z (1990) Etude de trois espèces de terfez:caractères culturaux et cytologie du mycélium isolé et associé à Helianthemum guttatum. Thèse de Doctorat. Université d'Oran Es-Senia, Algérie

Fortas Z, Chevalier G (1990). Effet des conditions de culture sur la mycorhization d' *Helianthemum guttatum* par trois espèces des genres *Terfezia* et *Tirmania* (truffes du désert). In: Atti Secondo Congr Intern sul Tartufo, 24–27 Nov 1988, Spoleto, Italie. pp. 197–203

Fortas Z, Chevalier G (1992) Effet des conditions de culture sur la mycorhization de l' *Helianthemum guttatum* par trois espèces de terfez des genres *Terfezia* et *Tirmania* d' Algérie. Can J Bot 70:2453–2460

Gutierrez A, Morte A, Honrubia M (2003) Morphological characterization of the mycorrhiza formed by *Helianthemum almeriense* Pau with *Terfezia claveryi* Chat. and *Picoa lefebvrei* (Pat.) Maire. Mycorrhiza 13(6):299–307

Gutierrez A, Moreno G, Morte A, Honrubia M (2004) *Terfezia olbiensis* Ness. in South-East Spain: a different species from *T. leptoderma* Tul. In: Actes du Premier Symposium sur les Champignons Hypogés du Bassin Méditerranéen. Rabat, 6–8 avril 2004. pp. 102–105

Hollos L (1933) Fungi of the region of Szekszàrd. Math Termtud Közl 37:205 (in Hungarian)

Honrubia M, Cano A, Molina-Ninirola C (1992) Hypogeous fungi from Southern Spanish semi-arid lands. Persoonia 14(4):647–653

Janex-Favre MC, Parguey-Leduc A, Riousset L (1988) L'ascocarpe hypogé d'une terfez française (*Terfezia leptoderma* Tul., Tubérales, Discomycètes). Bull Soc Mycol Fr 104(3):145–178

Khabar L, Najim L, Janex-Favre MC, Parguey-Leduc A (2001) Contribution à l'étude de la flore mycologique du Maroc: les truffes marocaines (Discomycètes). Bull Soc Mycol Fr 117 (3):213–229

Kagan-Zur V, Roth-Bejerano N (2009) Studying the brown desert truffles of Israel. Isr J Plant Sci 56:309–314

Kiraly I, Bratek Z (1992) *Terfezia terfezioides* a common truffle in Hungary. Micol Vegetazione Mediterr 7(1):43–45

Kovacs GM (2002) Study of mycorrhizae on the Great Hungarian Plain. Summary of Ph.D. dissertation. Univ. Szeged, Hungary

Kovacs GM, Bagi I, Vagvolgyi C, Kottke I, Oberwinkler F (2002) Studies on the root associations of the truffle *Terfezia terfezioides*. Acta Microb Immunol Hung 49(2–3):207–213

Langiu T (1979) La *Terfezia leonis* in Sardegna. Bol Gruppo Micol G Bresadola 22:32–35

Lawrynowicz M, Markovic M, Milenkovic M, Ivancevic B (1997) *Terfezia terfezioides* - a new hypogeous fungus from Balkan Peninsula. Acta Mycol 32:233–238

Leduc JP, Dexheimer J, Chevalier G (1986) Etude ultrastructurale comparée des associations de *Terfezia leptoderma* avec *Helianthemum salicifolium, Cistus albidus* et *C. salviifolius*. In: Mycorrhizae. Physiology and Genetics. Actes1er Symposium Européen sur les Mycorhizes, Dijon, 1–5 juillet 1985, Dijon. INRA, Paris. pp. 291–294

Maire R (1907) Contribution à l' étude de la flore mycologique de l' Afrique du Nord. Bull Soc Myc France 53:180–215

Malençon MG (1973) Champignons hypogés du nord de l' Afrique. I: Ascomycètes. Persoonia 7 (2):261–288

Mattirolo O (1900) Gli ipogei di Sardegna e di Sicilia. Malpighia 14:39–110

Mattirolo O (1905) Prima contribuzione allo studio della flora ipogea del Portogallo. Boll Soc Broteriana 21:86–105

Mattirolo O (1906) Seconda contribuzione allo studio della flora ipogea del Portogallo. Boll Soc Broteriana 22:227–246

Montecchi A, Sarasini M (2000) Fungi ipogei d' Europa. Associazione Micologica Bresadola. Centro Studii Micologici, Vicenza, Italie, 714 p

Morara M, Saitta A, Venturella G, Iotti M, Gargano ML, Zambonelli A (2009) I funghi ipogei dell' Emilia Romagna e della Sicilia a confronto. Annales Cofederationis Europaeae Mycologiae Mediterraneensis. pp. 53–67

Moreno G (1980) Notas sobre la vegetacion micologica centro occidental espanola. Acta Bot Malact 6:175–202

Moreno G, Diez J, Manjon JL (2002) *Terfezia boudieri* first records from Europe of a rare vernal hypogenous mycorrhizal fungus. Persoonia 17:637–641

Moreno G, Galan R, Montecchi A (1991) Hypogeous fungi from peninsular Spain II. Mycotaxon 42:207–238

Moreno GR, Galan R, Ortega A (1986) Hypogenous fungi from continental Spain. I Crypto Mycol 7(3):201–229

Moris JH (1827) Stirpium sardoarum Elenchus. Caralis 3:22

Morte MA, Cano A, Honrubia M, Torres P (1994) In *vitro* mycorrhization of micropropagated *Helianthemum almeriense* plantlets with *Terfezia claveryi* (desert truffle). Agric Sc Finland 3:309–314

Morte A, Navarro-Rodenas A, Nicolas E (2010) Physiological parameters of desert truffles mycorrhizal *Helianthemum almeriense* plants cultivated in orchards under water deficit conditions. Symbiosis 52:133–139

Moyen J (1889) Les champignons: traités élémentaires et pratiques de mycologie. J Rotschild, Paris, 762 p

Novak E, Zeller L. Ein neuer Fundort von *Mattirolomyces terfezioides* in Hungarn und die Verhältnisse seines Vorkommens.. Ann. Univ. Sci. Budapest, Sect. Biol. 2. pp. 197–200

Riousset L. & G., Dupré C., Chevalier G. (2004) Truffes et terfez communes à la France et au Maroc. In : Actes 1er Symposium sur les Champignons hypogés du bassin méditerranéen, 6–8 avril 2004, Rabat, Maroc, pp. 192–220

Roth-Bejerano N, Livne D, Kagan-Zur V (1990) *Helianthemum-Terfezia* relations in different growth media. New Phytol 114(2):235–238

Slama A, Neffati M, Khabar L, Boudabous A (2004) Etude taxonomique des truffes (Terféziacées) de la Tunisie méridionale. In: Actes 1er Symposium sur les Champignons Hypogés du bassin méditerranéen, 6–8 avril 2004, Rabat, Maroc. pp. 233–243

Szemere L (1965) Die unterirdische Pilze des Karpatenbeckens. Akademiai Kiado, Budapest. 319 p

Tulasne LR, Tulasne C (1851) Fungi Hypogaei. apud F. Klinchsieck. Paris. 222 p

Vittadini C (1831) Monographia Tuberacerum. Milan

Zitouni-Haouar FE-L, Fortas Z, Chevalier G (2012) Study of mycorrhizal associations between three species of desert truffles of *Terfezia* genus and various *Cistaceae* and conifers under controlled conditions. Mycorrhiza 2013

Chapter 10
Mediterranean Basin: North Africa

Lahsen Khabar

10.1 Introduction

Mediterranean countries, including those of North Africa, are the location of an abundant harvest of edible truffles known by several different names: "Terfass," "Torfàs," "Kama," "Kame," "Kholassi," "Zubaidi," "desert truffles," and "sand truffles." Because of their pale coloration they are often called "white truffles" (Maublanc 1946, 1952).

These names refer to a variety of edible mushrooms, which are less appreciated as compared to the true (forest) truffles (Maublanc 1946, 1952). They were classified along with forest truffles, into the Tuberales (Janex-Favre and Parguey-Leduc 1985; Parguey-Leduc et al. 1987a, b, 1988, 1989, 1990; Janex-Favre et al. 1988; Khabar et al. 1994) to form a distinct family Terfeziaceae (Trappe 1979). This order Tuberales was, however, abandoned and members were moved into of the order Pezizales (Korf 1973; Trappe 1979; Donadini 1983).

The genera *Delastria*, *Terfezia*, *Tirmania*, *Picoa*, *Balsamia*, and *Melanogaster* were placed in the Terfeziaceae family within this order.

The Bataille key (1922) includes thirteen species and four varieties. Chatin (1891a, b, 1896a, b, c), Maire (1907, 1933), Maire and Werner (1937), Malençon (1973), Alsheikh and Trappe (1983), and Khabar et al. (2001) added to the list several other species found in Morocco, North Africa, and the Middle East.

"Desert truffles" or "Terfass" are collected in semiarid and arid areas with hot climates. Countries around the Mediterranean, especially in North Africa and the Middle East, are renowned for their desert truffle yields. The following species have been reported from North Africa (Algeria, Morocco, Libya, Tunisia, and Egypt) by Chatin (1891a, b, 1896b), Patouillard (1894a, b), Maire (1906, 1907, 1933), Maire and Werner (1937), Malençon (1973), Khabar (1984, 1988), Fortas and Chevalier

L. Khabar (✉)
Team "Mycologie Forestière et Trufficulture", Department of Biology, Faculty of Sciences, Mohammed V-Agdal University, B.P. 1014, Avenue Ibn Battouta, Rabat, Morocco
e-mail: khabar@fsr.ac.ma; l.khabar@yahoo.fr

V. Kagan-Zur et al. (eds.), *Desert Truffles*, Soil Biology 38,
DOI 10.1007/978-3-642-40096-4_10,

(1990), El-Kholy and Assim (1991), Pacioni and El-Kholy (1994), Moawad et al. (1997), Ali et al. (1998), Khabar and Najim (1999), Khabar et al. (1994, 1999, 2001, 2005), and Slama et al. (2006): *Terfezia boudieri*, *T. goffartii*, *T. leptoderma* and *T. arenaria*, *T. leonis*, *T. leonis* var. *heterospora* and *T. mellerionis*, *T. claveryi* (="red Terfass" of Tafilalet), *Tirmania nivea* and *T. pinoyi* (="white Terfass" or "Zoubaïdi"), *Picoa juniperi*, and *P. carthusiana*. We note that many varieties and species have been revised or abandoned.

Some species of desert truffles are found outside of North Africa: in the Middle East (Saudi Arabia, Iraq, Iran, Kuwait, Syria, Israel, etc.) as well as in European countries in the Mediterranean basin (the south of Italy, Spain and France, Portugal, Greece) (Tulasne and Tulasne 1851; Moyen 1889; Chatin 1891a, b, 1896a, b, c; Patouillard 1894a, b; Fischer 1897; Mattirolo 1905, 1922; Bataille 1922; Maublanc 1927, 1946; Imai 1933; Rayss 1959; Bresadola 1960; Ceruti 1960; Trappe 1971, 1979; Malençon 1973; Calonge 1982; Calonge et al. 1985; Donadini 1979, 1986; Langiu 1979; Pacioni 1979; Awameh and Alsheikh 1980a, b; Binyamini 1980; Girel 1980; Castro and Freire 1982; Moustafa 1985; Moreno et al. 1986; Bokhary 1987; Bokhary and Parvez 1988; Rueda and De Rueda 1989; Daneshpazhuh 1991; Ewaze et al. 1991; Hashem and Al-Homaidan 1991; El-Kholy et al. 1992a, b, c; Honrubia et al. 1992; Pacioni and El-kholy 1994; Hashem and Al-Obaid 1997; Moawad et al. 1997; Kagan-Zur 1998; Diez et al. 1999).

Desert truffles are also found in Hungary (Kiraly and Bratek 1992; Kiraly et al. 1992; Bratek et al. 1996), Turkey (Agaoglu and Artik 1992; Afyon 1996), the Canary Islands (Korf and Zhuang 1991), India (Khare 1975), China (Bin-Cheng 1992), South Africa (Marasas and Trappe 1973; Ackerman et al. 1975; Taylor et al. 1995; Kagan-Zur et al. 1999a, b), North America (Gilkey 1947, 1954; Knighton 1976; Trappe and Sundberg 1977), South America (Trappe 1979), Germany (Boetticher 1987), Chihuahuan Desert in Mexico (Zak and Whilford 1986), and in the Balkan Peninsula (Lawrynowicz et al. 1997).

While forest truffles such as *Tuber aestivum* and *T. rufum Pico Vitt* Syn *uncinatum* which were collected by Malençon (1973) in the middle Atlas between elevation 1,600 and 2,000 m and *T. melanosporum* recently introduced at the chain of horsts and the massive of Debdou in eastern Morocco (1,700 m) are collected in regions of subhumid to humid climate and limestone soil, desert truffles are generally harvested in semiarid or arid zones in sandy loam or sandy soils which may be slightly acidic or basic.

Truffles of the genus *Tuber* develop under poplars, oaks, olive trees, pines, chestnut, and more (Chevalier et al. 1984). Desert truffles are collected under phanerogam herbaceous plants such as Cistus (species reported are *Cistus halimifolius*, *C. ladaniferus* var. *halimioides*, *C. salicifolius*, *C. monspeliensis*, and *C. salvifolius*) and *Helianthemum* spp., detailed below (Chatin 1891a, b; Awameh and Alsheikh 1978; Awameh et al. 1979; Awameh 1981; Alsheikh 1984; Dexheimer et al. 1985) or Pins (Chevalier et al. 1984; Korf and Zhuang 1991; Janex-Favre et al. 1988; Khabar and Najim 1999; Khabar et al. 1999, 2001).

However, the desert truffle *Terfezia leptoderma*, a rather common species in southern Europe and northern Africa, was collected under *Pinus radiata,* in San Leonardo, Italy (Janex-Favre et al. 1988); under *Pinus canariensis*, in the Canary Islands (Korf and Zhuang 1991); and under *Pinus pinaster* var. *atlantica*, in the Mamora forest, Morocco (Khabar and Najim 1999; Khabar et al. 1999, 2001).

Other tree species have also been reported as host plants of desert truffles; *Robinia pseudoacacia* associated to *Terfezia terfezoides* (=*Mattirolomyces terfezioides*) (Bratek et al. 1996) as well as some oak species (Chevalier et al. 1984).

But the most common desert truffles host plants are, in general, annual or perennial herbaceous *Helianthemum* spp. (Chatin 1891a, b, 1896a, b, c; Awameh 1981; Awameh and Alsheikh 1980a, b; Alsheikh 1984; Khabar 1984, 1988; Khabar and Najim 1999; Khabar et al. 1994, 1999, 2001; Dexheimer et al. 1985; Fortas and Chevalier 1990; 1992a, b; Roth-Bejareno et al. 1990; Cano et al. 1991; Kagan-Zur et al. 1994; Morte et al. 1994; 1995; Morte and Honrubia 1997). Species reported are the following: *Helianthemum tuberaria*, *H. guttatum* (L. Foureau), *H. salicifolium* (L). Mill., *H. ledifolium* (L). Mill., *H. salicifolium*, *H. almeriense,* and *H. sessiliflorum.*

The most common method of harvesting desert truffles is by observing the ground as it is often swollen and cracked at the base of the host plant.

In this chapter, we will include species recorded or collected by us in the countries of North Africa and give their geographical distribution.

10.2 The Species and Their Distribution

10.2.1 The Genus Delastria *(Tul. and C. Tul. 1843) Terfeziaceae (Montecchi and Lazzari 1993)*

10.2.1.1 *Delastria rosea* Tulasne 1843 (Figs. 10.1a, 10.2a, 10.3a, b)

(syn. *Terfezia rosea* (Tul.) Torrend 1907)

In Morocco, it is locally known as "bitter Terfass of pine." We harvest this species under pine (*Pinus pinaster* var. *atlantica*) on acid soils and under semiarid climate in the Mamora forest and in the green belt of Temara (near Rabat) the same places as *Tuber oligospermum* (Khabar et al. 2001), but it was also reported in Larache (Malençon 1973). Little buried, mature ascocarps are often exposed by the December rains. The harvest period is very short, from November to December. Little appreciated because of its bitter taste and unpleasant smell.

Fig. 10.1 Ascocarps of *Delastria rosea* (**a**), *Terfezia arenaria* (**b**), *T. boudieri* (**c**), *T. claveryi* (**d**), *T. leptoderma* (**e**), *Tirmania pinoyi* (**f**), and *T. nivea* (**g**)

10.2.2 The Genus Terfezia *(Tul. and C. Tul. 1851) Terfeziaceae (Montecchi and Lazzari 1993)*

10.2.2.1 *Terfezia arenaria* Trappe 1971 (Figs. 10.1b, 10.2b–d, 10.3c)

(syn. *Tuber arenarium* Moris 1829; *Terfezia leonis* Tul. 1851)

This species is very common in the countries of northern Africa. Generally, *Terfezia arenaria* is collected on acid soil in semiarid climate, under *Helianthemum guttatum*, as of the first week of March until May. It is highly prized by farmers, is

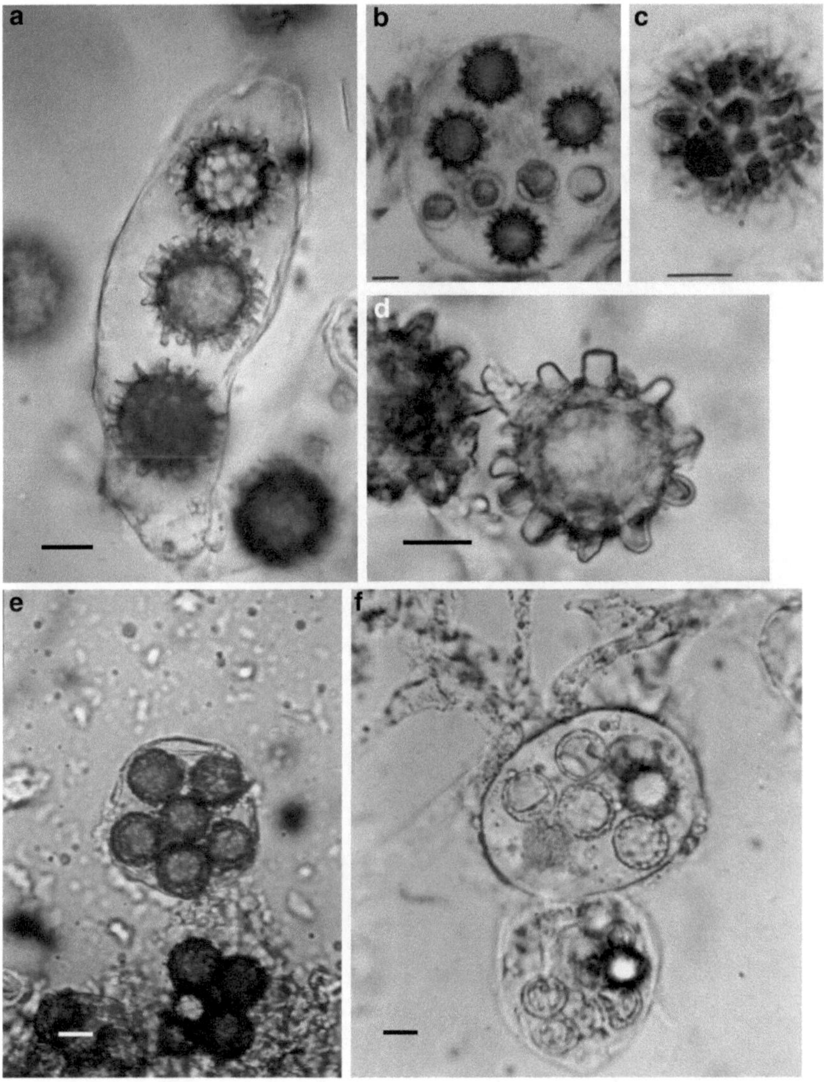

Fig. 10.2 (continued)

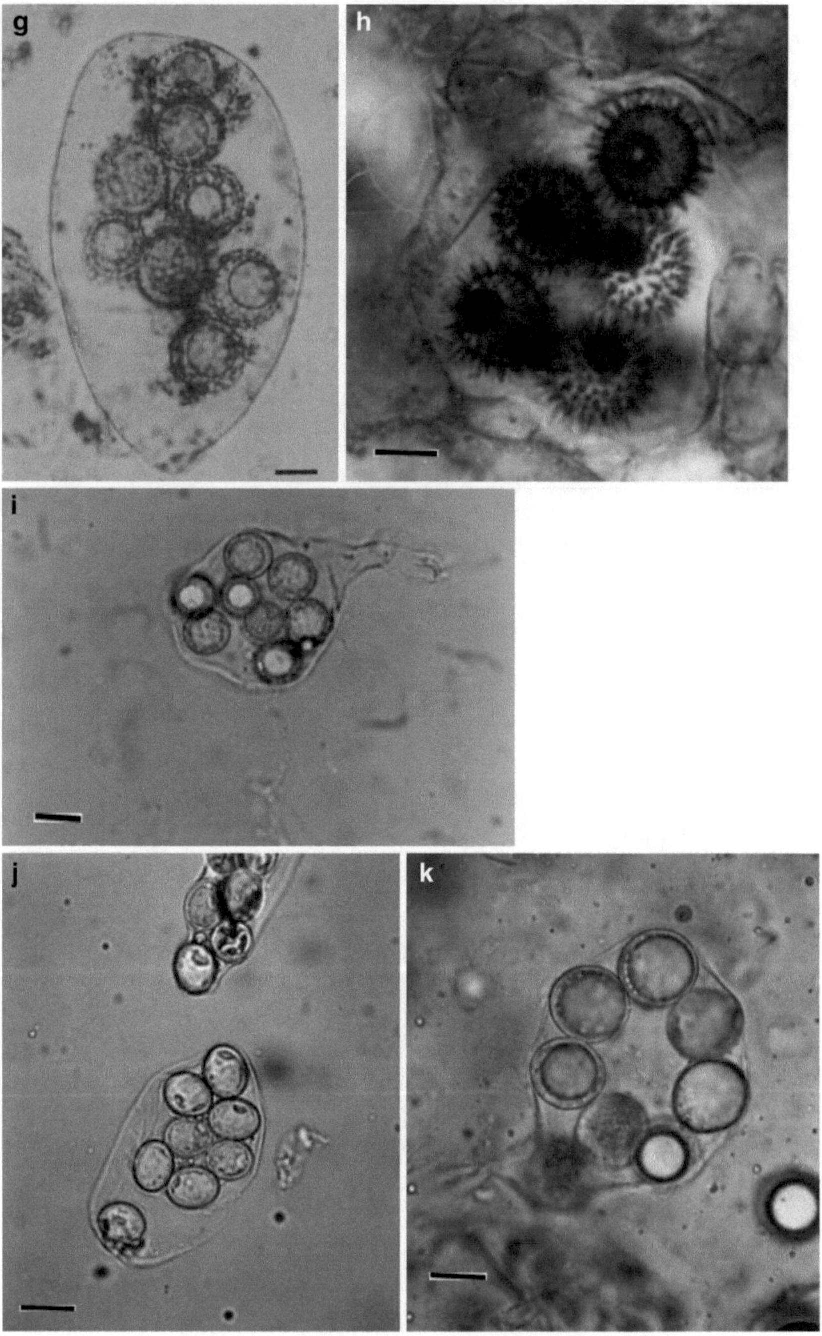

Fig. 10.2 Asci and ascospores (scale 10 μm); *Delastria rosea* (**a**); *Terfezia arenaria* (**b–d**); *T. boudieri* (**e**); *T. claveryi* (**f**); *T. leptoderma* (**g**, **h**); *Tirmania pinoyi* (**i**); *T. nivea* (**j**); and *Picoa juniperi* (**k**)

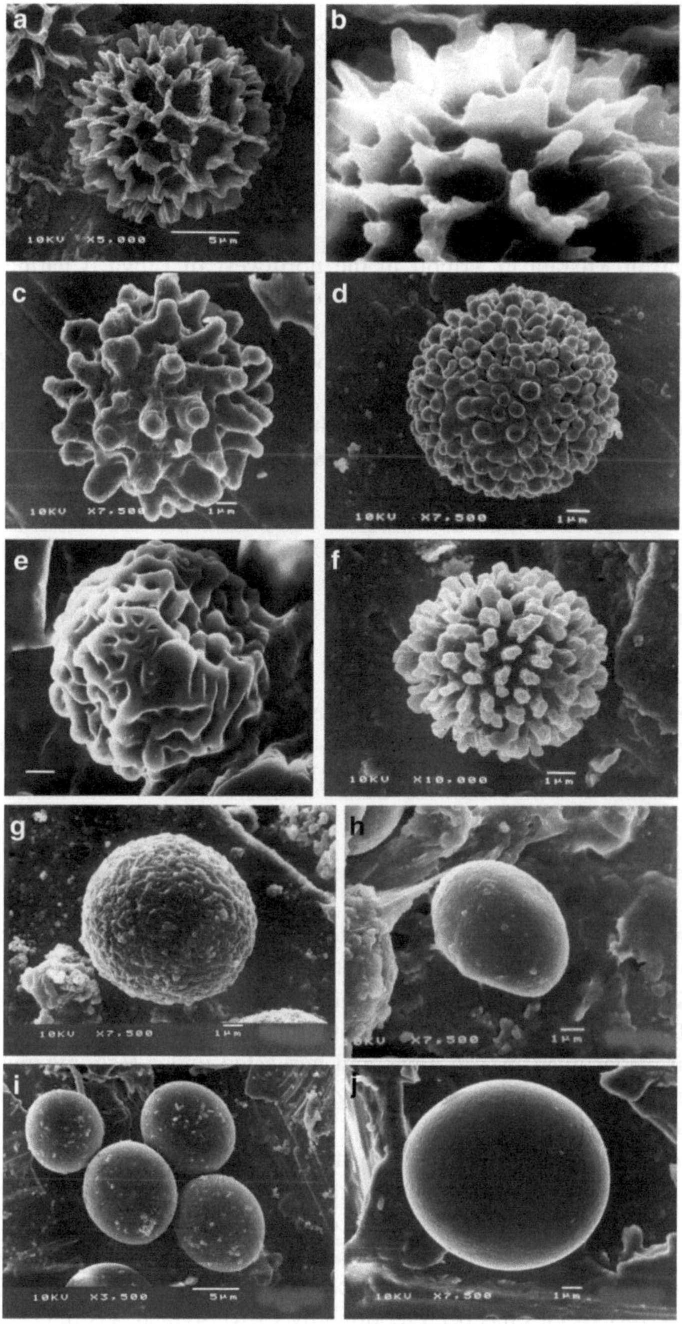

Fig. 10.3 Ascospores in scanning electron microscopy (SEM); *Delastria rosea* (**a**, b—detail); *Terfezia arenaria* (**c**); *T. boudieri* (**d**); *T. claveryi* (**e**); *T. leptoderma* (**f**); *Tirmania pinoyi* (**g**); *T. nivea* (**h**); and *Picoa juniperi* (**i**, **j**)

traded in local markets, and is even exported to Kuwait, to Saudi Arabia, and to Eastern countries generally (Khabar et al. 2001).

Distribution

In Algeria: this species is cited as being collected at Bou-Saada; Zeriguet by Fortas and Chevalier (1992a).

In Libya: at Misrata in Tripoli in the northwest of Libya (Chatin 1896a) in typically Mediterranean climate: hot and dry summers, cool winters, and some modest rainfall.

In Morocco: at the Moroccan Meseta between Rabat and Tangier (Malençon 1973); Mamora forest (Khabar et al. 2001)

In Tunisia: at Carthage (Patouillard 1894a, b). This fungus also grows in parts of the Northwest; Tabarka and Ain Draham (Khabar et al. 2005).

10.2.2.2 *Terfezia boudieri* Chat. 1891

(syn. *T. boudieri* var. *arabica* Chat. 1892; *T. deflersii* Pat. 1894) (Figs. 10.1c, 10.2e, 10.3d)

More southern than T. *arenaria.*

This species is harvested under *Helianthemum* spp., in desert plains (with arid to semiarid climate) of North Africa, in spring (March–April), generally on limestone soil.

Distribution

In Algeria: this species has been reported in southern Algeria (Biskra, Barika, Batna, Bou-Saada, and El Golea) by Chatin (1891a, b).

In Tunisia: this species is harvested from the second half of February throughout southern Tunisia, including Medenine, Tataouine, and Gafsa. It is called locally red terfez. It is widespread in arid areas and harvested under *Helianthemum lipii* or *H. sessiliflorum* and sometimes under *Rhanterium suaveolens* (Khabar et al. 2005; Slama et al. 2006).

In Morocco: this species is harvested at Had Hrara, region of Oualidia, about ten miles east of Safi area on the plain of Abda, not far from the Atlantic coast, also in Ain Beni Mathar, (region of Erfoud), on plateau of eastern Morocco and in southeast of Morocco. It is widespread in arid and calcareous soil under *Helianthemum sessiliflorum* and *H. ledifolium* (Khabar et al. 2001, 2005).

In Libya: this species has been reported by El-Kholy and Assim (1991).

In Egypt: this species has been reported by Ali et al. (1998) to grow both at the east and west of the Nile. Fruit bodies were also collected from the northwestern and northeastern parts of the Egyptian desert by Moawad et al. (1997) and were reported from the northwestern coast of Egyptian desert including northern coast of North Sinai, Sidi Barrani, and Sallum (El-Kholy and Assim 1991; El-Kholy et al. 1992a, b; Pacioni and El-Kholy 1994).

10.2.2.3 *Terfezia claveryi* Chat. 1891

(syn. *T. hafizi* Chat. 1892) (Figs. 10.1d, 10.2f, 10.3e)

This species is very common in the semidesert plains of southern Mediterranean. It is widespread in sandy soils and limestone in arid and sub-Saharan regions. It is harvested under several species of *Helianthemum*: perennial, annual, herbaceous, or hemicryptophytes.

Distribution

In Algeria: at Laghouat oasis, 380 km south of Biskra (Chatin 1891a, b).

In Tunisia: this species is harvested in Dbin, the region of Tozeur, from mid-February on slightly sandy soils (khabar et al. 2005; Slama et al. 2006).

In Morocco: reported along the parallel 32° North, near the centers of Ksar-es-Souk, Bou-Bernous, Boudenib, and Figuig by Malençon (1973); also collected in calcareous soil, arid, and Saharan regions, at Ain Beni Mathar, Tendrara, Bouarfa, and region of Erfoud, in southeastern Morocco (Khabar et al. 2001, 2005). *T. claveryi* is collected in the proximity of *Helianthemum lipii* and *H. apertum* and sold under the name "red Terfass of Tafilalet." Very common, especially after heavy March rains. Highly sought after, it is the subject of major commercial activity in the east of Morocco between March and May (Khabar et al. 2001).

In Libya: cited by El-Kholy and Assim (1991).

In Egypt: collected from the northwestern and northeastern parts of the Egyptian desert (Moawad et al. 1997; El-Kholy and Assim 1991; Pacioni and El-Kholy 1994).

10.2.2.4 *Terfezia leptoderma* Tul. 1851

(syn. *Terfezia olbiensis* Tul. 1851 The *Terfezia olbiensis* Tul. Cited by several authors (Chatin 1891a, b; Ceruti 1960; Mattirolo 1905) was regarded as an immature form of *T. leptoderma* (Malençon cited by Moreno et al. 1986). New research has restated *T. olbiensis* as a distinct independent species (see Chap. 3 by Bordallo and Rodríguez).

(syn. *Choiromyces leptodermus* Tul. 1845; *Terfezia fanfanii* Matt. 1900) (Figs. 10.1e, 10.2g, h, 10.3f)

Distribution

We collect *T. leptoderma* in the Mamora forest on acid soil under *Helianthemum guttatum* from the third week of February until May. We also collected it under pine (*Pinus pinaster* var. *atlantica*) from November to January. Often sold mixed with *Terfezia arenaria*. It is considered by farmers as an indicator of the arrival of the "real" Terfess (*Terfezia arenaria*) (Khabar et al. 2001).

10.2.3 The Genus Tirmania *(Chat. 1891)*

Species of the genus *Tirmania* are locally known as "white Terfass of Tafilalet" or "Zubaidi." They are very abundant at calcareous soils of desert areas in arid, sub-Saharan, and Saharan regions. They are harvested under *Helianthemum* spp. such as *Helianthemum hirtum*, *H. lipii*, *H. ledifolium*, *H. salicifolium* at the second week of December until the end of March.

Two species are known *Tirmania* to date.

10.2.3.1 *Tirmania pinoyi* (Maire) Malençon 1973 (syn. *Terfezia pinoyi* Maire 1906) (Figs. 10.1f, 10.2i, 10.3g) and *Tirmania nivea* (Desf. Ex. Fr.) Trappe 1971 (syn. *Tuber niveum* Desf. ex Fr. 1823; *Terfezia ovalispora* Patouillard 1890; *Tirmania ovalispora* (Pat.) Pat. 1892; *Tirmania africana* Chatin 1892; *Terfezia africana* (Chat.) Maire 1916; *Tirmania camboni* Chatin 1892) (Figs. 10.1g, 10.2j, 10.3h)

Distribution

In Algeria: cited at Biskra, Batna, and Barika (Chatin 1891a, b) and at Zériguet and west of Naama (Fortas and Chevalier 1992a, b.

In Morocco: very abundant in the southeast of Morocco; in the high plateau of eastern Morocco; and regions of Ain Beni Mathar, Bou-Bernous, Tendrara, Bouarfa, Erfoud, Figuig, and Rissani (Khabar et al. 2001). Also reported at the Hamada Daoura, southern Morocco, 29°N latitude by Malençon (1973). These species have been collected in calcareous soil, arid, and Saharan regions under *Helianthemum hirtum*, from the second week of December until the end of March (Khabar et al. 2001).

In Tunisia: the species of this genus are very common at Ben Gardane, Tataouine, and Gafsa in southern Tunisia. They grow near *Helianthemum lipii* and are harvested in the month of February (Khabar et al. 2005; Slama et al. 2006).

In Libya: (El-Kholy and Assim 1991).

In Egypt: collected from the northwestern and northeastern parts of the Egyptian desert (Moawad et al. 1997; El-Kholy and Assim 1991 and Pacioni and El-Kholy 1994).

10.2.4 The Genus Picoa *(Vitt. 1831)*

10.2.4.1 *Picoa juniperi* Vitt. 1831 (syn. *Picoa juniperina Tul. 1851)* (Figs. 10.2k, 10.3i, j)

In Morocco, locally called "Ed doukar," this species is very rare; it is harvested with *Terfezia claveryi* under *Helianthemum lipii* at the end of February (Khabar et al. 2001, 2005)

In Tunisia: this species is harvested in the region of Tozeur and Gafsa to the second half of February near *Rhanterium suaveolens* and *Helianthemum lipii* (Khabar et al. 2005; Slama et al. 2006).

10.2.4.2 *Picoa carthusiana* Tul. and C. Tul. 1851

In Tunisia: this species is called "zouber." It is harvested in the region of Tozeur and Gafsa to the second half of February near *Rhanterium suaveolens* and *Helianthemum lipii.* It is considered as indicator of sites *Terfezia* and *Tirmania* (Khabar et al. 2005; Slama et al. 2006).

In Egypt: collected by Pacioni and El-kholy (1994).

10.3 Comparisons and Conclusions

We note that the countries of northern Africa (Morocco, Algeria, Tunisia, Libya, and Egypt) have a high species diversity of desert truffles, and there are a dozen species distributed more or less throughout large areas with a specific distribution depending on the nature of the soil and climate. We note both analogies with other Mediterranean countries and those of Middle Eastern Asia. Thus the species recorded in other Mediterranean countries, Spain, southern France, and southern Italy (*Terfezia arenaria*, *T. leptoderma*) are found in the north where the climate is semiarid. By contrast, species which are rather xerothermophilous (*Terfezia claveryi*, *T. boudieri*, *Tirmania pinoyi,* and *T. nivea*) are harvested in the south where the climate is arid and relatively sub-Saharan. These species are encountered in some countries of East Asia (Kuwait, Saudi Arabia, Israel, Turkey, Iraq, Syria) (Chatin 1891a; b; Dickson 1955; Awameh et al. 1979, see also Chaps. 12, 9 and 11 by Ammarellou et al., Chevalier and Kagan-Zur and Akyuz). We also find that some species, *Terfezia leptoderma* and *T. arenaria*, prefer acidic soils while

Terfezia boudieri, *T. claveryi*, *Picoa juniperi* and *P. carthusiana*, *Tirmania pinoyi* and *T. nivea*, which are species xerothermophilous, prefer rather alkaline soils (limestone). Finally, both our observations and the consulted bibliographies confirmed the characteristic distribution of truffle species between Europe and North Africa, as claimed by Malençon (1973). So, these are the truffles of the genus *Tuber* dominate in humid and subhumid countries of Europe, while the genera *Tirmania* (*T. pinoyi* and *T. nivea*), Picoa (*P. juniper* and *P. carthusiana*), and *Terfezia* (*Terfezia claveryi* and *T. boudieri*) characterize arid and Saharan countries of the southern Mediterranean.

However, few species of *Terfezia* (*T. arenaria*, *T. leptoderma*, and *Delastria rosea*) are common to both southern European and North African countries.

References

Ackerman LGJ, Van Wyk PJ, Du Plessis LM (1975) Some aspects of the composition of the Kalahari truffle or n'abba. South Afr Food Rev 2(5):145–147

Afyon A (1996) Isparta yoresinde belirlenen bazi makroskobik mantarlar. Some macrofungi of Isparta (Turkey). Turk J Bot 20(2):161–164

Agaoglu Y, Artik N (1992) Chemical compositions of desert truffle (*Terfezia boudieri*) grown in Turkey. Micologiae e vegetazione. In: Tuber, proceeding of the international conference on Truffle, L'Aquila, Italy, 5–8 March 1992

Ali MI, El-Saidi MT, Moawad H, Abu-Shanab RAI (1998) Host range diversity and impact of vegetation on the desert truffles of Egypt. J Union Arab Biol, Cairo 5(B):43–58

Alsheikh AM (1984) Mycorrhizae of annual *Helianthemum* species formed with desert. In: Molina R (ed.) Proceedings of the sixth N. Am. Conf. on Mycorrhizae. Bend, OR, 25–29 June

Alsheikh AM, Trappe JM (1983) Desert truffles: the genus *Tirmania*. Trans Br Mycol Soc 81(1):83–90

Awameh MS (1981) The response of *Helianthemum salicifolium* and *H. ledifolium* to infection by the desert truffle *Terfezia boudieri*. Mushroom sciences XI proceeding of the eleventh international scientific congress on the cultivation of edible fungi Australia, 1980. pp. 843–853

Awameh MS, Alsheikh A (1978) Laboratory and field study of for kinds of truffle (Kamah), Terfezia and Tirmania species, for cultivation. Mushroom science X (Part II) proceedings of the tenth international congress on the science and cultivation of edible fungi, France, 1978. pp. 507–517

Awameh MS, Alsheikh A (1980a) Ascospore germination of black kame (*Terfezia boudieri*). Mycologia 72(1):50–54

Awameh MS, Alsheikh A (1980b) Features and analysis of spore germination in the brown kame *Terfezia claveryi*. Mycologia 72(3):494–499

Awameh MS, Alsheikh A, Al Ghawas S (1979) Mycorrhizal synthesis between *Helianthemum ledifolium*, *H. salicifolium* and four species of the genera *Terfezia* and *Tirmania* using ascospores and mycelial cultures obtained from ascospore germination. In: Fourth North American Conf. on Mycorrhizae, 24–28 June 1979. Colorado State University, Fort Collins, Colorado. p. 23

Bataille MF (1922) Flore analytique et descriptive des tuberoides de l'Europe et de l'Afrique du nord. Bull Soc Mycol Fr 37(4):155–207

Binyamini N (1980) Addenda to the Hypogeous Mycoflora of Israel. Nova Hedwigia 32:9–20

Bin-cheng Z (1992) Chinese hypogeous mycoflora: Terfeziaceae, new species and new. Micol Vegetazione Mediterr 7(1):39–42

Boetticher W (1987) Truffles and their use. Industrielle-Obst-und Gemueseverwertung (Germany, F.R.) 72(11–12): 464–466
Bokhary HA (1987) Desert truffles 'Al-Kamah' of the Kingdom of Saudi Arabia. I. Occurrence, identification and distribution. Arab Gulf J Sci Res B (Agric Biol Sci) 5(2):245–255
Bokhary HA, Parvez S (1988) Desert truffles 'al-kamah' of the Kingdom of Saudi Arabia. 2. Additional contributions. Arab Gulf J Sci Res 6(1):103–112
Bratek Z, Jakucs E, Boka K, Szedlay G (1996) Mycorrhizae between black locust (*Robinia pseudoacacia*) and *Terfezia terfezioides*. Mycorrhiza 6(4):271–274
Bresadola J (1960) Elaphomycetales et Tuberales. Iconographia Mycologica 28(supplement II)
Calonge FD (1982) Adiciones al catalogo de hongos hipogeos de Espana. Garcia de Orta, Sér Est Agron Lisboa 9(1–2):143–146
Calonge FD, Rocabruna A, Tabras M (1985) Nuevos datos sobre los hongos hipogeos de espana. Bol Soc Micol Castellana 9:45–54
Cano A, Honrubia M, Molina-ninirola C (1991) Mycorrhizae in semiarid ecosystems: synthesis of mycorrhizae between *Terfezia claveryi* chat., *Picoa* juniperi vit. and *Helianthemum almeriense* (Cistaceae). Third European symposium on mycorrhizas, mycorrhizas in ecosystems—structure and function Sheffield, 19–23 August
Castro ML, Freire L (1982) *Terfezia berberiodora* y *T. oligosperma* encontradas en Galicia, nuevas para Espagna. Collect Bot Esp 13(2):441–444
Ceruti A. (1960) Elaphomycétales et Tubérales. Ab. Dr. J. Bresadola (Tridenti). In: Bresadola G (ed.) Iconographia mycologica 28(supplementum II)
Chatin A (1891a) Terfas ou truffes d'Afrique et d'Arabie. Genre *Terfezia* et *Tirmania*. Bull Soc Bot Fr 38:59–64
Chatin A (1891b) Contribution à l'histoire botanique de la truffe : kamé de Damas (*Terfezia claveryi*). Bull Soc Bot Fr 38:332–335
Chatin A (1896a) Truffes (terfas) de Mesrata, en Tripolitaine. Bull Soc Bot Fr 43:139–143
Chatin A (1896b) Un terfes d'Espagne et trois nouveaux terfas du Maroc. Bull Soc Bot Fr 43:397–399
Chatin MA (1896c) Truffes (terfaz) de Grèce, *Terfezia* gennadii. Bull Soc Bot Fr 43:611–617
Chevalier G, Riousset L, Dexheimer J, Dupre C (1984) Synthèse mycorhizienne entre *Terfezia leptoderma* Tul. et diverses Cistacées. Agronomie 4:210–211
Daneshpazhuh B. (1991) Two new records of Ascomycetes (Truffles) from Iran. Proceedings of the tenth Plant Protection Congress of Iran 1–5 Sep 1991 Kerman (Iran Islamic Republic) (1991). p. 123
Dexheimer J, Gerard J, Leduc JP, Chevalier G (1985) Etude ultrastructurale comparée des associations symbiotiques mycorhiziennes *Helianthemum salicifolium-Terfezia claveryi* et *Helianthemum salicifolium-Terfezia leptoderma*. Can J Bot 63(3):582–591
Dickson V (1955) The wild flowers of Kowait and Bahrain. George allen and Unwin LTD, London, 103p
Diez J, Manjon JL, Moreno G (1999) Two species of the genera *Picoa* and *Tirmania* in the Iberian peninsula. Vème Congrès International Sciences et culture de la truffe et des autres champignons hypogés comestibles, 4–6 Mars 1999, Aix-En-provence, France
Donadini JC (1979) Les Champignons hypogés des Iles D'hyères (Ascomycetes et Basidiomycètes). Trav Sci Parc Nation Port-Cros 5:9–18
Donadini JC (1983) Etude des discomycètes I. Critères taxinomiques des pézizales et tubérales. Bull Soc Linn Provence 35:53–73
Donadini JC (1986) Les *Balsamia*cees sont des *Helvellacees*: cytologie et scanning de *Balsamia vulgaris* Vitt. et de *Balsamia platyspora* Berk. et Br. Bull Trimestriel Soc Mycol Fr 102(4):373–387
El-Kholy HK, Assim MMA (1991) Truffles in Egypt: survey and identification. Micol Vegetatione Mediterr 7(1):46
El-Kholy HK, Assim MMA, Ali Kamil AA (1992a) Conidia formation in desert truffles, their dna content and the possibility of Homothallism. Micol Vegetatione Mediterr 7(1):66

El-Kholy HK, Assim MM, Hamouda MH (1992b) Genetic possibilities for the difficulty of truffle ascospore. Micol Vegetatione Mediterr 7(1):65

El-Kholy HK, Assim MM, Hamouda MH (1992c) The histology and cytology of Egyptian truffles. Micol Vegetatione Mediterr 7(1):56

Ewaze JO, Al-Naama MM, Alabdul M, Hussain RK (1991) Biochemical changes during morphogenesis of *Picoa* juniperi vitt. Sporacarps. Crypt Bot 2(2/3):121–124

Fischer E (1897) Plectascineae. In: Engler et Prantl, Die natürlichen Pflanzenfamilien, Engelmann Leipzig. pp. 290–320

Fortas Z, Chevalier G (1990) Effet des conditions de culture sur la mycorhization d'*Helianthemum guttatum* par trois espèces des genres *Terfezia* et *Tirmania* (truffes du désert). Atti del Secondo congresso internazionale sul Tartufo. In: Bencivenga M, Granetti B (eds.) Comunita Montana del Monti Martani del Serano, Spoleto, Italy, 24–27 Novembre 1988. Pp. 197–204

Fortas Z, Chevalier G (1992a) Caracteristiques de la germination des Ascospores de *Terfezia arenaria* (Moris) Trappe, recolté en Algerie. Crypt Mycol 13:21–29

Fortas Z, Chevalier G (1992b) Effet des conditions de culture sur la mycorhization de l'*Helianthemum guttatum* par trois espèces de Terfez des genres *Terfezia* et *Tirmania* d'Algérie. Can J Bot 70(12):2453–2460

Gilkey HM (1947) New or otherwise noteworthy species of Tuberales. Mycologia 39(4):441–452

Gilkey HM (1954) Tuberales. North Am Flora 2:1–36

Girel R (1980) J'ai gouté les Terfas chantées par Juvénal. Bull Féd Mycol Dauphiné-Savoie 78:18–21

Hashem AR, Al-Homaidan AA (1991) Metal analysis of soil and *Tirmania pinoyi* from Saudi Arabia. Crypt Bot 2(2/3):118–120

Hashem AR, Al-Obaid AM (1997) Mineral composition of soil and wild desert truffles in Saudi Arabia. J King Saud Univ Sci 8(1):5–10

Honrubia M, Cano A, Molina-Ninirola C (1992) Hypogeous fungi from southern Spanish semi-arid lands. Persoonia 14(4):647–653

Imai S (1933) On two new species of Tuberaceae. Proc Imp Acad 9:183–184

Janex-Favre MC, Parguey-Leduc A (1985) Les asques et les ascospores du *Terfezia claveryi* Ch. (Tubérales). Crypt Mycol 6(2):87–99

Janex-Favre MC, Parguey-Leduc A, Riousset L (1988) L'ascocarpe hypogé d'une Terfez française (*Terfezia leptoderma* Tul., Tubérales, Discomycètes). Bull Trimestriel Soc Mycol Fr 104(3):145–178

Kagan-Zur V, Raveh E, Lischinsky S, Roth-Bejerano N (1994) Initial association between *Helianthemum* and *Terfezia* is enhanced by low iron in the growth. New Phytol 127(3):567–570

Kagan-Zur V (1998) *Terfezia*s, a family of mycorrhizal edible mushrooms for arid zones. Proceedings of the first international meeting on "Ecology, Physiology and Cultivation of Edible Mycorrhizal Mushrooms"

Kagan-Zur V, Kuang J, Stabak S, Taylor FW, Roth-Bejerano N (1999a) Potential verification of a host plant for the desert truffle *Terfezia pfeilii* by molecular methods molecular verification of *Terfezia* host. Mycol Res 103(10): 1270–1274

Kagan-Zur V, Taylor FW, Roth-Bejerano N (1999b) Identification of plant hosts of the Kalahari desert-truffle *Terfezia pfeilli*. Vème Congrès International Sciences et culture de la truffe et des autres champignons hypoges comestibles, 4–6 Mars 1999 , Aix-en-Provence, France

Khabar L (1984) Etude systématique et inventaire des espèces de *Terfezia* de la forêt de la Mamora (Règion de Salé). Mém. Pour le C.E.A Mycologie, Université Mohammed V, Fac. Sci. Rabat, Dépt. Biologie, 38 p. + planches

Khabar L (1988) Le genre *Terfezia* Tull. (Terfass) de la forêt de la Mamora (Région de Salé: étude systématique, écologique, morphologique, cytologique et ultrastructurale). Thèse de troisième cycle (mycologie). Département de Biologie, Faculté des Sciences, Université Mohammed V, Rabat, Maroc

Khabar L., Najim L. (1999) Les truffes au Maroc : biodiversité et circuit de commercialisation. Actes du Vème Congrès International Sciences et culture de la truffe et des autres champignons hypogés comestibles, 4–6 Mars 1999, Aix-En-provence, France. pp. 511–516

Khabar L, Najim L, Janex-Favre MC, Parguey-Leduc A (1994) L'ascocarpe de Terfezia leonis Tul. (Discomycètes, Tuberales). Crypt Mycol 15:187–206

Khabar L, Najim L, Janex-Favre MC, Parguey-Leduc A (2001) Contribution à l'étude de la flore mycologique du Maroc: les truffes marocaines, discomycètes. Bull Soc Mycol Fr 117(3):213–229

Khabar L, Najim L, Moreno G, Diez J., Manjon JL, Morte A, Gutierrez A, Honrubia M, Abourouh M (1999) Truffles in Morocco. XIII congress of european mycologists, 21–25 Sep 1999, Alcala de Henares, Madrid, Spain

Khabar L, Slama A, Neffati M (2005) Terfess common to morocco and Tunisia. Actes of IWEMM 4, 27 novembre au 2 décembre 2005, Murcia, Espagne

Khare KB (1975) *Terfezia terfezioides*. a new record for India. Curr Sci India 44(16):601–602

Kiraly I, Bratek Z (1992) Terfezia terfezioides a common truffle in Hungary. Micologia e Vegetazione Mediterranea. In: Tuber, proceeding of the international conference on truffle, l'Aquila, Italy, 5–8 March, 1992) 7(1):43–45

Kiraly I, Bratek Z, Albert L, Lukacs Z (1992) A homoki szarvasgomba (*Terfezia terfezioides*). The truffle (*Terfezia terfezioides*). Mikologiai Kozlemenyek 31:49–54

Knighton H (1976) Truffle excitement [*Terfezia gigantea*, Vermont]. McIlvainea (USA) 2(2): 74

Korf R (1973) Sparassoid ascocarps in pezizales and tuberales. Rep Tottori Mycol Inst 10:389–403

Korf RP, Zhuang WY (1991) A preliminary Discomycete flora of Macaronesia: Part 15, *Terfeziaceae*, and *Otideaceae*. Mycotaxon 40:413–433

Langiu T (1979) La *Terfezia leonis* in Sardegna. Boll Gruppo Micol G Bresadola, Trento 22:32–35

Lawrynowicz M, Markovic M, Milenkovic M, Ivancevic B (1997) *Terfezia terfezioides*—a new hypogeous fungus for Balkan peninsula. Acta Mycol 32:233–238

Maire R (1906) Notes mycologiques. Ann Mycol 4:329–335

Maire R (1907) Contribution à l'étude de la flore mycologique de l'Afrique du Nord. Bull Soc Bot Fr 53:180–215

Maire R. (avec la coll. de Codina J. et Fontquer P.) (1933) *Fungi Catalaunici*. Contributions à l'étude de la flore mycologique de la Catalogne. Treb Mus Cienc Nat, Barcelona, sér bot 15: 120

Maire R, Werner RG (1937) Fungi Marroccani. Catalogue raisonné des champignons connus jusqu'ici au Maroc. Mémoires de la société des sciences naturelles du Maroc no XLV 31 Décembre 1937. p. 17

Malençon G (1973) Champignons hypoges du Nord de l'Afrique. I. Ascomycètes. Persoonia 7(2):261–288

Marasas WFO, Trappe JM (1973) Notes on Southern African Tuberales. Bothalia 11(1–2):139–141

Mattirolo O (1905) Prima contribuzione allo studio della flora ipogea del Portogallo. In Bolm. Soc. broterniana 21: 86–105

Mattirolo O (1922) Osservazioni supra due ipogea della Cirenaica e considerazioni intorno ai generi *Tirmania e Terfezia*. Atti Acad. naz. Lincei Memerie (ser. 5) 13: 543–568.

Maublanc A (1927) Les champignons comestibles et vénéneux, 2ème édition, tome II. Paul Lechevalier, Paris

Maublanc A (1946) Les champignons de France. Les champignons de France, 3ème édition, Tome II, 1946. pp. 220–223

Maublanc A (1952) Les champignons de France 4ème édition tome II. Les champignons de France, 4ème édition, Tome II, 1952. pp. 232–233

Moawad H, Ali MI, El-Saidi MT, Abu-Shanab RAI (1997) Survey of edible hypogeous ascomycotina of Egypt. Afr J Mycol Biotechnol 5(1):1–12

Montecchi A, Lazzari G (1993) Atlante fotograficodi gunghi ipogei. Associazione Micol Bresadola Centro Studi Micologici—1993. A, Volta 46, 38100 Trento, 490 p.

Moreno G, Galan R, Ortega A (1986) Hypogeous fungi from continental Spain. Crypt Mycol 7(3):201–229

Morte A, Cano A, Honrubia M, Torres P (1994) In vitro mycorrhization of micropropagated *Helianthemum almeriense* plantlets with *Terfezia claveryi* (desert truffle). Agric Sci Finland 3(3):309–314

Morte A, Honrubia M (1997) Micropropagation of *Helianthemum almeriense*. Biotechnol Agric Forestry 40: 163–177. In: Bajaj YPS (ed.) High-tech and micropropagation VI. Springer, Berlin

Morte A, Honrubia M, Elliott TJ (1995) Improvement of mycorrhizal synthesis between micropropagated *Helianthemum almeriense* plantlets with *Terfezia claveryi* (desert truffle). Mushroom science XIV, Volume 2, Proceedings of the 14th international congress on the science and cultivation of edible fungi, Oxford, UK, 17–22 Sep 1995. pp. 863–868. Balkema AA, Rotterdam Netherlands

Moustafa AF (1985) Taxonomic studies on the fungi of Kuwait. III: Ascomycotina (Plectomycètes and Discomycètes). J Univ Kuwait (Sci) 12:79–100

Moyen J (1889) Les champignons : traités élémentaire et pratique de mycologie (page 671 et p. 691). Les champignons : traités élémentaire et pratique de mycologie. J Rotschild, Paris, 762 p.

Pacioni G (1979) The Sardinian hypogean mycological flora and some of is unknown members Funghi ipogei sardi poco conosciuti. Mic Ital 2:21–24

Pacioni G, El-kholy HK (1994) Truffles from Egyptian desert. Micol Vegetazione Mediterr 9(2):69–84

Parguey-Leduc A, Janex-Favre MC, Montant C (1987a) Formation et évolution des ascospores de *Tuber melanosporum* Vitt. (Truffe du Périgord, Discomycètes). Can J Bot 65:1491–1503

Parguey-Leduc A, Montant C, Kulifaj M (1987) Morphologie et structure de l'ascocarpe adulte du *Tuber melanosporum* Vitt. (Truffe du Périgord, Discomycètes). Crypt Mycol 8: 173–202

Parguey-Leduc A, Janex-Favre MC, Montant C (1988) L'ascocarpe du *Tuber melanosporum* Vitt. (Truffe noire du Périgord, Discomycètes): structure de la glèbe. I. les veines fertiles. 2ème Congrès International sur la Truffe, Spoleto, Italie, 24/27 novembre 1988: 101–109

Parguey-Leduc A, Janex-Favre MC, Montant C, Kulifaj M (1989) Ontogénie et structure de l'ascocarpe du *Tuber melanosporum* Vitt. (Truffe du Périgord, Discomycètes). Bull Soc Mycol Fr, t. 105, fasc.3: 227–246

Parguey-Leduc A, Janex-Favre MC, Montant C (1990) L'appareil sporophytique et les asques du *Tuber melanosporum* Vitt. (Truffe noire du Périgord, Discomycètes). Cryptog Mycol 11: 47–68

Patouillard MN (1894a) Les Terfez de la Tunisie. J Bot 10:153–156

Patouillard MN (1894b) Les Terfez de la Tunisie (2ème note). J Bot 10:181–183

Rayss T (1959) Quelques additions à la mycoflore d'Israel. Bull Res Council Israel 8(1):1–14

Roth-Bejareno N, Livne D, Kagan ZV (1990) *Helianthemum-Terfezia* relations in different growth media. New Phytol 114(2):235–238

Rueda Salguero JAO, de De Rueda Salguero JAO (1989) Silviculture and management of edible mycorrhizal fungi. Bol Soc Micol Madrid 13:175–188

Slama A, Fortas Z, Neffati M, Khabar L, Boudabous A (2006) Etude taxonomique de quelques Ascomycetes hypoges (Terfeziaceae) de la tunisie meridionale. Bull Soc Mycol Fr 122(2–3):187–195

Taylor FW, Thamage DM, Baker N, Roth-Bejerano N, Kagan-Zur V (1995) Notes on the Kalahari desert truffle, *Terfezia pfeilii*. Mycol Res 99(7):874–878

Trappe JM (1971) A synopsis of the carbomycetaceae and Terfeziaceae (Tuberales). Trans Br Mycol Soc 57(1):85–92

Trappe JM (1979) The orders, families and genera of hypogeous ascomycotina (truffles and their relatives). Mycotaxon 9(1):297–340

Trappe JM, Sundberg WJ (1977) *Terfezia gigantea* (Tuberales) in North America. Mycologia 69(2):433–437

Tulasne LR, Tulasne C (1851) Fungi hypogaei. Klinckseieck, Paris, 222 p

Zak JC, Whilford WG (1986) The occurrence of a hypogeous Ascomycete in the northern chihuahuan desert. Mycologia 78(5):840–841

Chapter 11
Asian Mediterranean Desert Truffles

Varda Kagan-Zur and Mehmet Akyuz

11.1 Short History

This chapter deals with the desert truffles encountered in the lands of Western Asia bordering on the Mediterranean Sea (Fig. 11.1). The focus will be on knowledge gathered in Turkey, Syria, Lebanon, and Israel.

Dwellers of arid regions have known about desert truffles for thousands of years: "The Middle Bronze Age Amorites collected them in the Syrian Desert, Egyptian pharaohs considered them exclusive royal delicacies, and Roman emperors shipped them home directly from Libya. Mishna and Talmud rabbis, who wrote interpretations of the laws of Judaism almost two thousand years ago, debated the origin of desert truffles and discussed ways to bless God for blessing people with them. The Islamic Prophet Muhammad said to his followers that desert truffles were the "manna" that God gave the Israelites in their travels through the desert" (from Shavit 2008).

11.2 Distribution of Desert Truffles

Very little is known about Lebanese and Syrian truffles. The single reference to a verified collection of desert truffles in Lebanon mentions the Chouf and Bekaa regions (Fig. 11.2) as places where desert truffles are to be found (Farhat 2006). Modad (2006)—who, however, did not collect any truffles during her

V. Kagan-Zur (✉)
Life Sciences Department, Ben Gurion University, POB 653, Beer-Sheva 84105, Israel
e-mail: zur@bgu.ac.il

M. Akyuz
Faculty of Science and Arts, Department of Biology, Bitlis Eren University, Bitlis 13000, Turkey

V. Kagan-Zur et al. (eds.), *Desert Truffles*, Soil Biology 38,
DOI 10.1007/978-3-642-40096-4_11, © Springer-Verlag Berlin Heidelberg 2014

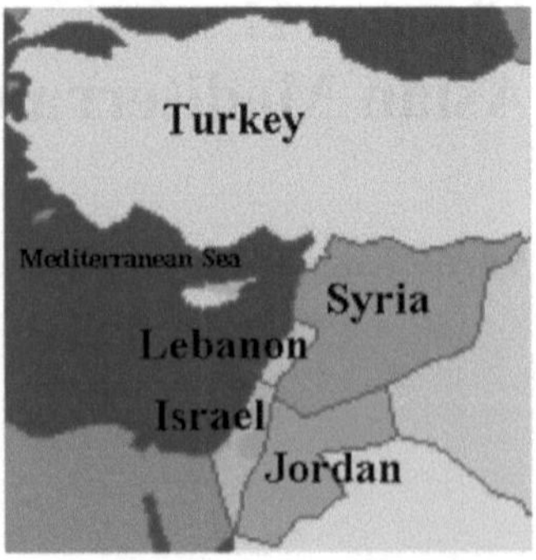

Fig. 11.1 West Asian countries bordering the Mediterranean Sea

research—also considers these to be potential truffle-producing regions based on the testimony of local inhabitants.

Regarding Syria, it seems that desert truffles were collected in ancient times at the foot of the Bishri Mountains (see Shavit 2008). Lonnquist (2004) states that the practice continues in the same area to this day among the nomadic Bedouins. Cavalcaselle (1998) mentions desert truffles being collected throughout the arid steppes of Syria (Fig. 11.2) by Bedouins, who sell them at markets in the surrounding countries of West Asia.

Rather more detailed information is available for both Turkey and Israel. In Turkey, while truffles may be found in most parts of the country, not all are desert truffles, and several species of forest truffles as well as false truffles are also encountered (Fig. 11.3). We note that the Turkish desert truffle region is part of the Al-Hamad, the stony plain that covers large swathes of Turkey, the Syrian Desert, Iraq, Jordan, and Saudi Arabia.

In Israel desert truffles are confined to undisturbed areas in the southern part of the country, namely, the Negev desert (Binyamini 1980; Kagan-Zur and Roth-Bejerano 2008b; Berseghyan and Wasser 2010). Indeed, while a variety of truffle genera belonging to different families occur from the south of the country all the way to the northern border (Binyamini 1980; Berseghyan and Wasser 2010), apart from a single documented collection of *Terfezia olbiensis* near Beit Shemesh (Alsheikh 1994), none of the species collected north of the Negev region are desert truffles.

The Negev occupies over 60 % of the country's area (Israel Land Fund 2012) and represents an extension of the North African desert, which stretches through the Sinai Peninsula and the Israeli Negev into Jordan. Desert truffles are found within the area demarcated by the lines of 250 and 50 mm annual rainfall (Fig. 11.4).

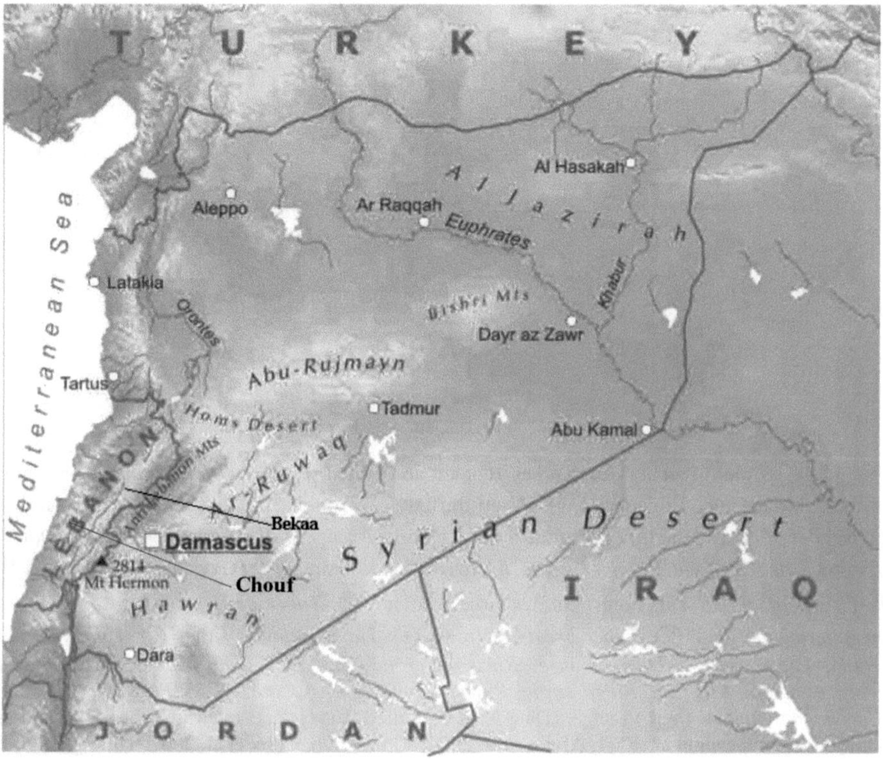

Fig. 11.2 Lebanon and Syria. Copied and edited from: http://www.worldmapsinfo.com/mapimage/syria.jpg

11.3 Soils

Generally speaking, the soil of the Bekaa Valley where desert truffles have been collected in Lebanon is composed of recent infill sediments, the main underlying rocks consisting of limestone (Wikipedia (d) (2012)). Even less is known about the Syrian soils supporting truffle production. In Turkey desert truffles are observed to inhabit sandy-clayey calcareous soils that become slimy when wet in the Anatolian climate region (Gücin and Dülger 1997; Akyuz et al. 2012). Most of the Negev soils were found to be deeper than 2 m and of the loess type (Kark 2003). Vast areas are, in fact, comprised of both light loess sand and quartz sand dunes (Ramat Negev Desert Agro-research Center report, and see Fig. 11.4), both of which are particularly well adapted to desert truffle proliferation. (For detailed soil analyses, see Chap. 4: Soil Properties, in this book.)

Fig. 11.3 Truffle distribution in Turkey (copied and edited from: http://en.wikipedia.org/wiki/File:BlankMapTurkeyProvinces.png). Desert truffles: ♀: *Tirmania pinoyi* (Maire) Malençon. ⦰: *Terfezia claveryi* Chatin. Ⓡ: *Terfezia arenaria* (Moris) Trappe. *Filled square*: *Terfezia boudieri* Chatin. *Filled inverted triangle*: *Picoa lefebvrei* (Pat.) Maire. Ⓟ: *Picoa juniperi* Vittad. [?]: *Terfezia leptoderma* Tul. Forest truffles (true truffles): ●: *Tuber aestivum* Vittad. ◉: *Tuber mesentericum* Vittad. ⊘: *Tuber brumale* Vittad. ⊛: *Tuber nitidum* Vittad. ©: *Tuber borchii* Vittad. False truffles: *Open square*: *Rhizopogon luteolus* Fr. *Filled triangle*: *Rhizopogon ochraceorubens* A.H. Sm. *Filled circle*: *Rhizopogon roseolus* (Corda) T.M. Fries. *Open circle*: *Rhizopogon vulgaris* (Vitt.) Lange. (Data were compiled from papers cited in Table 11.1, and in addition from: Gezer et al. 2004; Afyon et al. 2004; Aktaş 2006; Çelik et al. 2007; Türkoğlu 2008; Ekmekçiler 2009; Turkekul and Yıldız 2010; Türkekul and Zülfikaroğlu 2010; Subaşı 2010; Türkmenoğlu 2010; Erkuş, 2010; Kuyumcu 2011; Gezer et al. 2011; Kaya 2012; Doğan et al. 2012; Kırış et al. 2012; Sesli and Denchev 2012)

11.4 Climatic Conditions

Most if not all the Asian-Mediterranean regions that support desert truffle populations have certain climatic attributes in common: they are inland steppe areas, and they are characterized by semiarid to arid continental conditions with dry summers (no rain from June to September) and wet winters. Annual precipitation ranges between 300 and 600 mm in the relatively wet semiarid parts and 50 and 250 mm in the arid parts. Rains are often erratic, and drought years are not infrequent (see below for details and citations)

The three Anatolian regions in Turkey where desert truffles occur (central, southeastern, and eastern) present similar features. Both central and southeastern Anatolia have a semiarid continental climate with hot, dry summers and cold, snowy winters. The central region usually receives low precipitation (average yearly rainfall 382 mm). Temperatures range from −25 °C to 40 °C. The average humidity is 62 %. In the southeastern Anatolian region, temperatures reach 46 °C in summer and, unusually, a low of −12 °C in winter, with an average yearly rainfall of 576 mm.

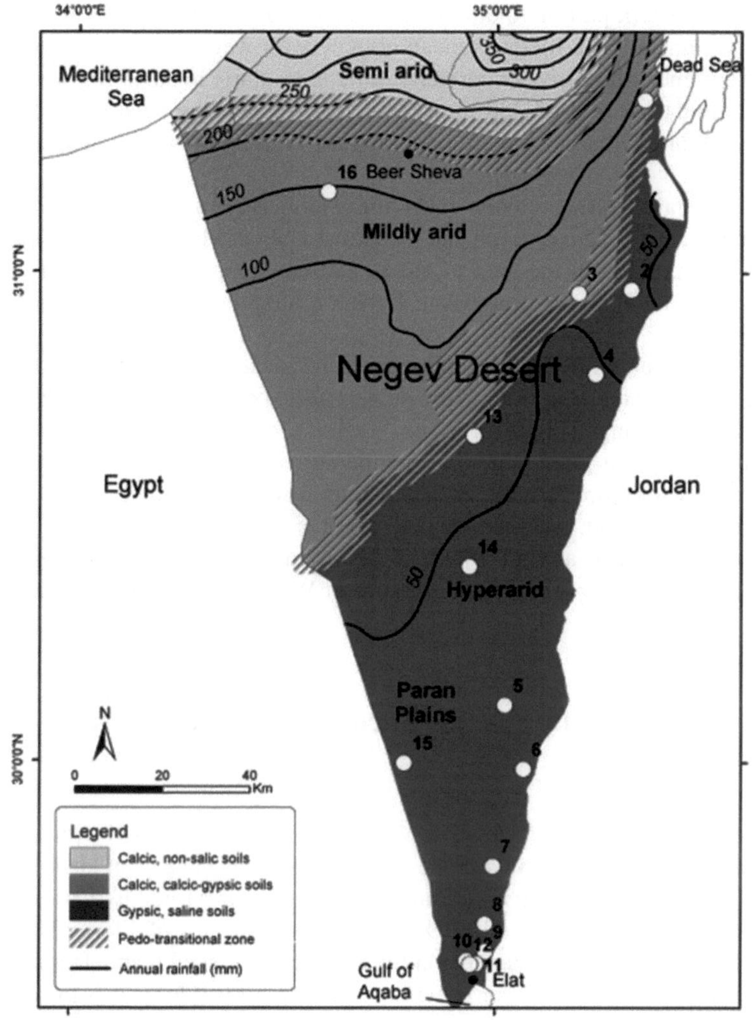

Fig. 11.4 Soil and rain map of the Negev (from: Amit et al. 2011; Permission no. GB 494 6272 12)

Since most of the eastern Anatolian region is remote from the sea and is characterized by a high elevation, it has a harsh continental climate with long winters and short summers. The winter is very cold and snowy, while in summer the weather is cool in the highlands and warm in the lowlands. The region experiences the lowest average winter temperature in all Turkey, namely, −25 °C, although temperatures here can, occasionally, drop below −40 °C. The summer average is about 20 °C (occasionally reaching up to 38 °C). However, the annual average is just 9 °C. Average yearly rainfall is 560 mm (compiled from the following: Wikipedia 2012, sites a, b, c; Turkey's Weather, Climate & Geography 2012).

Like the eastern part of Anatolia, the Lebanese Bekaa valley lies at a relatively high elevation. It has a Mediterranean climate with wet, often snowy winters and dry, hot summers. It receives limited rainfall, particularly in the north, which has an average annual rainfall of 230 mm versus 610 mm in the central valley; this is because Mount Lebanon blocks precipitation coming from the sea (Wikipedia 2012, site d).

In Syria rain-bearing clouds from the Mediterranean pass through the gap between the Jabal an Nusayriyah and the Anti-Lebanon Mountains, dropping rather limited precipitation in the area of Homs and the steppe region east of that city. Farther to the south, however, the Anti-Lebanon Mountains block the rains from the Mediterranean.

The Negev region of Israel is arid due to its location to the east of the Sahara Desert. Average annual rainfall in the desert truffle bearing areas of the Negev is in the 50–250 mm range (Fig. 11.4).

As in the Syrian Desert, midday temperatures in the Israeli Negev typically average 30 °C or more in summer. Highs may reach above 40 °C, while, typically for a desert climate, summer nights are far cooler (up to 20 °C). Winter temperatures average around 10–20 °C at midday and 0–7 °C at night; winter lows may reach below −5 °C (compiled from Wikipedia 2012 sites e, f).

11.5 Desert Truffle Species Encountered in the Study Region

In Lebanon so far only *Terfezia claveryi* Chatin has been identified with any degree of certainty (Farhat 2006). Several desert truffle species have been identified in the Syrian deserts: *Terfezia boudieri* Chatin (Awameh and Alsheikh 1980a), *Terfezia claveryi* (Awameh and Alsheikh 1980b), and two species of *Tirmania*, *T. nivea* (Desf.: Fr.) Trappe and *T. pinoyi* (Maire) Malençon (Alsheikh and Trappe 1983a). This is almost certainly an incomplete list of the desert truffle species to be found in Lebanon and Syria, which still await a comprehensive study. In the case of both Turkey and Israel the state of our knowledge is far more advanced. Figure 11.3 maps all the hypogeous fungi encountered and defined in Turkey; the desert truffles among them are summed up in Table 11.1.

Several desert truffle species are encountered in Israel, the most abundant by far being *Terfezia boudieri* (Fig. 11.5c). Their nomenclature and affiliations have undergone many changes over the years. *T. boudieri* was initially known as *Terfezia leonis* Tul. (e.g., Binyamini 1980). However, *T. leonis* proved to be incorrectly named and was, in fact, *T. arenaria* (Trappe 1971). Some authors still refer to the popular Negev brown truffle as *T. arenaria* (e.g., Berseghyan and Wasser 2010), but a new examination of specimens confirmed that this abundant brown truffle is indeed *Terfezia boudieri* (Holdengraeber et al. 2001). *Terfezia claveryi* is also

Table 11.1 Desert truffle distribution in Turkey

Desert truffle	Province	Citation
Picoa lefebvrei (Pat.) Maire (Fig. 11.5a)	Elazığ Şanlıurfa	Gücin et al. (2012)
Picoa juniperi Vittad.	Uşak	Türkoğlu and Yağız (2012)
Tirmania pinoyi (Maire) Malençon	Izmir	Ersel and Solak (2004)
Terfezia boudieri Chatin (Fig. 11.5c)	Batman	Demir et al. (2007)
	Karaman	Doğan and Öztürk (2006)
	Uşak	Akyuz et al. (2012)
	Elazığ	Türkoğlu and Yağız (2012)
	Konya	Gücin and Dülger (1997)
	Eskişehir	Gucin (1990)
	Kutahya	Yamaç (2012)
	Mardin	Yıldız and Ertekin (1997)
	Şanlıurfa	Yildiz et al. (2006)
	Gaziantep	Kasık et al. (2001)
	Diyarbakır	
	Niğde	
	Ankara	
	Aksaray	
	Kırşehir	
Terfezia claveryi Chatin	Kastamonu	Bekçi et al. (2011)
Terfezia leptoderma Tul.	Uşak	Castellano and Türkoğlu (2012)
Terfezia arenaria (Moris) Trappe (Fig. 11.5b)	Isparta	Afyon (1996)
	Malatya	Isıloğlu and Oder (1995)
	Konya	Kasık et al. (1998)
		Oder (1988)

Notes: (a) The genus *Picoa*, previously placed in the Helvellaceae, was transferred to the Pyronemataceae (Laessoe and Hansen 2007). Based on molecular studies, Ammarellou et al. (2011) proved *Picoa lefebvrei* (Fig. 11.5a) to be synonymous with *Phaeangium lefebvrei*, and therefore it too has been incorporated in the above family. (b) *Terfezia boudieri* Chatin (Fig. 11.5c), the most abundant and well studied truffle in Turkey, has been collected in a number of Turkish provinces (Fig. 11.3, Table 11.1; Akyuz et al. 2012). (c) The genera *Tirmania* and *Terfezia* were moved from the Terfeziaceae family, which was abolished, to the Pezizaceae family following the reorganization of the Pezizales (Laessoe and Hansen 2007; Kagan-Zur and Roth-Bejerano 2008a)

found in Israel, but very rarely (Berseghyan and Wasser 2010). A single specimen of *Terfezia olbiensis* was collected on 1979 near Beit Shemesh (close to Jerusalem), Israel (Alsheikh 1994).

The genus *Tirmania* is represented by *Tirmania africana* Chatin (Fig. 11.5d, Binyamini 1980; Berseghyan and Wasser 2010). *T. africana* is synonymous with *Tirmania nivea* (Desf.:Fr.) Trappe (Trappe 1971; Laessoe and Hansen 2007).

Terfezia boudieri, *Terfezia claveryi*, and *Tirmania africana* have all been placed in the Pezizaceae family.

In addition, *Phaeangium lefebvrei (Picoa lefebvrei)* Pat. (Pyronemataceae) is also occasionally encountered (Kagan-Zur and Roth-Bejerano 2008b). People rarely if ever trouble to collect these minute truffles, which are consumed mainly

Fig. 11.5 Representatives of desert truffles genera encountered in Asian countries bordering the Mediterranean. (**a**) *Picoa lefebvrei* (Pat.) Maire. With permission from "Trufomania" 2012. (**b**) *Terfezia arenaria* (Moris) Trappe. With permission from "Trufomania" 2012. (**c**) *Terfezia boudieri* Chatin. Photo by T. Turgeman, with permission. (**d**) *Tirmania africana* Chatin. From Kagan-Zur and Roth-Bejerano (2008a). With permission from "Fungi"

by birds and other animals (Alsheikh and Trappe 1983b). Ammarellou et al. (2011) proved by molecular means that this species is synonymous with *Picoa lefebvrei* (Pat.) Maire (Fig. 11.5a).

The truffle reported as *Terfezia oligosperma* by Berseghyan and Wasser (2010) is, in fact, *Tuber oligospermum* (Tul. & C. Tul.) Trappe (Trappe 1979). It belongs to the Tuberaceae and is not a desert truffle, being found under *Pinus* spp. in pine groves outside the desert truffle zone.

11.6 Host Plants

The majority of plant species supporting desert truffle mycorrhizas belong to the genus *Helianthemum* (Cistaceae). Although *Cistus* species are also known to form mycorrhizas with desert truffles (Comandini et al. 2006) and although we were able to obtain such mycorrhizas at will both in a greenhouse (Kagan-Zur and Roth-Bejerano unpublished) and in vitro (Zaretsky et al. 2006), in none of the Asian Mediterranean countries have cistus-desert truffle mycorrhizas been observed in the wild.

Fig. 11.6 The main *Helianthemum* species mycorrhized by desert truffles in the Asian Mediterranean deserts. (**a**) *Helianthemum sessiliflorum* (Defs.) Pers. Photo by T. Turgeman, with permission. (**b**) *Helianthemum salicifolium* (L.) Miller. Photo by M. Akyut

The genus *Helianthemum* includes both annual and perennial species, the latter consisting of small bushes (Fig. 11.6a). Desert truffles are known to form mycorrhizas with several *Helianthemum* species, some annual and others perennial. While each region is usually characterized by a single predominant symbiont, other species of the genus may also function as hosts. In the Negev region, for instance, *Helianthemum sessiliflorum* (Defs.) Pers. (Fig. 11.6a), a perennial bush, is the major host (Binyamini 1980), but other members of the genus Helianthemum—such as *H. kahiricum, H. stipulatum, H. ledifolium,* and *H. lippii*—have been observed to form associations with *T. boudieri* in rainy years (Danin and Arbel 1998). (No plant genus other than *Helianthemum* has been reported to harbor truffle mycorrhizas in Mediterranean Asia.)

While the plant symbionts of Turkish desert truffles have not been studied in depth, it has been reported that the main *Terfezia boudieri* symbiont in the Anatolian steppe is *Helianthemum salicifolium* (L.) Miller, an annual plant (Fig. 11.6b; Akyuz et al. 2012). *H. salicifolium* is an important *T. boudieri* symbiont in the Saudi Arabian peninsula, in addition to *H. ledifolium* (Awameh 1981). *T. boudieri* has also been found associated with *H. aegyptiacum* in the vicinity of the Ankara-Eskişehir highway (Anonymous 2012).

We may conclude that the truffle *T. boudieri*, at least, is not specific to any single species of *Helianthemum.*

Regarding two other species of desert truffle—*Terfezia leptoderma* (Castellano and Türkoğlu 2012) and *Picoa juniperi* (Türkoğlu and Yağız 2012)—the only information provided is that they were associated with geranium, helianthemum and cerastium plants, meaning, most probably, that these plants were found in the vicinity of the truffles. The most probable host out of the three is *Helianthemum* spp.; however, this question requires further elucidation.

No plant symbionts have been identified for the other desert truffles encountered in Turkey, such as members of the *Picoa* genus, *Tirmania pinoyi*, *Terfezia arenaria,* or *Terfezia claveryi*, no doubt because research has mainly concentrated on identifying the truffle species.

In the Negev area different desert truffles may enter into symbiosis with the same plant host; for example, *Tirmania africana* is found near *H. sessiliflorum* roots (Binyamini 1980; Berseghyan and Wasser 2010) and so is *Picoa lefebvrei* (Kagan-Zur and Roth Bejerano unpublished). However, no desert truffle is uniquely coupled with a single plant species and vice versa.

Nothing at all can be found in the accessible literature concerning the plant symbionts of either Lebanese or Syrian truffles.

11.7 Conclusions

Asian Mediterranean regions harboring desert truffles are either semiarid or arid and have dry summers and wet winters. Climatic conditions range from mild summers and harsh winters to hot summers and mild winters, depending on elevation.

Three genera of desert truffles are encountered in this region: *Picoa* (*juniperi* and *lefebvrei*), Pyronemataceae; *Tirmania* (*pinoyi* and *africana*), Pezizaceae; and *Terfezia* (*arenaria*, *boudieri*, *claveryi*, *leptoderma*, and potentially *olbiensis*), Pezizaceae. All verified host plants are members of the genus *Helianthemum*, Cistaceae.

References

Afyon A (1996) Some macrofungi determined of Isparta Province in Turkey. Turk J Bot 20:161–164

Afyon A, Yagız D, Konuk M (2004) Macrofungi of Sinop Province. Turk J Bot 28:351–360

Aktaş S (2006) Macrofungi of Amasya Province. Selçuk University, Graduate School of Natural and Applied Science Ph.D. thesis, Konya. 345 p. http://tez2.yok.gov.tr

Akyuz M, Kirbag S, Kursat M (2012) Ecological aspects of the arid and semi-arid truffle in Turkey: Evaluation of soil characteristics, morphology, distribution, and mycorrhizal relationships. Turk J Bot 36:386–391

Alsheikh AM (1994) Taxonomy and mycorrhizal ecology of the desert truffles in the genus Terfezia. Ph.D. thesis, Oregon State University, Oregon, USA, 239 p

Alsheikh AM, Trappe JM (1983a) Desert truffles: the genus *Tirmania*. Trans Br Mycol Soc 81:83–90

Alsheikh AM, Trappe JM (1983b) Taxonomy of *Phaeangium lefebvrei*, a desert truffle eaten by birds. Can J Bot 61:1919–1925

Ammarellou A, Smith ME, Tajick MA, Trappe JM (2011) The phylogenetic placement of *Picoa*, with a first report on *Picoa lefebvrei* (Pat.) Maire (=*Phaengium lefebvrei*) from Iran. Int J Environ Res 5:509–514

Amit R, Simhai O, Ayalon A, Enzel Y, Matmon A, Crouvi O, Porat N, McDonald E (2011) Transition from arid to hyper-arid environment in the southern Levant deserts as recorded by early Pleistocene cummulic Aridisols. Quat Sci Rev 30:312–323

Anonymous (2012) http://www.babamonk.com/2011/05/ekmek-aras-domalan-mantar.html#axzz29kcxBlRO

Awameh MS (1981) The response of *Helianthemum salicifolium* and *H. ledifolium* to the infection by the desert truffle *Terfezia boudieri*. Mushroom sci. XI. Part II. In: Proceedings of the eleventh international congress on the cultivation of edible fungi, Sydney, Australia, pp. 843–583
Awameh MS, Alsheikh AM (1980a) Ascospore Germination of Black Kamé (*Terfezia boudieri*). Mycologia 72:50–54
Awameh MS, Alsheikh AM (1980b) Features and analysis of spore germination in the Brown Kamé *Terfezia claveryi*. Mycologia 72:494–499
Bekçi H, Altınsoy B, Sarıkaya S, Onbasılı D, Çelik GY (2011) Antimicrobial activity of some macrofungi collected from Kastamonu Province. Kastamonu Univ J Forestry Faculty 11:187–190
Berseghyan GS, Wasser SP (2010) Species diversity of hypogeous Ascomycetes in Israel. Mycobiology 38:159–165
Binyamini N (1980) Addenda to the hypogeous mycoflora in Israel. Nova Hedwigia 32:9–19
Castellano MA, Türkoğlu A (2012) New records of truffle taxa in *Tuber* and *Terfezia* from Turkey. Turk J Bot 36:295–298
Cavalcaselle B (1998) Edible mushroom production in forest villages of Turkey, Syria and Jordan. http://www.fao.org/docrep/x5402e/x5402e10.htm
Çelik A, Uşak M, Gezer K, Türkoğlu A (2007) Macrofungi of Tavas (Denizli) District in Turkey. Pak J Biol Sci 10:4087–4091
Comandini O, Conta M, Rinaldi AC (2006) An overview of *Cistus* ectomycorrhizal fungi. Mycorrhiza 16:381–395
Danin A, Arbel A (1998) The fauna and flora of the Holyland. Carta, Jerusalem, 144 pp. (in Hebrew)
Demir S, Demirel K, Uzun Y (2007) Macrofungi of Batman Province. Ekoloji 16(64):37–42
Doğan HH, Aktaş S, Öztürk C, Kaşık G (2012) Macrofungi distribution of Cocakdere valley (Arslanköy, Mersin). Turk J Bot 36:83–94
Doğan HH, Öztürk C (2006) Macrofungi and their distribution in Karaman Province, Turkey. Turk J Bot 30:193–207
Ekmekçiler M (2009) Macrofungi of Akşehir (KonyaNYA) Province. Selçuk University Graduate School of Natural and Applied Science M.Sc. thesis, Konya. 62 p. http://tez2.yok.gov.tr
Erkuş İ (2010) Macrofungi of Nevşehir (Gülşehir-Derinkuyu) Province. Gaziosmanpaşa University, Graduate School of Natural and Applied Science M.Sc. thesis. Tokat. 72 p. http://tez2.yok.gov.tr
Ersel FY, Solak MH (2004) Contributions to the Macrofungi of İzmir Province. Turk J Bot 28:487–490
Farhat CC (2006) Analysis of indigenous nutritional knowledge, cultural importance and nutritional content of wild edible plants. American University of Beirut. Master of Science, Beirut, Lebanon. 112 p. https://scholarworks.aub.edu.lb/handle/10938/7204
Gezer K, Kaygusuz O, Soylu U, Ermiş A (2011) Çamlık mesire alanı (Denizli) Makrofungusları. J Fungus 2:15–24
Gezer K, Çelik A, Uşak M, Türkoğlu A (2004) Macrofungi of Tavas (Denizli) District. Afyon Kocatepe Univ J Sci 7:439–446
Gucin F (1990) Macrofungi found surroundings of Elazığ. Turk J Bot 14:171–177
Gücin F, Dülger B (1997) Yenen ve Antimikrobiyal Aktiviteleri Olan Keme Mantari (*Terfezia boudieri* Chatin) Üzerinde Araştirmalar. Ekoloji 23:27–33
Gücin F, Kaya A, Soylu MK UY (2012) *Picoa Vittad.* a new truffle genus record for Turkey. Biol Diver Conserv 3:23–25
Holdengraeber S, Martin F, Roth-Bejerano N, Kagan-Zur V (2001) Some lobed fruit bodies of *Terfezia boudieri* (a desert truffle) combine two ITS – RFLP patterns, suggesting independent origins. In: Federation Francaise des Trufficuteurs (FFT) eds. Proceedings of the fifth International Congress, Science And Cultivation of Truffles. Aix en Provence, France, 4–6 Mars 1999. FFT Publ. Paris, pp 78–81.

Israel Land Fund (2012) http://www.israellandfund.com/en-us/info/regions/the-negev.htm
Isıloğlu M, Oder N (1995) Macrofungi of Malatya Province. Turk J Bot 19:321–324
Kagan-Zur V, Roth-Bejerano N (2008a) Desert Truffles. Fungi 1(3):32–37
Kagan-Zur V, Roth-Bejerano N (2008b) Studying the brown desert truffles of Israel. Isr J Plant Sci 56:309–314
Kark R (2003) The first observation posts in the Negev, the sixtieth anniversary: 1943–2003. Ariel Publishing House (Hebrew)
Kasık G, Ozturk C, Akkoz C, Doğan HH (1998) Some macrofungi determined in SU Alaaddin Keykubat campus. Selcuk Univ Fen Edebiyat Fakultesi Fen Dergisi 15:87–99
Kasık G, Ozturk C, Toprak E (2001) Macrofungi of Niğde Province (Turkey). Herb J System Bot 8:137–142
Kaya A (2012) Macrofungal diversity of Adıyaman Province (Turkey). Mycotaxon 110:43–46
Kırış Z, Halıcı MG, Akata I, Allı H (2012) Macrofungi of Akdağmadeni (Yozgat/Turkey) and Gemerek (Sivas/Turkey). Biol Diver Conserv 5:53–58
Kuyumcu S (2011) Genetic characterisation of indigenous tuber aestivum samples and phylogenetic analyses of their rdna its regions. Nevşehir University, Graduate School of Natural and Applied Science M.Sc. thesis, June 2011. http://tez2.yok.gov.tr
Laessoe T, Hansen K (2007) Truffle trouble: What happened to the Tuberales? Mycol Res 111:1075–1099
Lebanese and Syrian map (2012) http://www.worldmapsinfo.com/mapimage/syria.jpg
Lonnquist M (2004) Jebel Bishri in Syria and nomadism in the end of the early Bronze Age. British Museum Colloquium
Modad NN (2006). Survey and identification of wild mushrooms in Lebanon. American University of Beirut, Master of Science. 329 p. https://scholarworks.aub.edu.lb/handle/10938/7355
Oder N (1988) Taxonomic investigations of important edible and poisonous mushrooms growing in the Konya center and some Districts of Konya. Selcuk Univ Fen Edeb Fak Derg 8:237–257
Ramat Negev Desert Agro-research Center Report. www.moprn.org/foreign-studies/docs/about-us.doc
Sesli E, Denchev CM (2012) Checklists of the myxomycetes, larger ascomycetes, and larger basidiomycetes in Turkey. Mycotaxon 106:65–67
Shavit E (2008) Medicinal mushrooms: truffles roasting in the evening fires. Fungi 1:18–23
Subaşı R (2010) Macrofungi of Sivas (Ulaş ilçesi) Province. Gaziosmanpaşa University, Graduate School of Natural and Applied Science M.Sc. thesis. Tokat. 86 p. http://tez2.yok.gov.tr
Trappe JM (1971) A synopsis of the Carbonycetaceae and Terfeziaceae (Tuberales). Trans Br Mycol Soc 57:83–92
Trappe JM (1979) The orders, families, and genera of hypogeous Ascomycotina (truffles and their relatives). Mycotaxon 9:297–340
Trufomania (2012) http://www.trufamania.com/Terfezia%20arenaria%20English.htm
Turkekul İ, Yıldız MA (2010) Macrofungi from Artova (Tokat) District. Kafkas Üniv Fen Bil Enst Derg 3:29–34
Türkekul İ, Zülfikaroğlu E (2010) Çamlibel ilçesi (Tokat) makromantar florası. SAÜ Fen Edebiyat Dergisi. pp. 55–63
Turkey, Map (2012) http://en.wikipedia.org/wiki/File:BlankMapTurkeyProvinces.png
Türkmenoğlu A (2010) Macrofungi of Anamur (Mersin). Selçuk University, Graduate School of Natural and Applied Science, M.Sc. thesis. 48 s. Konya. http://tez2.yok.gov.tr
Türkoğlu A (2008) Macrofungal diversity of Babadağ (Denizli, Turkey). Afr J Biotechnol 7:192–200
Türkoğlu A, Yağız D (2012) Contributions to the macrofungal diversity of Uşak Province. Turk J Bot 36:580–589
Turkey's Weather, Climate & Geography (2012) http://www.turkeytravelplanner.com/WhenToGo/ClimaticRegions.html
Wikipedia (a) (2012) http://en.wikipedia.org/wiki/Central_Anatolia_Region
Wikipedia (b) (2012) http://en.wikipedia.org/wiki/Southeastern_Anatolia_Region

Wikipedia (c) (2012) http://en.wikipedia.org/wiki/Eastern_Anatolia_Region
Wikipedia (d) (2012) http://en.wikipedia.org/wiki/Beqaa_Valley
Wikipedia(e) (2012) http://en.wikipedia.org/wiki/Geography_of_Syria#Climate)
Wikipedia (f) (2012) http://en.wikipedia.org/wiki/Negev#cite_note-weathermsn-1
Yamaç M (2012) Eskişehir İli Makrofungus Çeşitliliği. Eski Yeni Aylık Şehir Kültür Dergisi 4 (42):5–10
Yıldız A, Ertekin AS (1997) Contribution to the macrofungal flora of Diyarbakır. Turk J Bot 21:119–122
Yildiz A, Dündar A, Acay H, Akyüz M, Yeşil ÖF (2006). Nutritive value of *Pleurotus eryngii* and *Terfezia boudieri*. The Scientific and Technological Research Council of Turkey, Agriculture, Forestry and Veterinary Sciences Research Grant Group (TOVAG), Project No: TUBITAK-TOVAG-1040108. 43 pp. http://uvt.ulakbim.gov.tr/uvt/index.php?cwid=9&vtadi=TPRJ&ano=86645_10304a7a72f2cff9c8c9e633cc218a40
Zaretsky M, Kagan-Zur V, Mills D, Roth-Bejerano N (2006) Analysis of mycorrhizal associations formed by *Cistus incanus* transformed root clones with *Terfezia boudieri* isolates. Plant Cell Rep 25:62–70

Chapter 12
Non-Mediterranean Asian Desert Countries

Ali Ammarellou, Yun Wang, Gorbanali Nematzadeh, and Mohammadali Tajick

12.1 Introduction

Truffles are edible hypogeous fruit bodies produced by many genera of fungi belonging to the order Pezizales (Ascomycetes). Truffles can be classified as forest truffles, desert truffles, and semiarid truffles (Akyuz et al. 2012). Desert truffles are encountered in arid and semiarid zones of all continents except Antarctica (Lawrynowicz et al 1997; Al-Ruqaie 2002; Diez et al. 2002; Moreno et al. 2002; Ammarellou et al. 2007; Ammarellou and Saremi 2008; Trappe et al. 2008; Al-Laith 2010; Akyuz et al. 2012; Tao 1988; Trappe and Weber 2001). This chapter undertakes to sum up information concerning desert truffles in non-Mediterranean Asian countries (Fig. 12.1): Species distribution, common knowledge, and climatic and soil conditions are presented. The most extensive studies were carried out in Iran, specifically in Zanjan province (State of Tarom).

A. Ammarellou (✉)
Department of Medicinal Plants and Fungi, Research Institute of Modern Biological Techniques, University of Zanjan, Zanjan, Iran

Sari University of Agricultural Sciences & Natural Resources, Sari, Iran
e-mail: ammarellou@yahoo.com

Y. Wang
The New Zealand Institute for Plant & Food Research Ltd., Private Bag 4704, Christchurch 8140, New Zealand

Yunnan Institute for Tropical Crop Research, Jinghong, Yunnan, China
e-mail: wang.yun@plantandfood.co.nz

G. Nematzadeh • M. Tajick
Sari University of Agricultural Sciences & Natural Resources, Sari, Iran
e-mail: gh.nematzadeh@sanru.ac.ir; tajickma@yahoo.com

V. Kagan-Zur et al. (eds.), *Desert Truffles*, Soil Biology 38,
DOI 10.1007/978-3-642-40096-4_12, © Springer-Verlag Berlin Heidelberg 2014

Fig. 12.1 Map of studied countries for desert truffles distribution (http://www.mapsofworld.com/asia/#)

12.2 Iran

Iran is situated in Southwest Asia, the Middle East. Its geographic coordinates are 32°N, 53°E. It is bordered on the eastern side of Iraq and Turkey, western side of Afghanistan and Pakistan, and southern side of Azerbaijan, Armenia, and Turkmenistan. The largest lake of the world, the Caspian Sea, is at the north of Iran and the Persian Gulf and the Oman Sea are at the south. Mean annual precipitation in different regions of Iran is given in Fig. 12.2. A yearly rainfall range between 100 and 400 mm (Fig. 12.2) is suitable for truffle cultivation, provided other demands, such as the temperature range, soil type, and suitable host plants, are met. Areas, marked as grazing lands in Fig. 12.3, are the most prolific natural truffle sites in Iran. To date, no truffle cultivation is practiced and fruit bodies are, usually, collected from the wild by local shepherds.

There are several different climatic districts in Iran. The Caspian coastal plain remains humid all year round. The high-altitude inhabited areas on the west of Iran have a cold winter that is often reaching below the freezing point. Both the Central and Eastern Iran have seasonal climatic variations. In general, these latter areas are arid and semiarid during most parts of the year. The southern coastal plains of Iran have mild winters, but very hot and extremely humid summer. The temperature could exceed 48 °C during July in the interior part of Southern Iran (http://ivansahar.ir/iran-geography.htm).

In general, Iran has an arid climate in which most of the relatively scant annual precipitation falls from October through April. In most of the country, yearly precipitation averages 250 mm or less (Fig. 12.2). The major exceptions are the higher mountain valleys of the Zagros and the Caspian coastal plain, where precipitation averages at least 500 mm annually. In the western part of the Caspian,

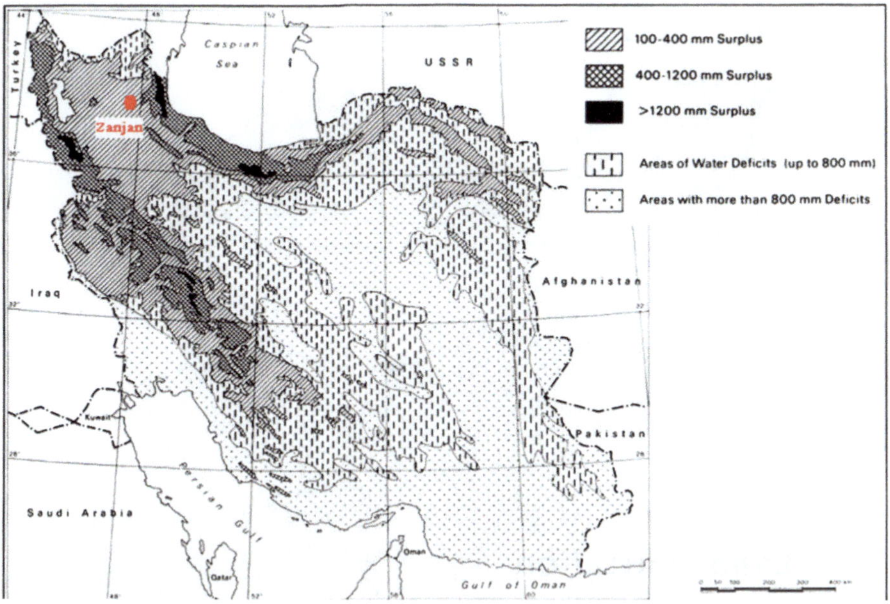

Fig. 12.2 Map of mean annual precipitation in different parts of Iran (www.iranicaonline.org)

rainfall exceeds 1,000 mm annually and is distributed relatively evenly throughout the year. This contrasts with some basins of the Central Plateau which receive 100 mm or less of precipitation annually. The Tarom region (province of Zanjan), enclosed by mountains, has a subtropical or Mediterranean climate and receives humid north air currents. In dry years the annual rainfall is only 150 mm, but in wet years it may be as high as 400 mm. Normally a high autumn rainfall induces good truffle production. Climate is of primary importance in the distribution of vascular plants, which are the hosts of desert truffles. In other words, climate affects truffles indirectly (Ammarellou et al. 2007).

12.2.1 Desert Truffle Species and Their Host Plants

Several species of delicious desert truffles can be found growing between the 27°N and 29°N lat., primarily in the Bandar-e Abbas, Shiraz, Tabriz, Qazvin, Kurdistan, and Zanjan provinces of Iran. They are naturally harvested during mid-winter and early spring, particularly from January to March. Experienced harvesters identify the location of the truffles from crevices that appear in the surface of the soil above the truffles. The local names for desert truffles (depending on local dialects) are al-kamah, al-chamae, kamaa, kame, al-faga and faga, donbalan, dombalan, or donbal (Bokhary et al. 1987; Mandeel and Al-Laith 2007; Akyuz et al. 2012). In

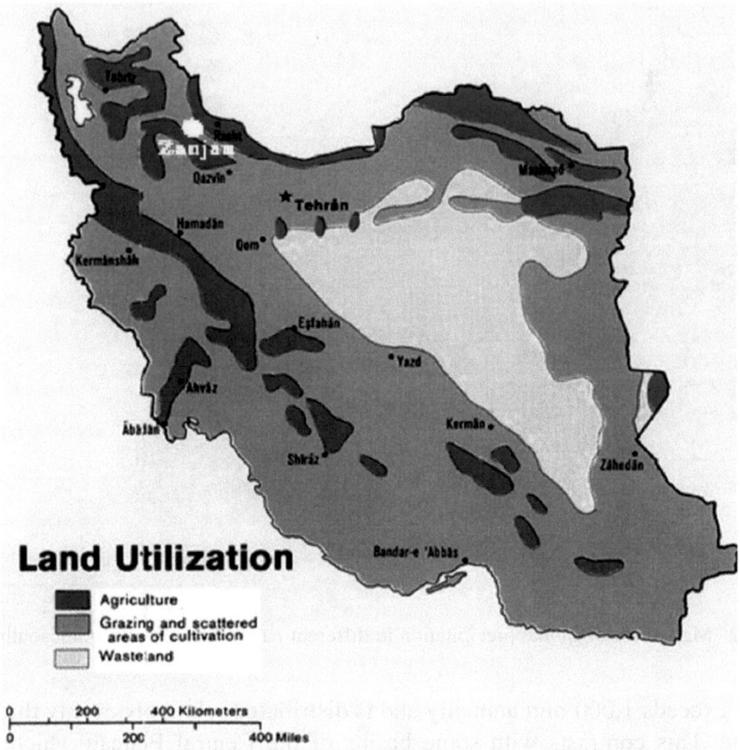

Fig. 12.3 Map of the land utilization of Iran and situation of Zanjan province (one of the most important areas of truffle growth) (www.mapcruzin.com)

Fig. 12.4 A view of the lands with truffle growth potential, early spring (Province of Zanjan, Tarom—Abbar region) (**a**) Photograph taken by Ali and truffle collection by indigenous peoples (Province of Kurdistan, Bijar) (**b**). Photograph taken by Ammarellou

the Tarom region (province of Zanjan) normally a high autumn rainfall induces good truffle production.

The truffles are mostly found in barren, uncultivated fields in mountainous areas (Fig. 12.4). The soils are generally rendzinas, chalky, alkaline, granular, and well drained. About 1 % of organic matter in the topsoil is produced by grasses.

Fig. 12.5 The primary desert truffles' plant symbionts in Iran. (**a**) *Kobresia bellardii* (bog sedge). (**b**) *Helianthemum* spp. Photograph taken by Ali Ammarellou

Terfezia boudieri Chatin was cytologically and morphologically identified in this region (Ammarellou et al. 2007, 2008). Although the most abundant desert truffle hosts around the Mediterranean basin are species of the genus *Helianthemum* (e.g., see Chaps. 11, 9, and 10 by: Kagan-Zur and Akyuz, Chavalier, and Khabar respectively). Our studies provided evidence that, on top of a *Helianthemum* sp., *Kobresia bellardii* (Cyperaceae) also forms ectomycorrhizal associations with *T. boudieri* (Ammarellou et al. 2007, 2008). These two plants are very different from each other. *Helianthemum* sp. is a dicotyledon plant, while *Kobresia bellardii* is a monocotyledon (Fig. 12.5).

Other species of Iranian desert truffles including *Terfezia claveryi* Chatin, *T. aphroditis*, *T. hanotauxii*, *T. hafizi*, *T. leonis* (a synonym of *Terfezia arenaria*, Laessøe and Hansen 2007), and *Tirmania pinoyi* were reported from Iran (Esfandiari and Petrak 1950; Daneshpajuh 1991).

All these desert truffles are seasonal and socioeconomically important mushrooms. They usually appear in arid and semiarid zones following the rainy season from March to April. They are an expensive delicacy in Iran. The cost of 1 kg of *Terfezia boudieri* and other similar desert truffles may reach as high as $ 10–30 depending on seasonal availability. As knowledge of their existence spread, along with recognition of their delicious taste and nutritional value—their consumption, and therefore, demand, was increased. This caused an increased and uncontrolled harvest resulting in a reduction of their production and propagation.

The increase in cost might reflect the above consideration of increased demand coupled with reduced production on top of the problem of erratic rainfall at different seasons in each province, city, and region.

12.2.2 Truffle Use

Most of the harvested desert truffles are used by the rural people who collect them. The collected truffles are consumed in two ways: (1) roasted over fire on skewers

Fig. 12.6 *Left*: Roasting terfezia in Iran. *Right:* Some of collected terfezia from the province of Kurdistan, Bijar, Spring 2011, Photograph taken by Ali Ammarellou

Fig. 12.7 The Iranian local market (Province of Zanjan, Zanjan—Alley Rasthchi) that sells terfezia and other mushrooms with seasonally medicinal plants. Photograph taken by Ali Ammarellou

(Fig. 12.6) and (2) fried with eggs. It seems that the first method of preparing truffles is considered to be tastier than the second.

Fresh mushrooms and desert truffles are important and attractive commodities for Iranian people and are sold, in season, along with fresh medicinal plants (Fig 12.7).

12.2.3 Prospects for the Future

The demand for desert truffles is increasing due to their delicious taste, high protein content, and biological activities. At the same time prime production areas for desert truffles in Iran (which include Zanjan, Kurdistan, Azerbaijanis, and other western parts) are gradually diminishing for several reasons: (1) Unfortunately, uncontrolled, extensive harvest leads to reduction of spore availability in the soil. (2) Development of agricultural and horticultural practices both reduced the sites available for truffle production and decreased the genetic diversity of plants in addition to causing destruction of other natural resources. Thus, the sites, the renewal of necessary host plants, and the loss of fungal inoculation materials reduced truffle production in cultivated areas to zero.

The Research Institute of Physiology and Biotechnology of Zanjan, at the University of Zanjan, has designed national programs for assessing methods and options leading, potentially, to mass production of these fungi via in vivo and

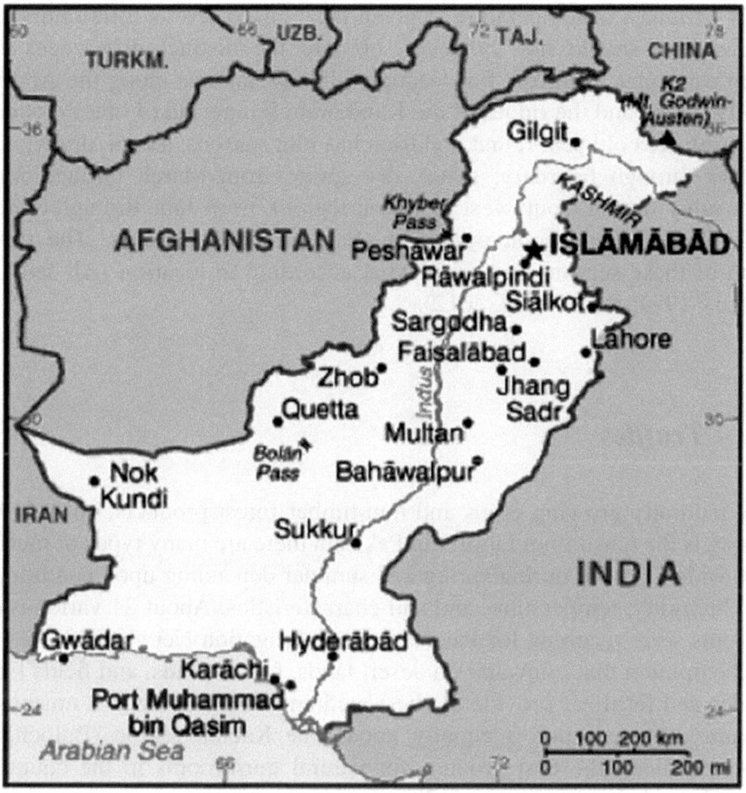

Fig. 12.8 Pakistan (map reference: http://www.mapsofworld.com/lat_long/pakistan-lat-long.html)

in vitro studies. Iran has both opportunities and a potential for improving desert truffle production (see Sect. 12.1).

12.3 Pakistan

Pakistan is situated in Southwest Asia, bordering the Arabian Sea on the south, between India on the east and Iran and Afghanistan on the west and China in the north. Its Geographic coordinates are 30°N, 70°E (Fig. 12.8).

12.3.1 Climate

Pakistan is located in the temperate zone. The climate is generally arid, characterized by hot summers and cool or cold winters, and wide variations between extremes of temperature at given locations. There is little rainfall. These generalizations should not, however, obscure the distinct differences existing between particular locations. For example, the coastal area along the Arabian Sea is usually warm, and the ridges of the Karakoram Range and of other mountains of the far north are cold year round. Pakistan has four seasons: a cool, dry winter from December through February; a hot, dry spring from March through May; the summer rainy season (southwest monsoon region), from June through September; and the retreating monsoon period of October and November. The onset and duration of these seasons vary somewhat according to location (Ali 1978, 1991; Chaudhary 1994; Salma et al. 2012).

12.3.2 Truffles

Among naturally growing crops and non-timber forest products, one of the most promising is the mushroom family. In Pakistan there are many types of mushrooms that are widely spread during spring and summer depending upon conditions such as rain, humidity, temperature, and soil characteristics. About 21 varieties of wild mushrooms were recorded for Pakistan (Spate Irrigation Network 2011). Farmers are of the opinion that rainwater on desert lands, fallow lands, and fields free from pesticides and fertilizer provide the best medium for the growth of mushrooms in the country. Spate water is equally good. The Kachhi region (Balochistan) is considered among the best grounds for natural mushrooms in the country (The Pakistan Spate Irrigation Network 2011).

Information regarding Pakistani truffles is limited. They are usually found in Barani, spate-irrigated areas, arid and wet mountains, in sandy loam soils. Spate irrigation fields and adjacent sites, where floodwater has once spread, are considered particularly suitable for truffles.

Season varies from area to area; they appear when the microclimatic requirements are fulfilled. In D. I. Khan, truffles appear in September–October, or during the growing season of sorghum and millets. In Murree hills, they are usually found during the rainy season of March, where, according to the legends of local rural people, lightning helps truffles to grow.

Two Pakistanian truffle researchers Majeed and Ahmad (2011) report that some of the spate-irrigated areas were visited by the NRD survey team, where underground mushrooms were unearthed. People at these places were interviewed to get information about the usage of natural mushrooms. Local names for truffles vary from place to place. In Ghora Gali of Murree hills, the local name is Gandair. In D. I. Khan it is called Zami Zung.

Murree hills people prefer to eat wild above ground mushrooms such as Morels. As far as local consumption of truffles is concerned people do not have any awareness of this underground fungus being a food and because of its strange aroma they usually throw it away. This attitude hinders truffle research (Majeed

Fig. 12.9 Truffle harvesting in sorghum and millets fields in D. G. Khan (Majeed and Ahmad 2011 with permission)

and Ahmad 2011). Murree hills people are, in fact, unaware of the value of truffles as a commercial commodity.

Some people at different places got some awareness about truffles being an expensive food in foreign countries; people who are aware of its value are reluctant to share the information in order to get maximal benefit out of its sale. But those who tried to make it a business are facing difficulty in identification of species, finding market locations, meeting export standards, understanding the shipment process, etc. Truffle sale is not a successful business enterprise to date.

In D. I. Khan and D. G. Khan truffles are mostly found in the crop fields (i.e., sorghum and millets) and other spate areas (Fig. 12.9). However, other areas are not fully explored (Majeed and Ahmad 2011).

Specifically, in D. I. Khan district of KPK truffles are found at Saggu, Kohawar, Shero Kohna, and Gundi Umar Khan. D. G. Khan truffles are reported from spate irrigation areas of Thana Bula Khan, Barkhan, Nushki, Narran, Kaghan, and Murree hills. A comprehensive survey may reveal additional truffle-producing areas.

12.3.2.1 Truffle Species

To date, there is no scientific identification of any truffle species. In D. I. Khan, the "Zami Zung," is a light brown truffle, good in taste, and smell like muddy ginger (looks like a species of the genus Terfezia). An off-white to white truffle species which resembles Italian white truffles (looks like a species of the genus Tirmania) is also found. Nationwide surveys to identify species available at different regions are called for.

12.3.2.2 Host Plants

In D. G. Khan and D. I. Khan districts, white truffles are found in large quantity in sorghum and millet fields (Fig. 12.9). In the Murree hills, they were found on the roots of pine trees but the information was not verified.

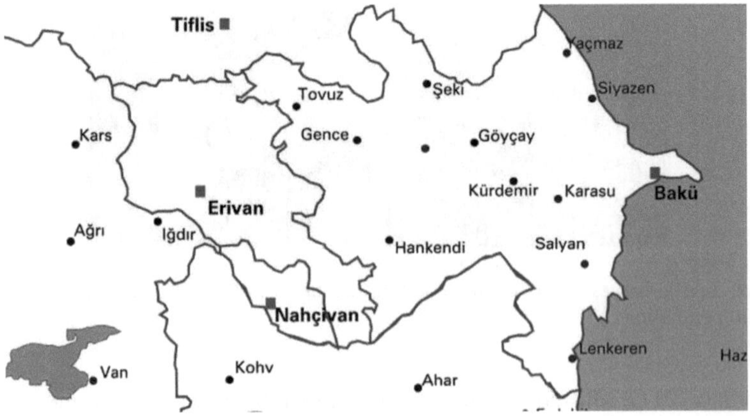

Fig. 12.10 Azerbaijan Republic, Image Source: http://www.forum.98ia.com/t212966.html

Harvest methods: The soil bulges out and cracks, indicating the presence of a truffle. As per local knowledge truffles appear in the morning, following a humid night with no wind. Collection is done early in the morning. The principle of "first come-first take" is the custom.

12.4 Nakhchivan (Azerbaijan) Autonomous Republic

12.4.1 Location and Climate

Nakhchivan Autonomous Republic (AR) is a part of the Azerbaijan Republic (Fig. 12.10). It is located in the southwestern part of the Lesser Caucasus Mountains. The region covers 5,500 km^2 and borders Armenia to the east and north, Iran to the south and west, and Turkey to the northwest. Its highest point is Gapudzhik peak (3,906 m) and the lowest point of the autonomous republic (600 m) is situated on the left bank of the Aras River, at the foot of the steep slope of Soyugdag ridge. The climate of the autonomous republic is of extreme continental type, with hot summers and severe winters. The average annual temperature is 10–14 °C. Areas located above 2,300–2,400 m have a mean annual air temperature below 4 °C. The maximum air temperature in the lower part of the republic is 18 °C in January and 41–43 °C in July–August. Relative humidity varies in different parts of the republic. In the city of Nakhchivan it is 74–76 % in December–February and 39–40 % in July–August. In the middle mountain zone it is 69–78 % and 52–55 % in December–February and July–August, respectively, similar to the foothills of the Lesser Caucasus. The bulk of precipitation falls in spring (March–May) and the minimum in July–August. In the lowland region the annual rainfall is 210–310 mm, in the mid-mountainous area it is 365–550 mm, and in the alpine area it is 660 mm.

Nakhchivan Republic is considered a separate climatic and physical-geographical region of Azerbaijan (Mirzeyev 1972; Seyidova and Huseinm 2012).

12.4.2 Desert Truffles

The flora of Nakhchivan Autonomous Republic is very diverse. Seyidova and Huseinm (2012) claim that 2,835 species of higher plants were defined in the region, and many studies concerning higher plants have been carried out.

5,020 fungal species were described in Azerbaijan so far (Azerbaijan National Report 2010). Based on this survey, conducted during the 2005–2010 period, 73 species of macromycetes were registered, 2 belonging to the division Ascomycota and 71 to the division Basidiomycota. In this scientific document, *Terfezia leonis* Tul. (a synonym of *Terfezia arenaria* (Moris) Trappe) was identified and reported. Specimens of this fungus were collected in Nehram village, Babek area, on 16 June 1963; Nehram village of Babek area, on 16 April 2010, *H.S.* 107; Darakend village, Sharur area, on 24 April 2010, *H.S.* 109; and Tananam village, Sharur area, on 6 May 2010, *H.S.* 122 (Seyidova and Huseinm 2012).

In addition, *Terfezia transcaucasica* (Tikhon), which is endemic to Azerbaijan, is found in the regions of Araz, Absheron, the Lesser Caucasus, and Karaback (Azerbaijan National Report 2010). It is considered to be at risk of extinction.

12.5 The Arab Peninsula

12.5.1 Geography

12.5.1.1 Saudi Arabia

Saudi Arabia occupies about 80 % of the Arabian Peninsula, lying between latitudes 16° and 33°N and longitudes 34° and 56°E (Fig. 12.11). Because the country's southern borders with the United Arab Emirates and Oman are not precisely defined or marked, the exact size of the country remains unknown. The CIA World Factbook's estimate is 2,250,000 km^2 (868,730 square miles) and lists Saudi Arabia as the world's 13th largest state. Saudi Arabia's geography is dominated by the Arabian Desert and associated semidesert and shrub land. It is, in fact, a number of linked deserts and includes the 647,500 km^2 (250,001 square mile) Rub' al Khali ("Empty Quarter") in the southern part of the country, the world's largest contiguous sand desert. There are virtually no rivers or lakes in the country, but wadis (dry riverbeds, where only superfluous rainwater flows when available) are numerous. The few fertile areas are to be found in the alluvial

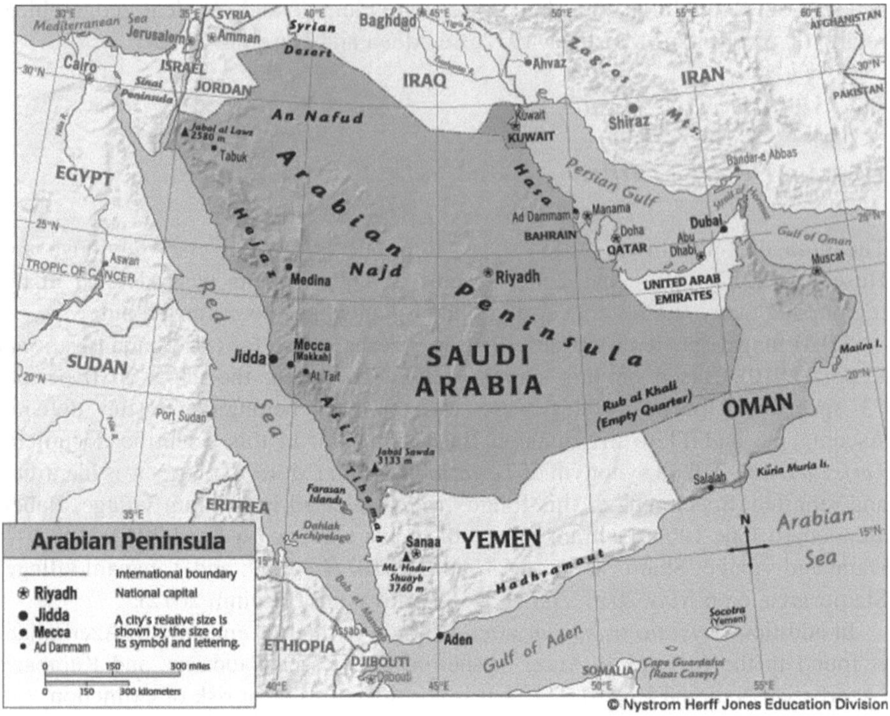

Fig. 12.11 The Arab Peninsula (reference http://farm4.static.flickr.com/3089/2483086462_d3f3c2d19d_o.jpg)

deposits in wadis, basins, and oases (Al-Rahmah 2001; Abou-Zeid and Altalhi 2006).

12.5.1.2 Bahrain

The Kingdom of Bahrain (2,600°N, 5,033°E) comprises an archipelago of 33 islands situated midway in the Persian Gulf close to the shore of the Arabian Peninsula. The islands are about 25 km from the east coast of Saudi Arabia and 28 km from Qatar. The total area of the islands is about 710 km^2. The largest island, accounting for 83 % of the area, is Bahrain, which is 572 km^2. Most of the island is low-lying, barren desert. Outcroppings of limestone form low rolling hills, stubby cliffs, and shallow dry riverbeds (wadis). There is a fertile strip 5 km wide along the northern coast. The interior contains an escarpment that rises to 134 m (the highest point on the island) to form Jabal ad Dukhan (Mountain of Smoke), named for the mists that often wreathe the summit. Most of the other islands are flat and sandy (Al-Shehabi 2005; Mandeel and Al-Laith 2007).

12.5.1.3 Qatar

Qatar is approximately 160 km long by 80 km wide, consisting mostly of low-lying, arid, stony desert.

12.5.2 Climate

12.5.2.1 Saudi Arabia

With the exception of the province of Asir (which is subject to Indian Ocean monsoons between October and March, averaging 300 mm of rain fall—60 % of the annual total), Saudi Arabia has a desert climate characterized by extreme heat during the day, an abrupt drop in temperature at night, and slight, erratic rainfall. On the coast, the desert temperature is moderated by the Red Sea and the Persian Gulf and temperatures seldom rise above 38 °C, but the relative humidity is high: between 85 and 100 %. A southerly wind is accompanied invariably by an increase in temperature and humidity and by a particular kind of storm known as a kauf. In late spring and early summer, a strong northwesterly wind, the shamal, blows; it is particularly severe in eastern Arabia and continues for almost 3 months. The shamal produces sandstorms and dust storms that can decrease visibility to a few meters.

A uniform climate prevails in Najd, Al-Qasim Province, and the great deserts. The average summer day temperature is 45 °C, but readings of up to 54 °C are common. Nights are cool. In the winter, the temperature seldom drops below 0 °C, but the almost total absence of humidity and the high wind-chill factor make a bitterly cold atmosphere. In the spring and autumn, temperatures average 29 °C. For the rest of the country, rainfall is low and erratic. The entire year's rainfall may consist of one or two torrential outbursts that flood the wadis and then rapidly disappear into the soil to be trapped above the layers of impervious rock. This is sufficient, however, to sustain forage growth. Although the average rainfall is 100 mm per year, whole regions may not experience rainfall for several years (Climate of Saudi Arabia 2013).

12.5.2.2 Bahrain

Bahrain has an arid desert climate characterized by very hot summers with high humidity while winters are relatively cooler (Airport Metrological Station, Muharraq, Bahrain). During the summer months, from April to October, the average temperature is 40 °C and can reach 48 °C during June and July (Mandeel and Al-Laith 2007). The combination of intense heat and high humidity makes this season uncomfortable. In addition, a hot, dry southwest wind, known locally as the qaws, periodically blows sand clouds across the barren southern end of Bahrain toward Manama in the summer. Winter temperatures (November–March) range between 10 and 20 °C. However, humidity often rises above 90 %. From December to March, prevailing winds from the southeast, the shamals, bring damp air over the islands. Daily temperatures are fairly uniform throughout the archipelago. Bahrain receives little precipitation. The average annual rainfall is $\leq$72 mm, usually confined to the winter months. No permanent rivers or streams exist on any of the islands. The winter rains tend to fall in brief, torrential bursts, flooding the shallow wadis that are dry the rest of the year (Mandeel and Al-Laith 2007).

12.5.2.3 Qatar

Average temperatures vary between 12 °C and 21 °C in January, and 35–49 °C from June to September. The climate during autumn and spring is moderate, while winter can be surprisingly cool. Average rainfall is 50–99 mm per year, although the actual amount varies considerably from one year to the next (Al-Thani 2010).

12.5.3 Desert Truffles

Four species of truffles have, so far, been identified in the Arab Peninsula (Al-Rahmah 2001): a white truffle, *Tirmania nivea* (Desf.) Trappe, locally called "Zubaidi"—the most appreciated of the four; two species of brown truffles, Terfezia claveryi Chatin and Terfezia boudieri Chatin, collectively called "Ikhlasi"—all three are members of the Terfeziaceae family; and *Picoa lefebvrei* (Pat.) Maire ("Heberi" or "Hoper") the bird truffle (consumed less by people and more by animals), belonging to the Pyronemataceae family. All form Mycorrhizal associations with species of the genus *Helianthemum* (Cistaceae) (Mandeel and Al-Laith 2007).

Although not specifically mentioned, the above holds for Kuwait, United Arab Emirates, and Oman, which are part of the Arab Peninsula, as well as for Iraq where some species of desert truffles were known and identified (Al-Shabibi et al. 1982; Moubasher 1993).

12.5.3.1 Saudi Arabia Desert Truffles

Desert truffles (locally known as AJ-Kamah or AI-Fag'a) are growing in the wild in particular regions of Saudi Arabia (Bokhary et al. 1987). They usually appear after the rainy season (February–March) and are reported to be a seasonal food component. In Saudi Arabia, as in several other deserts, fruit bodies have been collected and appreciated from ancient times to this day (Trappe 1990, see also Chap. 15 by Shavit). According to investigations at various Saudi Arabian markets the demand for truffles exceeds by far the supply. In Riyadh itself the annual offer varies between 1 and 1.2 t, whereas the demand is much higher. For the entire country the annual estimate is over than 50 t. Consequently, the price for 1 kg truffle remains high and may attain 60 US$ per kg or even more (Mandeel and Al-Laith 2007). The fruit bodies form close to the soil surface, and as they swell, they lift up the soil to form little cracked mounds recognizable to the trained eye (Awameh and Alsheikh 1979; Khanaqa 2006).

12.5.3.2 Bahrain

In general, the soil is poor in organic matter, slightly alkaline and saline. The limestone is covered by various densities of saline sand, capable of supporting only halophytic desert vegetation in the form of small trees and shrubs. Based on scientific reports Bahraini and non-Bahraini groups within the area actively consume truffles for food and other sociocultural practices and have a truffle heritage, which is yet to be adequately documented (Mandeel and Al-Laith 2007).

12.5.3.3 Qatar

Truffles are abundant in spring and autumn due to the humid climate as well as the richness of the flora at this time of the year (Sibounnavong et al. 2008). Truffles are important to the local community in Qatar due to their significant roles in human life. Traditionally, desert truffles are used as food, as a cash crop, and as medicine for a variety of ailments. Some of the different species of desert truffle that grow in arid and semiarid areas and that prefer high pH calcareous soils have been found in Qatar (Mandeel and Al-Laith 2007; Al-Thani 2010).

12.6 China

China has over 3,670,000 km^2 arid and semiarid land in the north and northwest regions, including over 600,000 km^2 of desert and semidesert (http://www.fao.org/docrep/w7539e/w7539e04.htm). Desertification will increase in China if global warming and poor land management continue. However, only a few Chinese desert truffle species have, so far, been reported. Extensive surveys in search of desert truffles are urgently needed. Hopefully, such surveys will discover additional Chinese desert truffle species.

12.6.1 Climatic and Soil Conditions, Vegetation

The climatic conditions in North China are semiarid and continental monsoon climate with dry and cold winter and hot and rainy summer. Annual precipitation ranges from 350 to 600 mm and is uneven (60–70 % in summer). Variability of the annual precipitation is rather large. Temperature ranges from −25 to 39 °C. The spring is dry and windy. The region receives extensive amounts of sunshine. In the autumn temperature drops and is suitable for fungal fructification (15–29 °C) (www.weather.gov.hk/wxinfo/. . ./china3_c.htm).

Geographically, pine-oak mixed broad-leaved deciduous forest should be the native vegetation in the north of China (Chen and Chen 2013). For example, in Liangshui, Heilongjiang, the natural vegetation was *Pinus koraiensis-Quercus mongolica* mixed broad-leaved deciduous forest, and in Hebei, Beijing, Shanxi, and Henan *Pinus tabulaeformis-Quercus* spp., especially in the lower mountains. In the higher mountains there are some spruce and fir trees. However, due to long-term large-scale deforestation the original vegetation has disappeared except in the native reserves and temples. *Populus* spp. and *Betula* spp. have become the dominant species. These areas are the sites where most collections of Chinese desert truffles were found. Brown forest soil dominates these forests (wenku.baidu.com/. . ./ac4f9b85b9d528ea81c77900) (ibid).

12.6.2 Chinese Desert Truffles

Since 1963 several species of desert truffle have been reported from China. Recent reexamination revealed most of the species to be misidentified. To date, only four species *T. terfezioides* (Matt.) Trappe (Tao 1988; Zhang 1990), *Terfezia* sp. (recorded from a collection of HMAS83766 in the Herbarium of Institute of Microbiology, Chinese Academy of Sciences, Beijing, China), *Terfezia parvocarpus* Zhang (Zhang 1990, 1992a, b), and *Picoa carthusiana* Tul & Tul (Tao 1988; Zhang 1990) are possibly present in China. *Terfezia terfezioides* has lately been transferred into the genus *Mattirolomyces* (Laessøe and Hansen 2007; Kovács et al. 2011), and *Picoa carthusiana* was transferred to *Leucangium carthusianum* (Li 1997).

12.6.3 Distribution

So far the Chinese truffles are mainly found in northern China: Beijing, Hebei, Shanxi, and Henan, N35°03′–N40°36′, E111°12′–E117°29′.

Mattirolomyces terfezioides (=*T. terfezioides*) has been found at Beijing: Luodaozhuang Village (N39°92′, 116°46′); Hebei, Xuanhua County (N40°35′,

Fig. 12.12 Distribution of desert truffles in China. (*Filled triangle*) *Mattirolomyces terfezioides* (=*T. terfezioides*), (*filled circle*) *Terfezia parvocarpus*, (*filled diamond*) *Terfezia* sp., and (*filled square*) *Leucangium carthusianum* (=*Picoa carthusiana*)

E115°03′) and Wanxian County (N38°83′, E115°13′); Shanxi, Taiyuan (N37°54′, E112°33′), Wenxi County (N35°22′, E111°12′), and Zhongtiaoshan Mountains (N35°03′, 33′); Henan, Yanshi County (N34°43′, E112°47′); and Shaanxi, Yunu Peak of Huashan Mountains (N34°31′, E110°03′). Fruit bodies develop around trees such as *Robinia pseudoacacia* L., *Pinus tabulaeformis,* as well as other species, as yet not proven plant species. They are also found in crop fields, such as maize (Tao 1988; Zhang 1990; Liu et al. 2002). Their biotrophic status is unknown (Laessøe and Hansen 2007) (Fig. 12.12).

Terfezia parvocarpus has been found at Hebei: Xinglong County, Mt. Wulingshan (N40°36′, E117°29′). It grows under trees of *Salix wallichiana* Anders (Zhang 1990).

Terfezia sp.: Heilongjiang, Liangshui (N46°32′, E131°52′). It grows in forests (recorded from the HMAS83766 collection in the Herbarium of Institute of Microbiology, Chinese Academy of Sciences, Beijing, China).

Picoa carthusiana (=*Leucangium carthusianum*) (Li 1997): Shanxi, Ningwu, Mt. Guanqinshan (N39°22′, E112°, 22′). It grows in *Picea wilsonii* forests and may form mycorrhizal association with spruce trees (Tao 1988; Zhang 1990).

12.7 Conclusions

The world population growth causes increasing demand for new sources of food, especially proteins. The nutritional value, delicious taste, and curative properties of desert truffles (e.g., see Chaps. 16 and 20 by Perez-Gilabert et al. as well as Shavit and Shavit) make desert truffles good candidates. They are already known and appreciated at certain countries, and, consequently, their prices are soaring. Efforts aimed at cultivation of desert truffles are still rather rudimentary (see Chaps. 21 and 22 by Morte et al. and Honrubia et al.). Cooperative global data collection leading to information on distribution and habitats of these fungi should be undertaken in order to be able to develop desert truffles as a good food source.

The habitats of desert truffles in various non-Mediterranean Asian countries were described above. We may conclude that all countries included in this chapter are capable of cultivating desert truffles, as they generally meet truffle requirements such as soil conditions (mainly calcareous loamy sandy soils; see Sects. 12.3 and 12.5), rainfall (amount and seasonal distribution), host plants, and fungal spore abundance. Based on evidence reported here, these fungi have several verified hosts (such as sedge, *Helianthemum* spp., sorghum, and millet and potentially more). Different truffle species are adapted to a variety of climatic conditions from cool mountain weather to hot lowland deserts. Although these truffles develop under rather restricted water availability (about 100–400 mm yearly average; see Sect. 12.1), fruiting will not occur in drought years. In other words, in areas suitable for truffle growth, good autumn rainfall is needed for truffle fruiting. This is common in all studied countries except Pakistan (in Pakistan truffles develop around and in spate-irrigated areas; see Sect. 12.3). Industry, urbanization, and other land use conversion factors are threatening the survival of these fungi in the wild. Two solutions are possible: international agreements to enforce protection and preservation of desert truffle natural habitats and the development of suitable agricultural practices for desert truffle cultivation.

References

Abou-Zeid AM, Altalhi AE (2006) Survey of some mushrooms in Al-Taif Governorate of Saudi Arabia. World J Agric Sci 2(1):01–05

Akyuz M, Kirbağ S, Kurşat M (2012) Ecological aspects of the arid and semi-arid truffle in Turkey: evaluation of soil characteristics, morphology, distribution, and mycorrhizal relationships. Turk J Bot 36:386–391

Al-Laith AAA (2010) Antioxidant components and antioxidant/antiradical activities of desert truffle (*Tirmania nivea*) from various Middle Eastern origins. J Food Comp Anal 23:15–22

Al-Rahmah AN (2001) Truffle of deserts and jungles. King Saud University publications, Riyadh, Saudi-Arabia, p 272 (in Arabic)

Al-Ruqaie IM (2002) Effect of different treatment processes and preservation methods on the quality of truffles: I. Conventional methods (drying/freezing). Pakistan J Biol Sci 5:1088–1093

Al-Shabibi MMA, Toma SJ, Hadad BA (1982) Studies on Iraqi truffles. I. Proximate analysis and characterization of lipids. Can Inst Food Sci Technol J 15:200–202

Al-Shehabi AA (2005) Wild flora of Bahrain (in Arabic). Kingdom of Bahrain, p. 299

Al-Thani R (2010) Survey of macrofungi (including Truffles) in Qatar. KBM J Biol 1:26–29

Ali SI (1978) The Flora of Pakistan: some general and analytical remarks. Notes Roy Bot Gard Edinburgh 36:427–439

Ali SI (1991) Introductory remarks. In: Ali SI, Ghaffar A (eds) Plant life of South Asia. Shamim, Karachi, pp 3–6

Ammarellou A, Saremi H, Gucin H (2007) Evaluation of morphology, cytology and mycorrhizal relationships of desert truffles (*Terfezia boudieri*) in Iran. Pakistan J Biol Sci 10:1486–1490

Ammarellou A, Saremi H (2008) Mycorrhiza between *Kobresia bellardii* (All.) Degel and *Terfezia boudieri* Chatin. Turk J Bot 32:17–23

Ammarellou A, Smith ME, Tajick MA, Trappe JM (2011) The phylogenetic placement of *Picoa*, with a first report on *Picoa lefebvrei* (Pat.) Maire (=*Phaeangium lefebvrei*) from Iran. Int J Environ Res 5(2):509–514

Awameh MS, Alsheikh A (1979) Laboratory and field study on four kinds of truffle (kamah), *Terfezia* and *Tirmania* species for cultivation. Mushroom Sci 10:507–517

Azerbaijan National Report (2010) Country Study on Biodiversity of Azerbaijan Republic. Fourth National Report to Convention of Biological Diversity. Baku

Bokhary HA, Suleiman AAA, Basalah MO, Parvez S (1987) Chemical composition of desert truffles from Saudi Arabia. C Inst F Sci Technol 20:336–341

Chaudhary QZ (1994) Pakistan summer monsoon rainfall's association with global and regional circulation features and its seasonal prediction. Proceeding of international conference on monsoon variability and prediction, Triest, Italy, 9–13 May, p 136–145

Chen Ling-zhi, Chen Qing-lang (2013). The forest diversity in China http://www.um.u-tokyo.ac.jp/publish_db/Bulletin/no37/no37006.html

Climate of Saudi Arabia (2013) http://ancienthistory.about.com/od/aneplacesgeography/qt/ClimSaudiArabia.htm

Daneshpajuh B (1991). Two new species of Hypogeous Ascomycetes in Iran. 10th Iran Plant Prot Cong Kerman. p. 123

Diez J, Manjon JL, Martin F (2002) Molecular phylogeny of the mycorrhizal desert truffles (*Terfezia* and *Tirmania*), host specificity and edaphic tolerance. Mycologia 94:247–259

Esfandiari E, Petrak F (1950) Pilze aus Iran-Sydowia. Horn 4:11–38

Khanaqa A (2006) Truffle production in the Kingdom of Saudi Arabia—potential and limitation. J Appl Bot Food Qual 80:14–18

Kovács GM, Trappe JM, Alsheikh AM, Hansen K, Healy RA, Vági P (2011) *Terfezia* disappears from the American truffle mycota as two new genera and *Mattirolomyces* species emerge. Mycologia 103(4):831–840. doi:10.3852/10-273. PMID 21262987

Laessøe T, Hansen K (2007) Truffle trouble: what happened to the Tuberales? Mycol Res 111:1075–1099

Ławrynowicz M, Marković M, Milenković M, Ivančević B (1997) *Terfezia terfezioides* a new hypogeous fungus for Balkan Peninsula. Acta Mycol 32:233–238

Li LT (1997) Ultrastructural studies of *Leucangium carthusianum* (hypogeous Pezizales). Int J Plant Sci 158:189–197

Liu B, Liu HY, Liu ZH (2002) Hypogeous fungi from China. Edible fungi of China 21–1: 3–4 and 2:14–15

Mandeel QA, Al-Laith AAA (2007) Ethnomycological aspects of the desert truffle among native Bahraini and non-Bahraini peoples of the Kingdom of Bahrain. J Ethnopharmacol 110:118–129

Majeed S, Ahmad S (2011). Truffles—1: an unexploited treasure in Pakistan—issues and options. Res Brief 3(19), NRD, PARC, Islamabad, Pakistan

Moubasher AH (1993) Soil fungi in Qatar and other Arab countries. University of Qatar Press, Doha, Qatar, p 566

Moreno G, Diez J, Manjon JL (2002) *Terfezia boudieri*, first records from Europe of a rare vernal hypogeous mycorrhizal fungus. Persoonia 17:637–641

Mirzeyev PS (1972) Nakhchivan MSSR-in Agroiglim Sedzhijjesi. Baku: Azerneshr ersoonia 17:637–641

Salma S, Rehman S, Shah MA (2012) Rainfall trends in different climate zones of Pakistan. Pakistan J Meteor 9(17):37–47

Seyidova H, Huseinm E (2012) Macrofungi of Nakhchivan (Azerbaijan) Autonomous Republic. Turk J Bot 36:1–8

Sibounnavong P, Cynthia CD, Kalaw SP, Reyes RG, Soytong K (2008) Some species of macrofungi at Puncan, Carranglan, Nueva Ecija in the Philippines. J Agric Technol 4 (2):105–115

Spate Irrigation Network, Pakistan. Practical notes #14. (2011) Desert Truffle Mushrooms in Spate Irrigation Areas, Pakistan

Tao K (1988) Taxonomic study of hypogeous fungi from China. M.Sc. thesis, Shanxi University, Taiyuan

Trappe, JM (1990). Use of truffles and false truffles around the world. In: Bencivenga M, Granetti B (eds.) Atti del secondo congresso internazionale sul tartufo. Spoleto, Italy 1988. Com. Mont. Dei Martini, Italy Pub. pp. 19–30

Trappe JM, Claridge AD, Arora D, Smit WA (2008) Desert truffles of the African Kalahari: ecology, ethnomycology, and taxonomy. Econ Bot 62:521–529

Trappe JM, Weber NS (2001) North American desert truffles: the genus *Carbomyces* (Ascomycota, Carbomycetaceae). Harvard Pap Bot 6:209–214

Zhang, BC (1990) Study on systematic position of tuber at tuberales and Chinese tuber species. Ph.D. thesis, Beijing Institute of Microbiology, CAS, Beijing

Zhang BC (1992a) Chinese hypogeous mycoflora: terfeziaceae, new species and new records. Micologia e Vegetazione Mediterranea 7:39–42

Zhang BC (1992b) Nuclear numbers in geneaceae and terfezia ascospores and their taxonomic value. Syst Ascomycet 11:31–35

Chapter 13
Ecology and Distribution of Desert Truffles in the Kalahari of Southern Africa

James M. Trappe, Andrew W. Claridge, and Varda Kagan-Zur

13.1 Introduction

The Kalahari basin of southern Africa covers some 930,000 km^2, including most of Botswana, the eastern third of Namibia, and Northern Cape Province of South Africa. Little is known of the hypogeous fungi of southern Africa: early reports were studded with misidentifications and inadequate documentation (Marasas and Trappe 1973), despite the long history of truffles as a highly preferred food of the !Kung San (Bushmen) indigenous people (Lee 1979; Silberbauer 1981). Verwoerd (1925) and Leistner (1967) examined several collections of hypogeous basidiomycetes in southern Africa but were hindered by the inadequate knowledge of the time; many collections were misidentified or introduced with exotic trees such as *Eucalyptus* spp. (Trappe unpublished data), and few were from the Kalahari. Moreover, the nature of these truffles was unclear in some circles as late as 1971: "It is uncertain whether this species is a root parasite or a fungus" (p. 107 in Lee and DeVore 1976).

J.M. Trappe (✉)
Department of Ecosystems and Society, Oregon State University, Corvallis, OR 97331-5752, USA

USDA Forest Service, Pacific Northwest Research Station, Forestry Sciences Laboratory, Corvallis, OR 97331, USA
e-mail: trappej@onid.orst.edu

A.W. Claridge
Office of Environment and Heritage, NSW National Parks and Wildlife Service, Nature Conservation Section, P.O. Box 733 Queanbeyan, NSW 2620, Australia

School of Physical, Environmental and Mathematical Sciences, University of New South Wales, Northcott Drive, Canberra, ACT 2600, Australia
e-mail: andrew.claridge@environment.nsw.gov.au

V. Kagan-Zur
Life Sciences Department, Ben-Gurion University, POB 653, Beer-Sheva 84105, Israel
e-mail: zur@bgu.ac.il

V. Kagan-Zur et al. (eds.), *Desert Truffles*, Soil Biology 38,
DOI 10.1007/978-3-642-40096-4_13,

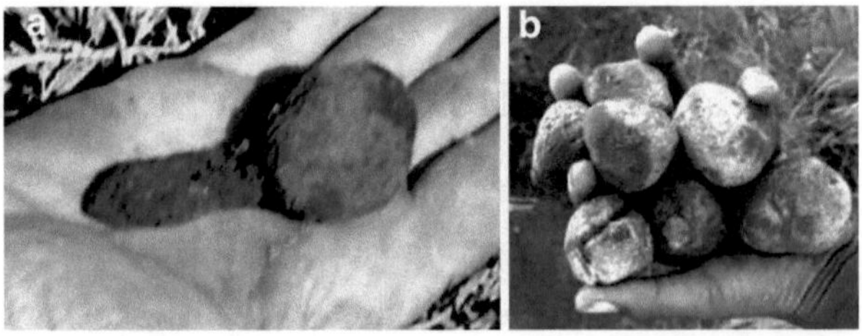

Fig. 13.1 (**a**) Kalahari truffle showing the basal "stalk" formed by a mass of entangled soil, roots, and mycelium. (**b**) Handful of mature Kalahari truffles (courtesy of Tim Turluck, Slow Foods Johannesberg, www.slowfood.co.za; all rights reserved)

As of now, three species of Kalahari truffles are known. One, *Kalaharituber pfeilii* (Henn.) Trappe & Kagan-Zur, is well known and widely distributed in the Kalahari (Fig. 13.1); *Eremiomyces echinulatus* (Trappe & Marasas) Trappe & Kagan-Zur and *Mattirolomyces austroafricanus* (Trappe & Marasas) Trappe, Kovács and Claridge, in contrast, are known only from a few collections each (Trappe et al. 2010a, b). Given the lack of systematic survey, more species could yet be found in the Kalahari. However, it seems to have a small truffle mycota compared to several desert regions elsewhere in the world (see other chapters in Part III of this volume).

13.2 The Kalahari

13.2.1 Landscapes

The Kalahari is a more or less featureless, gently undulating sand plain dominated by three types of surface: sand sheets, longitudinal dunes, and clay pans. All of it is higher than 900 m above sea level, but mountain ranges are absent. Bedrock is exposed only in low, vertical-walled hills (Thomas and Shaw 1991; Logan and Silberbauer 2013).

The southern part of the Kalahari has no perennial surface water except for widely scattered water holes (Thomas and Shaw 1991; Logan and Silberbauer 2013). Most rain disappears immediately into the sand or flows temporarily during and immediately after rains along short stretches of bedrock into claypans. The northern part, in contrast, has a complex drainage in which heavy rains from as far away as the Angola highlands flow southward in many streams that merge to form the perennial Okavango, Kwango, and several other rivers. These then break into numerous channels that may fill lakes, swamps, and pans. As Logan and Silberbauer (2013) phrases it, "Thus is created the paradoxical situation of an area with extensive excess of water in a region chronically short of water."

Despite the term "featureless," a relative term, the Kalahari abounds with diverse habitats.

13.2.2 Geology and Soil

Sand characterizes the Kalahari, but that sand is derived from diverse origins (Thomas and Shaw 1991). Mineralogical studies indicate its sands represent accumulation of materials derived from local sources, including the weathering products of pre-Kalahari rocks supplemented by other material transported over relatively short distances. In general, Kalahari sands are low in organic matter, relatively calcareous, and extremely dry (Logan and Silberbauer 2013).

The sand sheets of the eastern Kalahari apparently formed during the Pleistocene from about 12,000 to 2.6 million years before present and have occupied the same areas ever since (Logan and Silberbauer 2013); most were wind-formed. They change little in elevation, and the sand is generally deeper than 60 m. It is often red from iron compounds, a potentially meaningful trait in distribution of *Kalahari pfeilii*, the Kalahari truffle.

In contrast to the eastern Kalahari, the western Kalahari is entirely composed of north–south-oriented dunes at least 2.5 km long, a hundred meters or more broad, and 7–70 m tall. These dunes are separated from each other by broad, parallel depressions termed a *straat*, or street.

Pans are also common in parts of the Kalahari (Leistner 1967), having resulted from surface water flow in past times of greater precipitation. Where such flows disappeared into the sand in low places, the silt they carried formed pans (Logan and Silberbauer 2013). The pans of the southern Kalahari can be divided into three broad classes: white or calcareous, pink or slightly calcareous, and red or noncalcareous. These in turn may vary in physical and chemical characters from edge to center of the pan (Leistner 1967).

Thomas and Shaw (1991) otherwise provide a detailed description of the geology and climate of the Kalahari (also see Chap. 4 by Bonifacio and Morte).

13.2.3 Climate

The southwestern Kalahari meets the traditional definition of a desert as receiving less than 250 mm of rain per year. Moreover, it conforms to a more accurate definition in which potential evaporation is twice as great as the precipitation. As commonly the case in deserts, the southwest varies markedly between and within years in precipitation. The northeast, in contrast, cannot qualify climatically as a desert by these definitions. Nonetheless, it has no surface water; the rain drains instantly through its deep sands to produce an edaphic drought (Logan and Silberbauer 2013).

Two principles of Kalahari rainfall are variability and uncertainty (Thomas 2002). Pike (1971) reports for the Okavango Delta that more than 75 % of rain events are low in intensity, and half the storms produce less than 10 mm of precipitation per event. Annual rainfall varies notably as well. In many years rains do not start until December or January, and sometimes no rain falls after February or March. Silberbauer (1981) reported for the Ghanzi District of the central Kalahari that from 1961 to 1965 annual precipitation ranged from about 100 to 770 mm. Most rainfalls were of low intensity, and half of them delivered less than 10 mm, but most of the annual total was from localized, violent thunderstorms. The wet season usually begins in October or November and ends in April or May, but in many years the rain commences in December or January, and some years no rain may fall after February or March. The !Kung San divide the year into five seasons (Lee 1979): spring rains, main summer rains, autumn, winter, and the spring dry season. The spring rains usually begin in October or November and mostly consist of light thundershowers that hit in one place, miss others entirely, and generally last about an hour. These rains trigger plant growth. The main summer rains in December to March bring a season of relative plenty (Fig. 13.2), depending on their highly variable timing and intensity. The dry but still warm autumn season of April or May is followed by the dry but cool to cold winter months into August. The early spring dry season begins in late August, when temperatures begin to warm, and lasts until commencement of the spring rains in October or November. The Kalahari is huge, so variations on this theme are to be found in the wetter north and drier south.

The Kalahari's relatively high elevation and predominantly clear, dry weather produce large seasonal and diurnal temperatures. Summer day shade temperatures often reach 43–46 °C and drop to 21–27 °C at night. Winter night temperatures may drop to 12 °C (Logan and Silberbauer 2013).

13.2.4 Plants

The large diversity of Kalahari habitats on both macro- and microscales naturally results from the interaction of the diverse soils and climates over its 930,000 km^2. Most research on plant communities has been focused on the deserts of the central and southern Kalahari of Botswana, eastern Namibia, and western Northern Cape Province of South Africa (Leistner 1967; Lee 1979; Thomas and Shaw 1991). The documented collections of desert truffles of the Kalahari span these regions (Trappe et al. 2008a, 2010a).

Silberbauer (1981, Figs. 4–6) presents informative transect diagrams of the region's dune woodlands, scrub plain and pan, and thornveld. We will not repeat the details of these and other plant ecological features (Leistner 1967; Thomas and Shaw 1991, and others) for lack of specific information on habitats recorded for occurrence of desert truffles.

Fig. 13.2 Harvesting Kalahari truffles: lush growth of vegetation after intense summer rainstorms signals a good truffle fruiting season in the following winter

13.3 The Kalahari Desert Truffles

Compared to truffle-producing deserts elsewhere (see the other chapters in Part III of this volume), the Kalahari is poor in both species and genera. For example, the Kalahari has three genera, each with one species: *Eremiomyces echinulatus*, *Kalaharituber pfeilii*, and *Mattirolomyces austroafricanus*. A possible fourth also exists: specimens of an undescribed species of the genus *Tirmania* were collected by Kagan-Zur and Taylor on their Kalahari truffle collection expedition during 1990. Unfortunately, the specimens were lost, so scientific identification as well as herbarium specimen accession await further collections. In contrast, the other major desert region of the southern hemisphere, the Australian outback, has six genera and seven species (see Chap. 14 by Claridge et al.; Trappe et al. 2008b). The reason for this disparity is presently unclear, although the trend is in keeping with the general observation that the diversity of hypogeous fungi Australia-wide is unparalleled relative to other continents.

13.3.1 History of Discovery

The indigenous peoples of the Kalahari likely have used truffles for countless centuries, a presumption backed by the high regard they have for truffles as a preferred food to this day (Lee 1979; Silberbauer 1981). The first published record of Kalahari Desert truffles, however, appeared in 1897.

In the 1890s Joachim Count von Pfeil, German politician and explorer, obtained truffles in Damaraland, South-West Africa (now Namibia). He sent them pickled in vinegar in a large jar to the German mycologist Paul Christoph Hennings, who recognized them to be an undescribed species (Hennings 1897). He described and named them *Terfezia pfeilii* Henn. That name persisted for more than a century (see Trappe et al. 2008a for a brief historical review) and was accepted by Marasas and Trappe (1973), who added two additional new species to the truffles of the Kalahari: *Choiromyces echinulatus* Trappe & Marasas and *Terfezia austroafricana* Trappe & Marasas. The advent of molecular methods enabled a new, more precise way of evaluating phylogeny of fungi, and these were used to good effect by Ferdman et al. (2005). Their molecular data demonstrated that *C. echinulatus* and *T. pfeilii* each merited new and separate generic status. They described two new genera and transferred the two species accordingly: *Eremiomyces echinulatus* (Trappe & Marasas) Trappe & Kagan-Zur and *Kalaharituber pfeilii* (Trappe & Marasas) Trappe & Kagan-Zur.

Then, Trappe et al. (2010a, b) applied molecular techniques to *Terfezia austroafricana*, showing that species to belong to *Mattirolomyces*, a genus occurring in Mediterranean Europe. So, the new combination *M. austroafricanus* was formulated. With these nomenclatural corrections, the names should endure.

13.3.2 Taxonomy, Endemism, and Distribution

All taxa of Kalahari Desert truffles are described in detail, illustrated, discussed, and keyed by Trappe et al. (2008a, 2010a, b). The three genera/species of desert truffles so far discovered are all in the Ascomycota, order Pezizales, family Pezizaceae.

All three species are endemic to the Kalahari, but the genera *Eremiomyces* and *Mattirolomyces* are now known from elsewhere. Until a new species, *E. magnisporus* G. Moreno et al., was described from a semiarid habitat in central Spain (Alvarado et al. 2011), *E. echinulatus* was thought to represent both an endemic genus and species, but now only the species is endemic to the Kalahari.

Mattirolomyces austroafricanus is a species endemic to the Kalahari, but the genus *Mattirolomyces* was originally described by Fischer (1938) to accommodate *Choiromyces terfezioides* Matt. from a collection from Italy. Since then the genus *Mattirolomyces* has been found in many southern European localities (Kovács 2007), Australia (Trappe et al. 2010a), West Asia, and North America (Kovács et al. 2011). *Mattirolomyces austroafricanus*, originally described as a *Terfezia* species by Marasas and Trappe (1973), was revealed by phylogenetic analysis to belong in *Mattirolomyces* (Trappe et al. 2010a, b).

The currently known distributions of the Kalahari Desert truffles are summarized by Trappe et al. (2010a). *Eremiomyces echinulatus* is known from only three collections from 1961 to the present, one each from Botswana, Namibia, and Northern Cape Province of South Africa. Judging from these collections, it is widely distributed but infrequent. *Mattirolomyces austroafricanus* is represented by only two collections from Northern Cape Province, South Africa.

The common and much prized desert truffle from the Kalahari is *Kalaharituber pfeilii* (Fig. 13.1), represented from several collections from each of Botswana, Namibia, and Northern Cape Province of South Africa and reported without documentation from additional localities in these countries. Unfortunately, the specific localities and coordinates were not noted for any of the collections; rather, only broad regional areas are recorded (Ferdman et al. 2005), so a distribution map offers little more than is given in the first sentence of this paragraph.

13.3.3 Ecology: Key to Distribution

Little information on soils or climate has been documented for any of the collections we have examined or in the literature. In this section we summarize the review by Trappe et al. (2008a). Additional literature supplementing their review is cited in the text.

Analyses of truffle-bearing soils have been reported for only seven sites in Namibia and two in Botswana. *Kalaharituber pfeilii* tended to grow on compact, pink or sometimes white sands with a pH of 5.5–6.5, sometimes as high as 7.2. $CaCO_3$ is low, ranging from 0.3 to 3.1 %. Other observers reported the species in dips between sand dunes, but it also has been harvested in agricultural fields of various herbaceous plants (Kagan-Zur et al. 2001; Mshigeni et al. 2005).

Weather, particularly rainfall, is critical to the fruiting of Kalahari truffles as is true of all desert truffles, for example see Chap. 14 by Claridge et al. for Australian desert truffles. Thus, in his studies of the truffle-hunting !Kung San in northeastern Namibia and northwestern Botswana, Lee (1979) reported that *Kalaharituber pfeilii* did not fruit in the winter of 1964 but was common and easily obtained in the winter of 1968 (April and May) near water holes of the southern part of his study area. Silberbauer's (1981) visits with the G/wi-speaking tribesmen of central Botswana, who also preferred *K. pfeilii* over other foods when available, found it mostly in April and May and at least once in January, but the season was short. Mshigeni (2001) reported finding *K. pfeilii* in northern Namibia in June at the end of the truffle season. Depending on the region, the fruiting season may extend as late as June or July, the end of the rainy season when soil and air temperatures are cooling (Fig. 13.2).

The limited literature on plants associated with truffle fruiting, as reviewed by Trappe et al. (2008a), includes shrub–grass–forb communities, cultivated fields of melons, pearl millet and sorghum, shrubs of *Vachellia hebeclada* (DC) Kyal. & Boatwr., a mixed grassveld of *Aristida* and *Eragrostis* species with scattered *Vachellia* trees. (The name *Acacia* has been applied to African species, but molecular analyses show it to differ genetically from Australian members of *Acacia*, to which that generic name was first applied; the African species reported to associate with *K. pfeilii* now are placed in the new genus *Vachellia* [Kayalangalilwa and Bruce 2013].) None of these reports are specific enough to indicate a truffle mycorrhizal host: roots of desert plants can extend farther outward than is often realized, so a fruiting truffle is not necessarily forming mycorrhizae with its nearest plant. Only careful sampling and analysis of roots and experimentation can demonstrate actual mycorrhizal associates of the Kalahari truffle. Yet, knowing which plants form mycorrhizae is requisite to understanding where it occurs for purposes of sustainable management and harvest. Anything that threatens host plants threatens the truffle.

Identity of a several *K. pfeilii* hosts was confirmed by Taylor et al. (1995), who carefully examined the "stalk" of entangled rhizomorphs, hyphae, roots, and soil from which the truffle arises (Fig. 13.1a). This work provided strong evidence of mycorrhiza formation with roots of diverse herbaceous and woody plants: probable mycelium of *K. pfeilii* formed ectomycorrhizal structures such as Hartig nets within the roots. This study was followed by application of DNA analysis by Kagan-Zur et al. (1999) to even more convincingly demonstrate colonization of wild watermelon roots by *K. pfeilii*, this time not with a Hartig net but with structures similar to those formed by *Terfezia* and *Tirmania* spp. with desert annuals on the Arabian Peninsula (see Chap. 5, this volume). These two seminal papers provide the first scientific approach to determining the mycorrhizal hosts of *K. pfeilii* and thus form the basis for testing hypotheses about the role of plant hosts in its distribution.

Meerkats, baboons, bat-eared foxes, and hyenas have been observed to eat Kalahari truffles, presumably then dispersing the spores as do animals for hypogeous fungi around the world (Trappe and Claridge 2005). Mycophagy is also probable among any rodent species and other ground-dwelling desert animals.

13.4 The Outlook for Kalahari Desert Truffles

In 2008, a wildfire in Botswana's Central Kalahari Game Reserve, home of the Basarwa (another name for bushmen) people, raged over some 40,000 km^2 in 4 weeks (Earth Observatory 2008). Fire has always been a part of the Kalahari, both from lightning strikes and human activity. Paradoxically, wet years experience the worst fire potential, because the wet years induce fast seed germination and plant growth that rapidly dries to a high fuel load when the dry season begins. Native societies used prescribed burning, probably from ancient times (Thomas and Shaw 1991). How fire affects truffle production in the Kalahari inevitably depends on how fire affects the truffle host plants, particularly the fire sensitive ones. Probably the intense fire noted at the beginning of this paragraph is deleterious, because many plants may be killed. However, the entire topic of fire effects on truffle production, positive and negative, needs much careful research (Claridge and Trappe 2004).

Overgrazing, on the other hand, is well recognized as seriously detrimental to soils, water, and plants. Thomas (2002) elegantly describes evolution of the grazing industry in the Kalahari over time. A few highlights of his paper follow, although a full appreciation demands reading the original. Cattle have been grazing the Kalahari for a millennium or more, but until the last century limited water availability restricted use by cattle in many areas. Now boreholes dot the Kalahari landscape, enabling cattle to survive in formerly unsuitable areas; the cow is now the Kalahari's dominant mammal. With boreholes together with fencing, especially since the 1970s, the cattle populations have grown to impinge seriously on native wildlife, and concentrated grazing is opening the landscape to invasion of undesirable exotic plants and damaging truffle grounds (Khonga 2012).

Climate change will affect the Kalahari, including truffle-producing areas, potentially before the end of the twenty-first century. Precipitation may increase, but aridity may also increase because of higher temperatures and higher evaporation. As truffles have been observed to fruit in the depressions between dunes, the predicted increased mobility of dunes in this century due to increased erodibility and wind energy (Thomas et al. 2005) could bury significant areas of truffle production. Significant dune movement is predicted to occur by 2039 in the southern dune field and in the eastern and northern dune fields by 2069. All dune fields from South Africa to Angola and Zambia will likely be activated by the end of the century.

The serious problems for sustainable production, present and future, are the consequence of environmental factors and human interventions. Restoring and maintaining the health of the truffle population demands answers from research, which has been sporadic at best over the past century. Thus it is gratifying that the Botswana Agricultural College at Gaborone has established the Mahupu Project, named for the native word for truffle. It aims to address "the dwindling yields of the Kalahari desert truffle due to drought, overgrazing, and overexploitation." The project is assessing (1) various tree species and cucurbits as symbionts of the truffle fungus, (2) effect of fencing to reduce overgrazing, and (3) use of supplemental irrigation to enhance truffle yields in the field (Khonga 2012).

References

Alvarado P, Moreno G, Manjón JL, Sánz MA (2011) *Eremiomyces magnisporus* (*Pezizales*), a new species from central Spain. Mycotaxon 118:103–111

Claridge AW, Trappe JM (2004) Managing habitat for mycophagous (fungus-feeding) mammals: a burning issue? In: Lunney D (ed) Conservation of Australia's forest fauna, 2nd edn. Roy Zool Soc New S Wales, Mosman, pp 936–946

Earth Observatory (2008) http://earthobservatory.nasa.gov/IOTD/view.php?id=35403. Accessed 10 Dec 2012

Ferdman Y, Aviram S, Roth-Bejerano N, Trappe JM, Kagan-Zur V (2005) Phylogenetic studies of *Terfezia pfeilii* and *Choiromyces echinulatus* (*Pezizales*) support new genera for southern African truffles: *Kalaharituber* and *Eremiomyces*. Mycol Res 109:237–245

Fischer E (1938) Tuberineae. Engler and Prantl, Nat. Pflanzenfam. 2 ed., 5bVIII, p. 38

Hennings P (1897) Fungi camerunensis. I. Engler Bot Jahrb Syst Pflanzengesch Pflanzengeogr 22:72–111

Kagan-Zur V, Kuang J, Tabak S, Taylor FW, Roth-Bejerano N (1999) Potential verification of a host plant for the desert truffle *Terfezia pfeilii* by molecular methods. Mycol Res 103:1270–1274

Kagan-Zur V, Taylor FW, Roth-Bejerano N (2001) Identification of plant hosts of the Kalahari desert-truffle *Terfezia pfeilii*. In: Fédération française des trufficulteurs (eds.) Proceedings of the fifth international congress, science and cultivation of truffle. 4–6 Mars 1999, Aix-en-Provence, pp. 82–86

Kayalangalilwa B, Boatwright J (2013) Phylogenetic position and revised classification of *Acacia s. l.* (Fabaceae: Mimosoideae) in Africa, including new combinations in *Vachellia* and *Senegalia*. Bot J Linn Soc 172:500–523

Kovács GM (2007) Identification of host plants and description of sclerotia of the truffle *Mattirolomyces terfezioides*. Mycol Progr 6:19–26

Kovács GM, Trappe JM, Alsheikh AM, Hansen K, Healy RA, Vági P (2011) *Terfezia* disappears from the American truffle mycota as two new genera and *Mattirolomyces* species emerge. Mycologia 103:831–840

Khonga EB (2012) Mahupu project. http://www.bca.bw/csp/research_projects/mahupu-project_csp.php. Accessed 10 Dec 2012

Lee RB (1979) The !Kung San–men, women, and work in a foraging society. Cambridge Univ Press, Cambridge, UK

Lee RB, DeVore I (1976) Kalahari hunter-gatherers: studies of the !Kung San and their neighbors. Harvard Univ Press, Cambridge, MA

Leistner OA (1967) The Plant Ecology of the Southern Kalahari. Bot Surv S Africa Mem 38. The Government Printer, Pretoria

Logan RF, Silberbauer GB (2013) Kalahari. http://www.britannica.com/EBchecked/topic/309972/Kalahari. Accessed 19 Aug 2013

Marasas WFO, Trappe JM (1973) Notes on southern African Tuberales. Bothalia 11:139–141

Mshigeni KE (2001) The cost of scientific and technological ignorance with special reference to Africa's rich biodiversity. Univ. of Namibia, Windhoek

Mshigeni KE, Mtango D, Massele A, Mgonda Y, Lyamuya AE, Elineema M, Chang ST (2005) Intriguing biological treasures more precious than gold: the case of tuberous truffles, and immunomodulatin *Ganoderma* mushrooms with potential for HIV/AIDS Treatment. Discov Innov 17:1–7

Pike JG (1971) The development of the water resources of the Okavango Delta. Botswana Notes Record, Special Ed 1:35–40

Silberbauer GW (1981) Hunter and habitat in the central Kalahari desert. Cambridge Univ Press, London

Story R (1958) Some plants used by the Bushmen in obtaining food and water. Bot Surv S Africa Mem 38. The Government Printer, Pretoria.

Taylor D, Thamage M, Baker N, Roth-Bejerano N, Kagan-Zur V (1995) Notes on the Kalahari Desert Truffle, *Terfezia pfeilii*. Mycol Res 99:874–878

Thomas DSG (2002) Sand, grass, thorns, and…cattle: the modern Kalahari environment. In: Sporton D, Thomas D (eds) Sustainable livelihoods in Kalahari environments. Oxford Univ. Press, Oxford, pp 21–38

Thomas DSG, Knight M, Wiggs GFS (2005) Remobilization of southern African desert dune systems by twenty-fist century global warming. Nature 435:1218–1221

Thomas DSG, Shaw PA (1991) The Kalahari environment. Cambridge Univ Press, Cambridge, UK

Trappe JM, Claridge AW (2005) Hypogeous fungi: evolution of reproductive and dispersal strategies through interactions with animals and mycorrhizal plants. In: Dighton J, White JF, Oudemans P (eds) The fungal community—its organization and role in the ecosystem. Taylor & Francis, Boca Raton, FL, pp 613–623

Trappe JM, Claridge AW, Arora D, Smit WA (2008a) Desert truffles of the African Kalahari: ecology, ethnomycology, and taxonomy. Econ Bot 62:521–529

Trappe JM, Claridge AW, Claridge DL, Liddle L (2008b) Desert truffles of the Australian Outback: ecology, ethnomycology, and taxonomy. Econ Bot 62:497–506

Trappe JM, Kovács GM, Claridge AW (2010a) Comparative taxonomy of desert truffles of the Australian Outback and the African Kalahari. Mycol Progr 9:131–143

Trappe JM, Kovács GM, Claridge AW (2010b) Validation of the new combination *Mattirolomyces austroafricanus*. Mycol Progr 9:145

Verwoerd L (1925) A contribution to our knowledge of the South African Hymenogastraceae. S Afr J Sci 22:161–166

Chapter 14
Ecology and Distribution of Desert Truffles in the Australian Outback

Andrew W. Claridge, James M. Trappe, and David J. Paull

14.1 Introduction

Australia's ten deserts collectively encompass nearly 1.4 million km^2 (van Oosterzee 2000). Adding to that the extensive semiarid lands characterizing much of the rest of the continent, about 70 % of continental Australia is classified as arid or semiarid, making it the driest-inhabited continent (Geoscience Australia 2012). Yet conservative estimates award Australasia the greatest richness of hypogeous fungi (hereafter truffles) and the greatest number of endemics of all the world's regions (Mueller et al. 2007). Most of that Australian diversity comes from the 30 % of the continent covered by forests and woodlands subject to a temperate climate. For example, Claridge et al. (2000) sampled truffles on 136 plots of 0.1 ha each in forests of eastern Victoria and south-eastern New South Wales. These plots averaged nine species, with some having up to 17 species (with later sampling included,

A.W. Claridge (✉)
Office of Environment and Heritage, NSW National Parks and Wildlife Service, Nature Conservation Section, P.O. Box 733, Queanbeyan, NSW 2620, Australia

School of Physical, Environmental and Mathematical Sciences, University of New South Wales, Northcott Drive, Canberra, ACT 2600, Australia
e-mail: Andrew.Claridge@environment.nsw.gov.au

J.M. Trappe
Department of Ecosystems and Society, Oregon State University, Corvallis, OR 97331-5752, USA

USDA Forest Service, Pacific Northwest Research Station, Forestry Sciences Laboratory, Corvallis, OR 97331, USA
e-mail: trappej@onid.orst.edu

D.J. Paull
School of Physical, Environmental and Mathematical Sciences, University of New South Wales, Northcott Drive, Canberra, ACT 2600, Australia
e-mail: dpaull@adfa.edu.au

V. Kagan-Zur et al. (eds.), *Desert Truffles*, Soil Biology 38,
DOI 10.1007/978-3-642-40096-4_14,

the most productive plots had 25–31 species). The area represented by these plots is about 31,000 km^2, which, with additional sampling since the year 2000, has yielded overall about 60 genera and 300 species (Claridge and Trappe unpublished data). In contrast, Australia's 1.4 million km^2 of arid landscapes have so far produced only six genera and seven species of hypogeous fungi (Trappe et al. 2008b, 2010a). Thus, some individual 0.1 ha plots of the south-eastern forests sampled by Claridge et al. (2000) had more than four times the species of the 1.4 million km^2 arid areas of the Outback. To put it another way, the Outback arid area is 45 times larger than the south-eastern Australian forests studied by Claridge et al. (2000), but the forested areas have 10 times the genera and 43 times the species of the arid zone.

To some extent, the disparity described above can be partially explained by the limited formal survey so far carried out for truffles in Australia's deserts. Notwithstanding differing efforts, the disparity between species richness in Australia's forests and its deserts also evidences the restrictions imposed by desert environments on fungal evolution and distribution, especially truffle species that are presumably mycorrhiza formers with desert plants. The distribution of truffles, then, depends on where their suitable host plants occur. In Sect. 14.2, we summarize features of the Australian desert environment and the information, or lack of it, as needed to understand what governs the distribution of its truffles. In Sect. 14.3, we review what is known of the Outback desert truffles themselves, and in Sect. 14.4, we consider what may be the future outcomes of global warming and changing land use patterns on their distribution.

14.2 The Australian Deserts

14.2.1 Landscapes

Australian desert landscape formations are, for the most part, relatively flat to undulating and include large and extensive systems of sand ridges, interdunal swales, salt pans, alluvial flood plains and mostly ephemeral watercourses (Shepherd 1992, 1995; van Oosterzee 2000). However, in the centre and north-west of the continent, these flat landscapes are punctuated by large island mountains, or inselbergs, like Uluru (formerly Ayers Rock) and Kata Tjuta (formerly The Olgas). Other tall peaks occur within the McDonnell Ranges, also in Central Australia, with elevations up to 1,500 m above sea level (i.e. Mount Zeil), and the Hamersley Ranges in north-western Australia. Taken together, Australia's desert mountains cover about 580,000 km^2.

To the south-east of the McDonnell Ranges, the Simpson Desert comprises the largest set of sand ridges in the world, with individual dunes 10–35 m high, approximately 500 m apart, running parallel for several hundred kilometres in a NNW–SSE direction. A similarly vast sand belt occurs in the Great Victoria Desert, with dunes running parallel to one another in a W–E direction spanning the Western

Australia–South Australia border (Shepherd 1995). In the Tirari and Sturt Stony Deserts, by way of contrast to the large sand dune systems of the Simpson and Great Victorian Deserts, there are large areas of stony pavements known as gibber plains. These are formed from the remains of weathered rock or silcrete, slowly eroded over a long period of time by wind and water (Shepherd 1992).

Major river systems and associated floodplains include the Finke, which stems from the McDonnell Ranges and cuts a transect south-east to the western edge of the Simpson Desert, where it eventually flows to Lake Eyre, the fifth largest terminal lake in the world. Lake Eyre is also fed, in extreme flood events, by the major channel country formed by the Cooper, Diamantina and Georgina Rivers to the north-east of the Simpson and Sturt Stony Deserts.

The distribution of desert truffles within each of these major landforms is largely unknown. However, from the limited collections so far recorded, it would seem that preferred habitat mostly occurs in interdunal swales, along floodplains and associated watercourses. This most likely ties in with the distribution of woody trees and shrubs that are the putative hosts for these fungi.

14.2.2 Geology and Soil

The various soil formations of Australian deserts are derived from ancient bedrocks, mostly comprised of crystalline granites, gneisses and granite-derived sedimentary rocks such as sandstone at least 1,000 million years old and up to 4,200 million years old (van Oosterzee 2000). By area, the western two-thirds of the continent are formed from these Archean to Middle Proterozoic shields. Repeated faulting and sagging of these over millennia has resulted in the development of massive inland basins. Over time, these were filled with sediment mostly laid down by shallow seas that covered the Australian land mass for much of its geological history. Uluru itself is comprised of sandstone, laid down in an inland sea in the Cambrian Period approximately 500 million years ago. Subsequent earth uplifting and folding tipped the rock vertically, as evidenced by the layers now seen as a result of adjacent erosion of softer sediments. By way of contrast, adjacent Kata Tjuta is comprised of sedimentary conglomerate with iron oxide impurities.

While soil structure across the desert country is highly variable, the soils are universally infertile due to lack of replenishment by either volcanic or glacial activity. In the Simpson Desert, for example, coarse sands dominate the dune fields, while interdunal areas carry mostly sand with a higher clay content. Where coarse sand dominates the soil profile, water is usually free-draining down to the underlying water table. Such soil microhabitat is unlikely to be favourable for sustained truffle production. Higher clay contents can be found in soils along floodplains and watercourses, where finer sediments are deposited and redeposited over time. These microhabitats have a greater ability to retain moisture following rainfall. In cases where there has been ongoing deposition of fine sediments, clay pans form such as that occurring through the Channel Country to the north-west of the Simpson and

Sturt Stony Deserts. In cases where these sediments have high levels of nutrients such as calcium, magnesium and sodium, subsequent evaporation results in soils with very high salt levels—again, very unlikely to support desert truffles.

14.2.3 Climate

The climate of Australian desert regions is characterized by extremely low and intermittent rainfall and extremes of temperature. In this respect, it does not differ much from desert regions elsewhere around the world. On average, Australian deserts receive less than 200 mm of rainfall per annum and, in some cases, less than half of that amount (Shepherd 1995; van Oosterzee 2000). Effective rainfall is further reduced by the high evaporation rates caused principally by high daytime temperatures and low humidity: with 2,500–4,000 mm per annum being lost. When rain does fall, the timing is best described as unpredictable. Thus, Australian deserts may receive well above average rainfall in 1 year, then nothing the next. Winter rainfall events tend to occur from major frontal systems travelling across southern Australia, from west to east, while summer rainfall generally originates from localized thunderstorms or rain-bearing depressions representing the tail-end of cyclonic or tropical activity moving south from northern Australia. Temperatures also fluctuate greatly. In some locations maximum daytime temperatures in summer may reach 50 °C, whereas in winter frosts are common with overnight minimum temperatures as low as −6 °C (van Oosterzee 2000).

The unpredictability of rainfall and extremes of temperature in these deserts contrasts greatly with the more moderate climates of coastal and near-coastal mainland Australia. In these latter environments, annual rainfall is for the most part substantially higher and far more predictable. There, fruiting patterns of truffles generally follow a seasonal pattern relating to seasonality of rainfall: in southern Australia, there are definite peaks following autumn and winter precipitation (Claridge et al. 1993, 2000), while further north truffles mainly fruit in spring and early summer (Vernes et al. 2001). In Australian deserts, it is highly unlikely that truffles fruit with any similar sort of regularity, but do so rather in response to individual rainfall events (see Sect. 14.3.3 below). This increases the challenge of further cataloguing the desert truffle flora, because the timing of surveys needs to be flexible.

14.2.4 Plants

Australian deserts and surrounding semiarid lands play host to a huge diversity of plant species, many of which are endemic to these landscapes. For example, the Simpson Desert alone is host to over 800 plant species (van Oosterzee 2000). Perhaps the most characteristic vegetation formation of the desert country, and certainly the most widespread, is the spinifex grasslands variously dominated by a mix of different plant species in the genera *Triodia* and *Plectrachne*. These occupy approximately 22 % of the entire continental land mass and are commonplace on the sand plains and dune fields of the five major deserts: Great Sandy, Great Victoria, Gibson, Tanami and Simpson (van Oosterzee 2000). Spinifex grasslands flourish under conditions of low rainfall and infertility and are highly flammable due to their inherent dryness and volatile resins. As a consequence, fire intervals of less than 5 years are commonplace. For the most part, spinifex grasslands grow on stable slopes of sand dunes or between interdunal corridors. Within these land formations, scattered shrubs may occur, within genera such as *Acacia* (i.e. mulga, *A. aneura*), *Allocasuarina* (i.e. desert oak, *A. decaisneana*), *Eremophila* and *Grevillea*. On dune crests with looser sand, spinifex is typically replaced by other plant species, for instance, desert canegrass (*Zygochloa paradoxa*) in the Simpson Desert.

Less sandy corridors between dune fields and ephemeral watercourses, containing a higher clay content, often give rise to woodlands or tall open shrublands dominated by species such as the coolibah (*Eucalyptus microtheca*), Georgina Gidgee (*Acacia georginae*), *A. aneura* and various species of *Hakea*. In places where permanent water occurs, ribbons of River Red Gum (*E. camaldulensis*) occur. Elsewhere, chenopod shrublands dominated by saltbush (*Atriplex*), bluebush (*Maireana*) and other succulents such as *Chenopodium*, *Rhagodia* and *Sclerolaena* occupy sites rich in nutrients such as calcium, magnesium and sodium. Among these perennial shrublands, which occur over an area of around 434,000 km^2, scattered trees such as *Allocasuarina* or Rosewood (*Heterodendrum olefolium*) occur.

As stated elsewhere in this paper, the mycorrhizal status of Australia's desert truffles is unknown. However, if we presume that most species are likely ectomycorrhizal, the logical putative plant hosts would be the species within the genera *Acacia*, *Allocasuarina* and *Eucalyptus*, as is seen in temperate landscapes elsewhere in Australia (Claridge et al. 2000; Jumpponen et al. 2004). Closer examination of the trophic status of desert truffles by techniques, such as isotopic analysis, would provide further clarification in this regard (Hobbie et al. 2001).

14.3 The Australian Desert Truffles

In Sect. 14.1, we noted six genera and seven species which have so far been recorded for the Outback deserts that cover about 1.4 million km^2. Thus, the Outback deserts are genus-rich (see Sect. 14.3.2) compared to the Kalahari with three genera, *Eremiomyces*, *Kalaharituber* and *Mattirolomyces* (see Chap. 13 by Trappe et al.; Trappe et al. 2008a); the North African and West Asian with four, *Mattirolomyces*, *Picoa*, *Terfezia* and *Tirmania* (Alsheikh 1994; Alsheikh and Trappe 1983; Ammarellou et al. 2011; Kovács et al. 2011a, b); and the North American with three, *Carbomyces*, *Mattirolomyces* and *Stouffera* (Trappe and

Weber 2001; Kovács et al. 2011a, b). Data are not available for deserts in East Asia or South America.

14.3.1 History of Discovery

Archaeological discoveries reveal that the Australian Aborigines have inhabited and explored these arid and semiarid landscapes for between 20,000 and 30,000 years, depending on location (Flood 2004). We can reasonably surmise that they knew of and treasured desert truffles for food over many centuries, probably millennia. The early Euro-Australian explorers of the mid-nineteenth century encountered and were often aided by Aboriginal tribes who had adapted well to living in these landscapes (Lindsay 1893; Murgatroyd 2002). However, the earliest desert truffle herbarium collection we could find was in the National Herbarium of Victoria from the Elder Exploring Expedition of 1891, collected in the Great Victoria Desert in June of that year. The report of that expedition does not mention truffles, but specimens were sent to M.C. Cooke in England, who with G. Massee described it (Cooke 1892) as *Stephensia arenivaga* Cooke & Massee. McLennan (1961) subsequently and correctly regarded the species as deserving a new genus, which she described and named *Elderia* in honour of the sponsor of the expedition and to which she transferred the species as *Elderia arenivaga* (Cooke & Massee) McLennan. The collection was large enough to split: specimens or parts of specimens have been lodged in the Royal Botanic Gardens, Kew, England; State Herbarium of South Australia, Adelaide; National Herbarium of Victoria, South Yarra, Australia; and Herbarium, Department of Botany and Plant Pathology, Oregon State University, Corvallis, Oregon, USA.

14.3.2 Taxonomy, Endemism and Distribution

All taxa of Australian desert truffles are described in detail, illustrated, discussed and keyed by Trappe et al. (2008a, 2010a). Five of the six genera of Australia are in the Ascomycota, Pezizales: *Elderia*, *Mattirolomyces*, *Mycoclelandia* and *Ulurua* in the Pezizaceae, and *Reddellomyces* in the Tuberaceae. It is the only Tuberaceae member known to have been found in a desert region, though not in a typical site; see below. The sixth genus, *Horakiella*, is in the Basidiomycota, Boletales, Sclerodermataceae, and as such is a 'false truffle'.

Elderia (one species, *E. arenivaga*), *Mycoclelandia* (two species, *M. arenacea* [Trappe] Trappe & G.W. Beaton and *M. bulundari* [G.W. Beaton] Trappe & G.W Beaton) and *Ulurua* (one species, *U. nonparaphysata* Trappe, Claridge & Kovács) are endemic to Australian deserts. *E. arenivaga* (Fig. 14.1), represented by four collections, is widespread, with collections from the Great Sandy, Great Victoria, Simpson and Tanami Deserts, thus spanning the continent from about 21–27°S

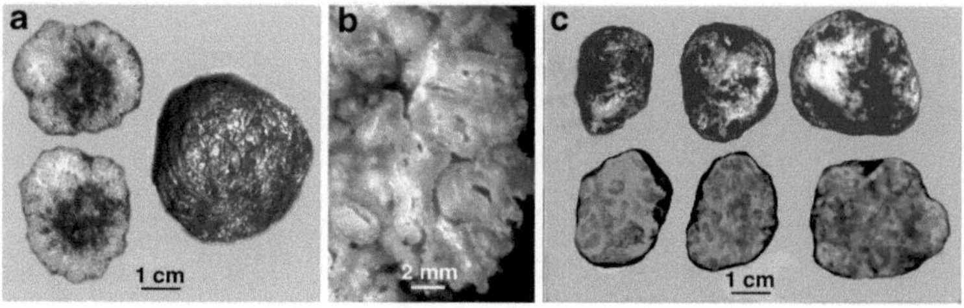

Fig. 14.1 Images of fruiting bodies of Australian desert truffles: (**a**) *Mycoclelandia arenacea* ascocarps, cross sections on *left*, surface view on *right*; (**b**) *Mycoclelandia arenacea*, cross section *close-up*; (**c**) *Reddellomyces westraliensis* ascocarps, surface views *above*, cross sections *below*

latitude and 128–134°E longitude so far as presently known (Fig. 14.2). The five collections of *Mycoclelandia arenacea* (Fig. 14.1) have been found only in the Great Sandy Desert, about 22–25°S latitude and 128–131°E longitude. *M. bulundari* is represented so far by four collections, two each from the Great Sandy and Tanami Deserts, about 20–26°S latitude and 127–131°E longitude. *Ulurua nonparaphysata* is known from only a single collection in Uluru–Kata Tjuta National Park (Ayers Rock) in the semiarid zone between the Great Victoria and Great Sandy Deserts, so the extent of its likely distribution cannot be estimated.

The other two genera are more widely distributed. *Reddellomyces westraliensis* (Fig. 14.1) has been found in dry habitats from south-eastern Western Australia, southern South Australia and New South Wales, in addition to a single collection in desert areas in the Australian Northern Territory south of Alice Springs. However, it occurred under *Eucalyptus* spp. in a moist swale along a creek 3 weeks after heavy rains, when the water had temporarily risen in the creek to flood the swale. *Mattirolomyces* is a cosmopolitan genus of dry to arid habitats of southern Europe, the Kalahari of southern Africa, Asia and North America (Alsheikh 1994; Kovács 2007; Kovács et al. 2011a, b; Trappe et al. 2010a,b). The sole Australian representative of this genus, *M. mulpu*, is known from one collection in the Tanami Desert and one in the eastern Great Sandy Desert, both in the Northern Territory; it is endemic to the Outback deserts.

Horakiella watarrkana Trappe & Claridge (Sclerodermataceae) is the sole Australian hypogeous Basidiomycete species found so far in Australia arid/semi-arid habitats and is represented by one collection from Watarrka National Park at 24°17′S latitude, 131°33′E longitude at the eastern fringe of the Great Sandy Desert in the Northern Territory. The only information about it is from the collection note, which gives only the date, place name and collector's name.

In brief, 17 collections offer the entire available documentation of truffle distribution in the 1.4 million km^2 of deserts in the Australian Outback. As shown in Fig. 14.2, these collections span a vast area, but what lies between and outside the map points remains to be determined. Even with a road system substantially expanded over the last century, huge areas of inaccessible arid land hinder mycological exploration. Add to this, the scant and erratic rainfall patterns may leave much of the area too dry for fungal fruiting in many, perhaps most years. The best

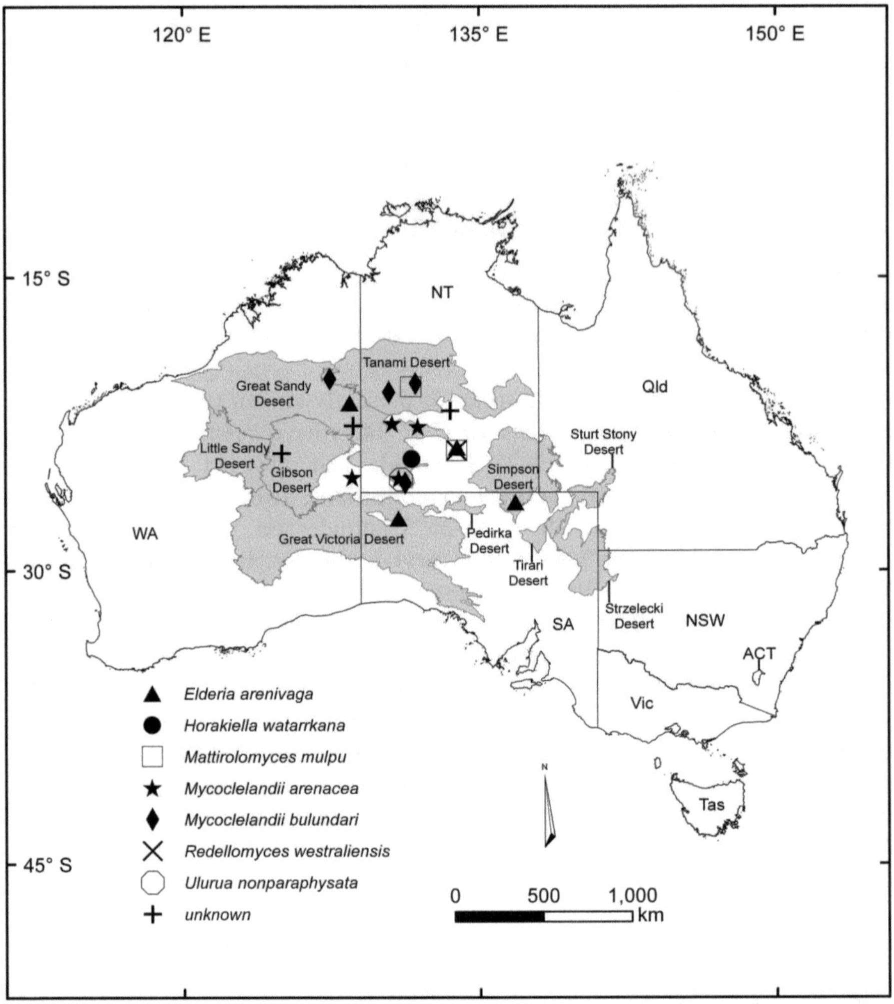

Fig. 14.2 A map of the deserts of Australia, adapted from the Interim Biogeographic Regionalisation of Australia Version 6.1. Approximate locations of truffle collections indicated. Unvouchered collections indicated as "unknown"

sources of information about fungal fruiting patterns would be the Aborigines whose ancestors inhabited these lands and handed down knowledge of desert truffles over so long a time. However, these traditions seem to be fading as Aboriginal lifestyles change (Trappe et al. 2008b).

14.3.3 Ecology: Key to Distribution

Only a few of the 17 available collections of Australian desert truffles are accompanied by data on weather, habitat and plant associations. What little is known about these features of Australian deserts in relation to truffle fruiting has been brought together by Trappe et al. (2008b). In summary, the substrates appear to be sands, variously termed as sandy soil, sand plains, dunes or at the foot of dunes. In one case, red sands are specified.

Rainfall patterns are particularly important for truffle fruiting, as would be expected in a desert environment (Claridge et al. 2000). The Australian Bureau of Meteorology (2012) records provide possible insights on the timing of truffle fruiting in deserts, at least at sites near government weather stations. We have yet to analyse these data formally, but for three desert sites in 2000, truffles were collected 6–8 weeks following 3–5 consecutive days of rain totalling 155–350 mm. These rainstorms were preceded and followed by many weeks of dry weather. Similar soil wetting rains occurred in 3 of the 10 years preceding 2000 and 4 of the 10 years following 2000. Large monthly or yearly total rainfall is less important than concentrated rain events, because high evaporation can quickly dry soils after scattered, low-intensity showers in desert environments. In any event, rainstorms likely to induce truffle fruiting occurred during less than half of the 10 years preceding and following 2000, the year represented by actual truffle collections.

Dates of the truffle collections at the three sites discussed above were also revealing. The collections were in May and June in 2000, as the autumn to early winter temperatures were nearing the coolest in the year. The mean daily temperatures for those months averaged ca. 19 °C, the highest daily temperatures averaged 26 °C and the lowest, ca. 15 °C.

None of the Australian desert truffles have yet been demonstrated to form mycorrhizae, but judging from desert truffles elsewhere (see Chap. 5 by Roth-Bejerano et al.), they are likely to be mycorrhiza formers. This would strongly influence distribution. Potentially ectomycorrhizal woody plants that occur on some habitats in the truffle-producing areas include *A. aneura*, witchetty bush (*A. kempeana*), *A. decaisneana*, Desert Bloodwood (*Corymbia terminalis*) and Blue Mallee (*Eucalyptus gamophylla*) (Parks Australia 2012). Other hosts or other types of mycorrhizae cannot be discounted, but none have been studied in this respect.

14.4 The Outlook for Australian Desert Truffles

Fire, crop production, grazing, feral animals (especially rabbits and camels) and urbanization could all potentially affect the distribution and abundance of Australian desert truffles. In the past, traditional burning practices by Aborigines usefully maintained biodiversity and habitat mosaics (Bowman 1998). These burns were usually done in winter to maintain the desired small burn size and low burning intensity (Bird et al. 2005; Bliege et al. 2008). With the fracturing of Aboriginal communities following European human settlement of the continent, the application of these patchy prescribed fires has diminished. Lightning storms are an integral part of the desert environment, so wildfires are commonplace. In contrast to traditional prescribed burning, these wildfires are often large and intense. Depending on vegetation type, habitat recovery following wildfire may take 25 or more years (Edwards and Allan 2009). Data on habitat requirements or mycorrhizal hosts of Australian desert truffles are too scant, however, to even speculate on effects of prescribed burning or wildfire on truffle distribution and production in either the short or long term.

For the same reasons, effects of crop production, grazing, feral animals or invasive exotic plants on short- or long-term truffle production cannot be forecast. All have in the past and often now adversely affect desert plant communities (Hatton et al. 2011). Urbanization of truffle grounds will certainly depress or eliminate truffles when practised on truffle-producing areas, as would crop agriculture. However, the 1.4 million km^2 of Outback desert contains only a few small towns and villages, and low and erratic water availability limits the area occupied by development. Consequently, these activities likely have negligible impact on truffle production now and for the foreseeable future. Overgrazing by livestock and feral animals in some areas could adversely affect truffle production by changing composition of plant communities or deteriorating soil quality.

Truffle harvesting by Aboriginal desert inhabitants or visitors seems unlikely to adversely affect truffle productivity for the same reasons truffles in other regions and continents withstand harvesting: their fruiting bodies have evolved to be dug up as a spore dispersal strategy. Moreover, the desert areas are huge and the human populations are low, large tracts are difficult of access, and the Aboriginal tradition of truffle hunting appears to be waning (Trappe et al. 2008b). We are unaware of desert truffles being harvested for sale in Australia as they are in the African Kalahari (Trappe et al. 2008a).

One factor that may have contributed to a decline in desert truffle distribution since European human settlement is the loss of primary dispersal agents such as medium-sized mammals. For example, populations of ground-dwelling marsupials such as the golden bandicoot (*Isoodon auratus*), bilby (*Macrotis lagotis*) and burrowing bettong (*Bettongia lesueur*) have all greatly declined in most arid zone areas across Australia, while others such as the desert bandicoot (*Perameles eremiana*) and broad-faced potoroo (*Potorous platyops*) have become extinct altogether (Johnson et al. 1989; Johnson 2006). Changing fire regimes and the introduction of feral predators such as the red fox (*Vulpes vulpes*) and feral cat (*Felis catus*) have all been implicated as causes of these declines. At least some of these marsupial species are known to consume truffles or were in all likelihood partially mycophagous, playing a vital role in truffle spore dispersal (Claridge and May 1994). Resurrecting populations of marsupials such as bandicoots, bettongs and potoroos in desert environments may therefore be an important conservation management initiative.

The aforementioned examples of drivers of environmental change (harvesting excepted) may each affect desert truffle sustainability in localized areas in ways unpredictable: we know too little about the diversity and ecology of those truffles to forecast their response to disturbance and environmental change. Climate change, however, will have large-scale effects on Australian desert ecosystems within this century (Hatton et al. 2011). Those effects will generally include increased temperatures and changes in precipitation. Because the Outback deserts encompass so large an area, considerable variability in climate change is expected from both north to south and east to west. It is beyond the scope of our paper to go into details of these predictions, especially because we cannot yet predict effects on truffles. We recommend the *State of the Environment 2011* report published by the Australian Department of Sustainability, Environment, Water, Population and Communities (Hatton et al. 2011) for many fascinating predictions in different scenarios of climate change.

Acknowledgements We thank Dr. David Albrecht for providing the images of *Mycoclelandia arenacea* shown in Fig. 14.1.

References

Alsheikh AM (1994) Taxonomy and mycorrhizal ecology of the desert truffles in the genus Terfezia. Ph.D. dissertation, Oregon State University

Alsheikh AM, Trappe JM (1983) Desert truffles: the genus *Tirmania*. Trans Br Mycol Soc 81:83–90

Ammarellou A, Smith ME, Tajick MA, Trappe JM (2011) The phylogenetic placement of *Picoa*, with a first report on *Picoa lefebvfei* (Pat.) Maire (=*Phaeangiumlefebvrei*) from Iran. Int J Environ Res 5:509–514

Bird DW, Bliege Bird R, Parker CH (2005) Aboriginal burning regimes and hunting strategies in Australia's western desert. Human Ecol 33:444–464

Bliege Bird R, Bird DW, Codding BF, Parker CH, Jones JH (2008) The "fire stick farming" hypothesis: Australian Aboriginal foraging strategies, biodiversity, and anthropogenic fire mosaics. Proc Natl Acad Sci 105:14796–14801

Bowman DJMS (1998) Tansley Review No. 101. The impact of Aboriginal landscape burning on the Australian biota. New Phytol 140:385–410

Bureau of Meteorology (2012) Climate data online. http://www.bom.gov.au/climate/data/index.shtml. Accessed 15 Oct 2012

Claridge AW, Robinson AP, Tanton MT, Cunningham RB (1993) Seasonal production of hypogeal fungal sporocarps in a mixed-species eucalypt forest stand in south-eastern Australia. Aust J Bot 41:145–167

Claridge AW, May TW (1994) Mycophagy by Australian mammals. Aust J Ecol 19:251–275

Claridge AW, Cork SJ, Trappe JM (2000) Diversity and habitat relationships of hypogeous fungi. I. Study design, sampling techniques and general survey results. Biodivers Conserv 9:151–172

Cooke MC (1892) Australian fungi. Grevillea 21:35–39

Geoscience Australia (2012) Deserts. http://www.ga.gov.au/education/geoscience-basics/landforms/deserts.html. Accessed 12 Oct 2012

Edwards GP, Allan GE (eds) (2009) Desert fire: fire and regional land management in the arid landscapes of Australia. Desert Knowledge Cooperative Research Centre, Alice Springs

Flood J (2004) Archaeology of the dreamtime: the story of prehistoric Australia and its people. J.B. Books, Marlestone, South Australia

Hatton T, Cork S, Harper P, Joy R, Kanowski P, Mackay R, McKenzie N, Ward T, Wienecke B (2011) State of the Environment 2011. http://www.environment.gov.au/soe/2011/report/index.html. Accessed 29 Oct 2012

Hobbie EA, Weber NS, Trappe JM (2001) Determining mycorrhizal or saprotrophic status of fungi from isotopic evidence: implications for element cycling and fungal evolution. New Phytol 150:601–610

Johnson CN (2006) Australia's mammal extinctions: a 50,000 year history. Cambridge University Press, New York

Johnson KA, Burbidge A, McKenzie NL (1989) Australian Macropodoidea: status, causes of decline and future research and management. In: Grigg G, Jarman P, Hume I (eds) Kangaroos, wallabies and rat-kangaroos. Surrey Beatty and Sons, New South Wales, pp 641–657

Jumpponen A, Claridge AW, Trappe JM, Lebel T, Claridge DL (2004) Ecological relationships among hypogeous fungi and trees: inferences from association analysis integrated with habitat modeling. Mycologia 96:510–525

Kovács GM (2007) Identification of host plants and description of sclerotia of the truffle *Mattirolomyces terfezioides*. Mycol Progr 6:19–26

Kovács GM, Balázs TK, Calonge FD, Martín MP (2011a) The diversity of *Terfezia* desert truffles: new species and a highly variable species complex with intrasporocarpic nrDNA ITS heterogeneity. Mycologia 103:841–853

Kovács GM, Trappe JM, Alsheikh AM, Hansen K, Healy RA, Vági P (2011b) *Terfezia* disappears from the American truffle mycota as two new genera and *Mattirolomyces* species emerge. Mycologia 103:831–840

Lindsay DM (1893) Journal of the elder exploring expedition, 1891. http://pds.lib.harvard.edu/pds/view/12803090. Accessed 18 Oct 2012

McLennan EI (1961) Australian tuberales. Proc Roy Soc Vic 74:111–117

Mueller GM, Schmit JP, Leacock PR, Buyck B, Cifuentes J, Desjardin DE, Halling RE, Hjortstam K, Iturriaga T, Larsson K-H, Lodge DJ, May TW, Minter D, Rajchenberg M, Redhead SA, Ryvarden L, Trappe JM, Watling R, Wu Q (2007) Global diversity and distribution of macrofungi. Biodivers Conserv 16:37–48

Murgatroyd SP (2002) The dig tree. Broadway, New York

Parks Australia (2012) Flora. http://www.environment.gov.au/parks/uluru/nature/flora.html. Accessed 15 Oct 2012

Shepherd M (1992) The Simpson Desert: Natural History and Human Endeavour. The Royal Geographic Society of Australasia. Adelaide, South Australia

Shepherd M (1995) The Great Victoria desert. Reed Books, Chatswood

Trappe JM, Claridge AW, Arora D, Smit WA (2008a) Desert truffles of the African Kalahari: ecology, ethnomycology, and taxonomy. Econ Bot 62:521–529

Trappe JM, Claridge AW, Claridge DL, Liddle L (2008b) Desert truffles of the Australian Outback: ecology, ethnomycology, and taxonomy. Econ Bot 62:497–506

Trappe JM, Kovács GM, Claridge AW (2010a) Comparative taxonomy of desert truffles of the Australian Outback and the African Kalahari. Mycol Progr 9:131–143

Trappe JM, Kovács GM, Claridge AW (2010b) Validation of the new combination *Mattirolomyces austroafricanus*. Mycol Progr 9:145

Trappe JM, Weber NS (2001) North American desert truffles: the genus *Carbomyces* (Ascomycota, Carbomycetaceae). Harvard Pap Bot 6:209–214

Van Oosterzee P (2000) The Centre: the natural history of Australia's desert regions. J.B. Books, Marlestone, South Australia

Vernes K, Castellano M, Johnson CN (2001) Effects of season and fire on the diversity of hypogeous fungi consumed by a tropical mycophagous marsupial. J Anim Ecol 70:945–954

Part IV
Fruit-Body Attributes

Chapter 15
The History of Desert Truffle Use

Elinoar Shavit

15.1 Introduction

The nineteenth-century writer Alexandre Dumas must have had desert truffles in mind when he said that to tell the story of the truffle was to tell the story of civilization (Dumas and Colman 1958). The records of the use of desert truffles by indigenous populations from antiquity until the present follow the story of civilization, beginning with the cuneiforms (inscribed clay tablets) left by the Bronze Age Amorites, to the Bedouins who now gather desert truffles in the same places (Sasson 2004; Lönnqvist 2004). The Bedouins have the longest recorded history of desert truffle use and have demonstrated a rich desert truffle culture (Mandaville 2011; Shavit 2008).

Indigenous people have used desert truffles wherever the truffles grow, but most of these cultures were strictly oral and have not left written records. This is the case with the indigenous people of the African Kalahari and the Aborigines of Australia. Despite their obviously long history of desert truffle use, little is known about either the truffles or how they were used by the indigenous people on both continents (Marasas and Trappe 1973; Trappe et al. 2008a, b).

The ancient desert truffle cultures of the Bedouins of the Middle East, the Khoisan (Bushmen in English, Basrawa in Setswana, the official Botswana language) of the African Kalahari, and the Aborigines of the Australian Outback represent entirely separate desert truffle cultures on three continents. However, they seem to share numerous similarities, including the methods by which the truffles are collected, cooked, and used (Trappe et al. 2008a, b).

Desert truffles have well-developed markets in the Middle East, and even though the quantities of the truffles in the wild have been diminishing, their popularity has

E. Shavit (✉)
FUNGI Magazine, Concord, MA, USA

North American Mycological Association, Watsonville, CA, USA
e-mail: elinoar.shavit@gmail.com

V. Kagan-Zur et al. (eds.), *Desert Truffles*, Soil Biology 38,
DOI 10.1007/978-3-642-40096-4_15, © Springer-Verlag Berlin Heidelberg 2014

been rising. Present-day Bedouins embrace their desert truffle culture, though fewer families now rely on the truffles for sustenance. The Kalahari truffle (*Kalaharituber pfeilii*) is a local favorite, its popularity has been growing, and it has limited local markets and some success in the European markets (Dixon 2006; Trappe et al. 2008b). Ecotourism that has developed in this region of Africa now includes excursions to learn about the culture of the indigenous people of the Kalahari and to gather and cook Kalahari truffles with the Khoisan. In contrast, there does not seem to be either a market for the native desert truffles in Australia or a strong interest in developing one (Trappe et al. 2008a; Latz 1995; Morse 2005). Some efforts to adapt the Kalahari concept of cultural ecotourism to the Australian Outback to help preserve the traditions of the Aborigines and their use of desert truffles are still in their initial stages (Latz 1995; Morse 2005). These efforts mainly concentrate on the hope that the successful introduction of European forest truffle farms to Australia will generate interest in the country's native desert truffles as well.

15.2 Some Characteristics of Culinary Desert Truffles

Desert truffles are hypogeous Ascomycetes that grow in arid and semiarid areas on every continent other than Antarctica (Kagan-Zur 2001). The desert truffles with the longest recorded history of use are species of *Terfezia* and *Tirmania*, which are popular mainly in the Middle East, Mediterranean basin, and Northern Africa (Diez et al. 2002; Kagan-Zur 2001). These desert truffles form mycorrhizas on the roots of plants in the Cistaceae family (Kagan-Zur 2001). They produce fruit bodies after the rainy season, and their fruiting is affected by the amount and distribution of rainfall (Mandeel and Al-Laith 2007; Trappe et al. 2008a). Desert truffle fruit bodies ripen just beneath the surface of the sand. They push up a thin crust as they swell, forming typical cracked bumps on the surface and often emerging onto it. It is easy to learn to recognize these telltale signs (Trappe 1990; Jabbur 1995). Desert truffles are traditionally gathered in the early morning when the slanted angle of the sun's rays makes the protruding truffles easier to spot (Mandeel and Al-Laith 2007).

15.3 Early Evidence of Societies with a Desert Truffle Culture

15.3.1 The Bronze Age Amorites

The oldest society with a recorded history of desert truffle use is the Bronze Age Amorites. The Amorites, believed to be Semitic, nomadic, pastoral people, arrived to the area of the Middle Euphrates (mostly in present-day Syria) in the third millennium BCE (Lönnqvist 2008; Gottlieb et al. 1998). These nomads clashed with the Mesopotamians but gradually settled in and built a long-lasting, powerful empire (Gottlieb et al. 1998). Evidence that the Amorites had a well-established desert truffle culture is found in the second millennia BCE cuneiform of the Sumerian mythopoetic hymn, *The Marriage of Mart'u*, which depicts the integration phase of the nomadic Amorites into the Sumerian society (Sasson 2004; Chiera 1934). The poem is written in Akkadian, a Semitic language akin to Arabic, Hebrew, and Aramaic. In the poem, the Sumerians describe the Amorites as a primitive people who live in drafty tents, dress in skins, neglect to bury their dead, eat raw flesh, and... *dig for truffles at the foothills* (Chiera 1934).

Among over 20,000 cuneiforms discovered during excavations in the Bronze Age Amorite city of Mari (present-day Tell Hariri in Syria), which portray all aspects of life in the prosperous city during the reign of its last ruler Zimri-Lim (1775 BCE), were numerous mentions of desert truffles (Sasson 2004). The records indicate that many baskets of desert truffles were shipped to the palace from their gathering areas in the regions of the Euphrates and the Khabur Rivers and the steppes (Sasson 2004; Saggs 1965). Correspondence cuneiforms between the king and his governors, who were responsible for supplying the palace with truffles, indicate that these truffles were gathered following the spring rain showers. Cuneiforms (*ARM 27; ARM 54*) mention two distinct kinds of desert truffles that were gathered in the area of the Middle Euphrates: the favored ***kam'u*** (***kam'a-tum***) and the less appreciated ***gib'u*** (or ***gibbi***) (Sasson 2004; Oppenheim 1971). They indicate that the two kinds of truffles were gathered in the same areas, at the same time of year, and that expert truffle gatherers were needed to locate and tell them apart (Cuneiforms # ARM 27; ARM 54; NABU 1989/58, in Sasson 2004).

Considered a delicacy, desert truffles were served at the Amorite palace banquets (Saggs 1965). The cuneiforms deal with the details of amassing large amounts of the best desert truffles for the ruler (Oppenheim 1971). Failing to supply King Zimri-Lim with the best truffles (*kam'u*) by offering him the lesser quality ones could have dire consequences, as is evident from the anguish in the reply cuneiform sent by Yaqqim-Addu, Governor of Saggaratum, to the Amorite king:

> *Ever since I reached Saggaratum five days ago, I have continuously dispatched truffles to my lord. But my lord wrote to me: You have sent me bad truffles! But my lord ought not to condemn with regards to these truffles. I have sent my lord what they have picked for me!* (Sasson 2004).

The Amorite tradition of offering the desirable desert truffles to rulers as a gift is a tradition still practiced among the Bedouins today (Mori 2007). Mandaville (2011) writes that he has observed that when a Bedouin finds a particularly large white desert truffle (*Tirmania nivea*), he will gift it to a local authority figure.

While no specific truffle preparation instructions have been found among the Amorite cuneiforms, they offer a wealth of information on raw food preservation techniques, which include slicing and drying, pickling in salt and vinegar, and salting. These methods are still used by Bedouins to preserve desert truffles (Sasson 2004; Mandaville 2011).

15.3.2 The Desert Truffle-Gathering Areas of the Amorites

The area of Jebel Bishri, the biblical Mountain of the Amorites, has been a sociopolitical border between nomadic, pastoral people and village agriculturalists for millennia. The nomadic groups in the region today consist of Bedouin tribes, and both nomadic and settled *fellahin* villagers gather large quantities of desert truffles for food and as a source of supplementary income (Lönnqvist and Törmä 2004). Jabbur (1995), who grew up in a Syrian village at the fringe of the desert, tells that the Bedouins and the villagers would enthusiastically await the arrival of the desert truffle season, when all skirmishes were temporarily put on hold and people from both groups would gather truffles from early February to late April. The Bedouins would live almost exclusively on desert truffles when they were in season, also shipping them to the cities where they were sold at a high price (Fig. 15.1).

Travelers to the Syrian Desert attested to the importance of this large desert truffle-producing area to generations of nomadic herders and settled populations. The Italian aristocrat Ludovico di Varthema (1470–1517) describes how in the desert truffle season many caravans, often 30 camels long, would carry the truffles (locally called *kam'a*) gathered in this prolific area to the Damascus market, where they would all be sold within a few days (De Varthema et al. 1863). Burckhardt (1831) observed a particularly good desert truffle season in the *hammad*, the great plain between Damascus, Baghdad, and Basrah. He writes that the vast number of semi-buried truffles created such a dense array of bumps on the surface of the sand that the camels would have to step on them. Truffles from this area were transported by Bedouin caravans to markets in Mosul, Baghdad, and Damascus (Burckhardt 1831).

15.4 Desert Truffles in the Old Testament: Did the Biblical *Manna* Refer to Desert Truffles?

Desert truffles were among the gourmet foods served to the Egyptian pharaohs, but very few details concerning the truffles have survived (Trappe 1990). Better descriptions of the kind of desert truffles that the pharaohs of Egypt may have consumed, along with an ancient version of traditional truffle preparations still popular in North Africa and the Middle East, can be found in the Bible (Pegler 2002; Shavit and Volk 2007). Two sections in the Bible, *Exodus* (16) and *Numbers* (11:7–9), describe the daily food that God would provide to the Israelites in the desert. The *daily meat,* in the form of quail, would fly into their camp at dusk, and the

Fig. 15.1 Desert truffles in the Damascus market. Photographer: Chadwan Al Yaghchi. Courtesy of Chadwan Al Yaghchi, Amateur cook and food blogger at http://www.syrianfoodie.blogspot.com, all rights reserved

daily bread would be found on the desert floor as soon as the morning dew would rise. If not for an interpretation that was artificially inserted into the translation of the original text, the Israelites' *daily bread* would have been known as *desert truffles* and not as *manna* (Brueggemann 1999; Bradlaugh 2011; Shavit and Volk 2007).

The term *manna* does not exist in the original text in Exodus 16:15, and the verbatim translation of it reads:

> *When the Israelites saw **it**, they said to one another, "**What is it**? [man hu]" For they did not know what **it** was.* (The Holy Bible, 1989).

The King James Version of Exodus 16:15, on the other hand, includes the interpretation:

> *And when the children of Israel saw **it**, they said to one another; **it is manna**: for they wist not what it was* (The King James Bible 2012).

Man hu means ***what is it?*** in the Hebrew and Aramaic languages (Steinberg 1960). It is neither an independent term nor does it designate an object. Were it not for the erroneous conjunction of *man* (what) and *hu* (is it), the two parts that make up the Hebrew question *what is it?*, the term *manna* would not exist (Brueggemann 1999; Bradlaugh 2011; Shavit and Volk 2007).

The remaining text in *Exodus* (16: 13–35) and *Numbers* (11: 7–8) proceeds to answer the question, *what is it?* posed by the puzzled Israelites, offering them expert instructions for first-time users of desert truffles. For example: what should the Israelites look for? Round objects the color of coriander seeds. Where and when should they gather them? Off the sand in early morning after the dew rises.

What would these objects taste like? The essence of oil and slightly sweet. How should the objects be prepared? By roasting, baking, boiling in a pot, and pounding. What should the Israelites watch out for? Spoilage: the truffles should be consumed immediately and not be kept overnight, because truffles putrefy quickly in the desert heat and infest with maggots. What religious instruction should be followed? The Israelites should bring a pot of them as an offering of gratitude to God. Desert truffles are gathered in the early morning not only because the slanted rays of the rising sun make the bumps easier to spot, but also because the cooler temperature is easier on the gatherer and prolongs the shelf life of the gathered truffles (Mandeel and Al-Laith 2007).

In the seventh century, Islam's Prophet Muhammad said to his followers that

> *Truffles are manna which Allah the Glorious and Exalted sent down upon the people of Israel, and its juice is a medicine for the eyes* (Hadith Sahih Muslim).

15.4.1 Quail and Desert Truffles Along Egypt's Desert Shores

Exodus (Chap.12) tells that the Israelites escaped during Passover (springtime) to the desert by the shores of the sea (The Holy Bible 1989). This coincides with two relevant spring phenomena: the arrival of the migratory quail and the fruiting of the desert truffles (Feeney 2002; El-Din 2005). Hunting the migratory quail is still popular along the Mediterranean shores of the Egyptian Desert (El-Din 2005). So is the gathering of desert truffles, albeit on a much smaller scale due to the loss of desert truffle habitats over the past century (Feeney 2002; Morte et al. 2008). A young Bedouin bird hunter and truffle gatherer described how he learned to hunt with his father, saying,

> *At the same time as hunting for birds and gazelle we would gather a basket of terfas and roast them in the ashes of our nightly coffee fires* (Feeney 2002) (Fig. 15.2).

15.5 Discussions of Desert Truffles in the Early Classics: The Greeks, Romans, and the Talmud

15.5.1 Discussions of the Origin and Nature of Desert Truffles in the Early Greek Classics

Truffles seem to be shrouded in mystery, and they puzzled the ancient Greek scholars. Theophrastus of Eresus (371–287 BCE), known as the father of botany, placed both mushrooms and truffles in the department of *plants without roots* (Helttula 1996). Theophrastus called the truffle in his *Historia plantarum*

Fig. 15.2 *Tirmania nivea* and quail in the desert. Photographer: Ahmed Al-Aseeri. Courtesy of Sultan Al-Aseeri and Ahmed Al-Aseeri, truffle researchers from Qatar, all rights reserved

(as repeated by Athenaeus) a *natural phenomenon of great complexity*, due to its mysterious appearance inside the soil without essential plant parts, such as roots, stems, branches, leaves, fruits, fibers, or veins (Ainsworth 1976 on Theophrastus *HP* 1.6.5, 1.6.9). However, Theophrastus also wrote of a common belief that truffles might be grown from seeds that were carried in the soil that is brought down each year by the swelling rivers and deposited on their banks, where truffles are later gathered (Athenaeus 2008). Theophrastus referred to truffles by different names depending on their habitats. The *iton* grew in Thrace, but the *misy*, which were the best truffles, grew in Cyrene (Libya). They tasted sweet and were said to smell like meat. Theophrastus also mentioned the *hydnon*, which was the term commonly used to refer to truffles in the ancient Greek language (Ramsbottom 1953).

15.5.1.1 Hydnophyllon: A Plant that Grows with Hydnon (A Truffle)

The first-century Greek grammarian, Pamphilus, discussed a plant called ***hydnophyllon***, which he described as a kind of grass that grew on top of truffles and by which the truffles were discovered (Athenaeus 2008; Ramsbottom 1953). This confirms that the ancient Greeks knew of the connection between truffles and certain plants with which the truffles tend to grow. The identity of *hydnophyllon* was first suggested in *Historia generalis plantarum*, attributed to d'Aléchamps, in which *hydnophyllon* is identified as both *Tuberaria major* and *T. minor*. According to *Historia generalis plantarum*, these plants grew in the vicinity of Castile where they were locally known as *yerva turmera* (truffle plant). Linnaeus united the two species into *Cistus tuberaria*, which was later transferred to *Helianthemum* (Ramsbottom 1953). A number of popular species of Terfezia, like *T. arenaria* and *T. fanfani*, grow in Spain's region of Extremadura, which is located within ancient Castile, where they are associated with the plant *Tuberaria guttata* (Trufmania 2008; Moreno et al. 2002). This suggests that *hydnon* may have referred to some species of *Terfezia*.

15.5.1.2 Pliny: Truffles Do Not Grow, They Appear in the Soil

Pliny the Elder (23–79 CE), in his influential work *Naturalis Historia,* followed Theophrastus and considered truffles a great wonder of nature (Ramsbottom 1953). In regard to the origin of the truffle (***tubera***), Pliny wrote that *among the most wonderful of all things is the fact that anything can spring up and live without a root* (*Nat. Hist.* 19.3 in Ramsbottom 1953). According to Pliny, truffles were generally found in dry, sandy places that were overgrown with shrubs and were so large that they could reach the size of a quince. He mentioned two kinds of truffles: those full of sand and thus *injurious to the teeth* and those free of impurities. He noted that some truffles were red or black and others white inside. Like Theophrastus, Pliny thought that the best truffles came from Africa, adding that the best truffles in Greece grew in Elis (southern Greece) (Ramsbottom 1953; Pliny in Bostock and Reily 1856). Pliny believed that truffles were made of soil elements and thus were nothing more than balled-together lumps of earthy substance (*Nat. Hist.* xixi, sect.11 in Ainsworth 1976).

15.5.1.3 Influence or Parallel Ideas: The Origin of the Truffle in Other Cultures

Lively debates on the origin and nature of desert truffles, some echoing the ideas of Theophrastus and Pliny, took place in Galilee in the first century (Shemesh 2010; Shavit 2008). These debates are recorded in the Mishna (the collection of the tractates of the systematically codified Jewish Oral Law), as well as in the Jerusalem and the Babylonian Talmuds (compiled and finally sealed in the fourth and sixth centuries, respectively) (Shemesh 2010; Shavit and Volk 2007). In these deliberations, the scholarly rabbis concluded that truffles did not grow from seed or root or with nourishment from the soil. Rather, the truffles *materialized in their final shape in one night, wide and round like dumplings,* and the soil eventually *spit them out* (Talmud Jerusalem; Talmud Babylonian; Shemesh 2010; Shavit 2008). A similar belief was expressed by a Khoisan Kalahari-truffle gatherer, who said that the reason he sold his truffles so cheaply was that the he did not toil to grow the truffles himself. Rather, desert truffles were God's given gift from the soil, emerging from it entirely on their own (Mshigeni 2001). Most likely unrelated to Greek, Jewish, or Khoisan beliefs, vernacular terms in China reflect similar confusion regarding tuberous fungi, as is evident in the name *wu-niang teng* (no-mother plant) given to a certain forest truffle (Hall et al. 2007).

15.5.2 The Belief That Thunderstorms Affect the Formation of Truffles

Pliny wrote that a good spring crop of desert truffles depended on soaking autumn showers, adding that thunderbolts had a particular contribution to the formation of the truffles (Pliny, *Nat. Hist.* XXV.9.67115; XIX.13.1, in Ramsbottom 1953). The relevance of thunder to the formation of truffles was the subject of a lively debate, organized by Plutarch (46–120 CE) during a meal of particularly large Grecian truffles, in Elis (Plutarch, *Mor.* VII 4 Q 2 in Clement and Hoffleit 1969; Helttula 1996). One of Plutarch's fellow debaters suggested that perhaps strong thunderclaps merely shake loose the already formed truffles inside the soil and expose them. Others thought that the contribution of strong thunderstorms was in the highly fertile rains that such storms brought down (due to the mixture of heat with the rainwater), which in turn boosted the growth of truffles (Plutarch, *Mor.* VIII 4 Q 2 in Clement and Hoffleit 1969; Klotz and Oikonomopoulou, 2011; Shavit and Volk 2007). The belief that thunderstorms, particularly thunderbolts, are especially important to the formation of spring truffles is still prevalent among desert truffle gatherers. Kagan-Zur (2001) mentions legends connecting the appearance of truffles to thunderstorms, noting similarities in such stories from truffle gatherers in Morocco and the Israeli Negev. Similarly, Feeney (2002) was told by a Bedouin that the force of the thunderclaps influences the number and size of the forming truffles. MacFarquhar (2004) writes that he was told by a professional truffle gatherer that when thunder cracks the soil, truffles appear. The present-day researcher Al-Thani, in a study published in 2010, names thunder among the three most important factors in the formation of truffles (heavy rains, thunder, and soil type) (Al-Thani 2010). Kalahari truffles are regarded as *eggs of the lightning bird* in the traditional Khoisan oral mythology, because they appear so close after spring's thunderstorms (Trappe et al. 2008b). A similar idea is conveyed by the Chinese, who named the truffle-like medicinal fungus *Omphalia lapidescens* (formerly *Polyporus mylittae*) from southern China, Lei Wan—*thunder ball fungus* (Hall et al. 2007).

15.6 Truffles Extravagance in Imperial Rome

The popularity of desert truffles in the ancient world reached its apex in imperial Rome. Desert truffles became a pricy, elite, gourmet delicacy (Helttula 1996). Roman emperors were well known for their lavish dining and love for truffles, and there are records of massive amounts of truffles sold in Roman markets. Their popularity and high price finally caused the Emperor Diocletian to include truffles among the foodstuffs subjected to price controls (Faas 2003; Helttula 1996). Truffles cost the same as the best cut of liver and salted pork and twice the price of the best cut of meat (Helttula 1996). The scandalous 14-year-old Emperor

Marcus Aurelius Antoninus (third century), nicknamed *the Golden Lunatic*, outdid other Roman emperors when he regularly served Libyan truffles cooked with sow's udders at his lavish parties, alongside flamingo brains and peacock tongues flavored with cinnamon (Hay 1911).

The Roman poet Juvenal (second century) considered truffles a decadent foodstuff and thought that the obsession that the wealthy Romans had with them was ridiculous. In his satires, he often used this obsession to ridicule the rich (Brothwell and Brothwell 1998). Referring to the particular popularity of the truffles that were shipped from Libya, an important supplier of much needed grain to Rome, Juvenal mocks,

> *If the spring will bring pearls of thunder, we will have the desired truffles. Keep your corn to yourself, O Libya!...Unyoke your oxen if only you send us truffles!* (Juvenal, *Satire V.*, in Page et al. 1920).

The fourth-century book of Roman recipes, known as Apicius' *De re coquinaria*, contains six recipes for truffle preparations that echo the Middle Eastern traditional desert truffle cooking methods (Helttula 1996; Brothwell and Brothwell 1998). The recipes begin with the removal of the truffles' outer skin followed by boiling the truffles with salt. The cooking techniques that follow include threading the truffles on skewers and grilling them smothered with fat (a preparation Iddison (2011) describes from Turkey), boiling the truffles in aromatics and sauces (a popular Bedouin method of cooking truffles), or wrapping them in fat and grilling them in the fire, as some of the Khoisan do (Feeney 2002; Iddison 2011; Dixon 2006) (Fig. 15.3).

15.6.1 Evidence Supporting the Claim That the Truffles Discussed in Early Classic Greek and Roman Writings Were Desert Truffles

Until the fourth century, Greek and Roman scholars' descriptions of truffles were more consistent with desert truffles than with European forest truffles (Helttula 1996; Hall et al. 2007). These descriptions include characteristics such as the truffles' sandy fruit bodies and habitats, their springtime fruiting season, their prevalence in geographic locations like Libya and North Africa, their mild bland taste, and the high-heat cooking methods that are used to prepare them (Helttula 1996; Brothwell and Brothwell 1998). While boiling, roasting, or frying may be the traditional cooking methods for desert truffles, the heat-sensitive aromatic compounds that the European forest truffles are famous for would be destroyed by the exposure to high-heat cooking (Alcock 2006; Diaz et al. 2009). In addition, Helttula (1996) did not find any reference in the classic Greek and Roman literature, up to the fourth century, either to the use of pigs or to the truffles' alleged aphrodisiac properties, both of which are typical of the European forest truffles.

Fig. 15.3 ***Mufaraket Kemeh*** (desert truffles with lamb). Photographer: Chadwan Al Yaghchi. Courtesy of Chadwan Al Yaghchi, Amateur cook and food blogger at http://www.syrianfoodie.blogspot.com, all rights reserved

Helttula (1996) concluded that it was clear that the truffles discussed in the early Greek and Roman writings were desert truffles.

15.7 The Evolution of Vernacular Terms for Desert Truffles in Historical Context

The widespread use of a vernacular term for truffles, along with the many phonetic variations of this term that have evolved in different languages and geographies, attests to the term's long lineage (Wasson and Wasson 1957). In early cultures with recorded languages, the ancient vernacular terms for desert truffles can be traced back for many centuries. The terms ***kam'a*** and ***faq'a*** are the two ancient terms from which the majority of the vernacular terms for desert truffles in the Middle East and North Africa seem to have been derived (Shavit and Volk 2007; Right and Patrick 2002; Bajiki 2011; Iddison 2011).

The term ***kama*** traces its roots to the third millennium BCE term ***kam'u***, first encountered in the Sumerian myth of *The Marriage of Mart'u* (Sasson 2004; Shavit 2008). *Kam'a* is the classic Arabic term for truffles. It refers to truffles in general, like the English word *truffles,* but it is also used to indicate desert truffles in particular (Gucin and Dulga 1997; Hussain and Al-Ruqaie 1999). In the ancient Amorite truffle-gathering areas of the Middle East, desert truffles are often referred to by variations of the term *kam'a*. In southern Turkey and northern Syria, truffles are called *kama*, ***keme,*** or ***kemeyeh***. In Israel, desert truffles are called *kam'a* in Arabic and ***kmeh'in*** in Hebrew. The term *kmeh'in* is also used in the Aramaic of the Talmud in reference to desert truffles and has a recorded history of use dating back to the first century (Shemesh 2010; Shavit and Volk 2007). In Iraq, desert truffles are called *kama*, *keme*, ***kima***, or ***chima*** (Feeney 2002; Iddison 2011). Some of the vernacular terms for plants in the Cistaceae family, with which desert truffles are associated, indicate an early knowledge of this association. For instance, an Arabic vernacular name for an annual species of Helianthemum is ***jaraid ach-chima*** (roughly *the Helianthemum of the desert truffle*) (Mandeel and Al-Laith 2007).

Faq'a (the root ***f/p-q/g-a***) or ***paqu'a*** is another ancient vernacular term for desert truffles, popular in the Middle East (Bajiki 2011; Bryant 1923; Shavit and Volk 2007). This term has a broader use, and it often refers to mushrooms in Arabic and Hebrew (Hussain and Al-Ruqaie 1999; Shavit and Volk 2007). The meaning of its root is *to burst*, *emerge*, or *break out*. Its noun refers to a "bundle" in Arabic, Hebrew, and even in Zulu, a language in the family of the southeastern Bantu, spoken in parts of Southern Africa (Bajiki 2011; Bryant 1923; Shavit and Volk 2007). Derived from this root, in Kuwait and Oman, desert truffles are called ***faqqa***; in Saudi Arabia ***faq***, ***fag***, ***fuga*** (for *Tirmania nivea* in Kuwait), ***faqah***, and ***figaa***; in Hebrew and in the Arabic dialect spoken near Jenin, the common terms are ***paqua*** and ***faqqua***, respectively (Shavit and Volk 2007; Iddison 2011). It is noteworthy that another vernacular term for desert truffle in use in the Arabian Peninsula also occurs in the Zulu language of southern Africa: in Oman, desert truffles are called ***kumba***, a term in the Zulu language indicating a bundle or a ball-like mass (Bryant 1923; Iddison 2011).

15.7.1 Vernacular Terms Based on Morphologic Characteristics

Vernacular terms for desert truffles abound among the numerous different indigenous languages spoken wherever desert truffles grow (Trappe et al. 2008a, b). One such term reflects a unique observation made by the Pitjantjatjara people of central Australia. They believe that the calls made by the gray shrike-thrush (*Colluricincla harmonica*) following heavy rains bring out the desert truffles. The Pitjantjatjara people call the bird *wititara* and know that when they hear its call following the rains, it will soon be time to gather ***witita***, the desert truffle (Trappe et al. 2008a).

Often based on morphological characteristics such as color or attributed virtues, some Bedouin vernacular names for desert truffles are abbreviated descriptions, such as the term ***zbedi*** for *Tirmania* spp., which is short for [*al-fag*] *az-zbedi* (the truffle that is white like cream) or ***khlas***, standing for [*al-fag*] *al-khlas* (the true, best kind of truffle) (Mandaville 2011). The same vernacular term, however, may indicate different species in different places, like the term *ikhlasi* for both *Terfezia claveryi* and *T. boudieri* (Mandaville 2011; Wang and Marcone 2011). In Spain, some vernacular terms for truffles are based on their suggestive shape, like the term *criadilla de tierra* (meaning *earth's testicles*), which refers to both *Terfezia arenaria* and *T. fanfani* (Trufmania 2008).

15.7.2 *The Evolution of the Terms Truffle and Terfezia*

In Latin, the truffle is called ***tuber***, meaning *to swell*. According to Wasson and Wasson (1957), tuber took the form of *tufer* in an Umbrian dialect. Its variant form *tufera*, which was used by Anthimus (early sixth century) and is found in the compound noun *terra tufera* (ground tuber), evolved into the French ***truffe***, the Spanish ***trufa***, and the English ***truffle*** (Alcock 2006; Helttula 1996). It is theorized that the Italian ***tartufo***, in its diminutive form *taratouffli*, evolved into the German ***kartoffel*** (potato) (Wasson and Wasson 1957). In some variations, this name spread eastward throughout the Slavic lands and Europe; until today much of Europe calls the humble potato by the word originally reserved for the truffle (Wasson and Wasson 1957).

The term Terfezia evolved from the North African Berber term ***Terfas*** (pl.), which in some variations like ***terfess*** and ***terfez*** are the vernacular terms for desert truffles in the countries of North Africa and in Egypt (Helttula 1996; Feeney 2002).

15.8 Historical Records of the Use of Desert Truffle by Travelers

15.8.1 *The Use of Desert Truffles Along the Ancient Caravan Routes*

A wealth of information regarding the use of desert truffles is available in the published diaries of explorers who travelled with the caravans that transported goods along the ancient caravan routes that crisscrossed the deserts of Asia, the Middle East, and the Sahara. These caravans, often thousands of camels long, were an essential part of world trade in ancient times, linking China, India, Central Asia, and parts of Africa to the markets of the Middle East and Europe (Ross 2011). Al-Idrisi (twelfth century), who travelled extensively in North Africa, observed that desert truffles were found in large quantities in the area of Awdaghust (present-day Tegdaoust, on the Senegal River in south-central Mauritania), an early Muslim outpost on the Trans-Saharan trade route (Gritzner 1981). The truffles were called by the Arabic term *kama* and were an important component of the local diet. The local population cooked the truffles with camel's meat, believing the dish to be the best food in the world. Al-Idrisi, who tasted it, concurred (Gritzner 1981). A century later, the large quantities of desert truffles gathered by the local population in the Sahara were also noted by the Tangier-born geographer Ibn Battuta (1304–1370 CE), who travelled in the West African Sahara as far as present-day Mali (Ibn Battuta and Gibb 1929).

Some of the desert truffles gathered in North Africa were so large that the famous Granada-born diplomat and geographer Leo Africanus (1494–1554 CE) wrote in his diary that a rabbit could make its burrow in them (Africanus and Brown 1896). He observed that the truffles were regarded by the local population as "great dainties" and were prepared by roasting on coals and serving smothered in fat (Africanus and Brown 1896).

Impressed by their similar look, Nachtigal noted in 1869 that the large Libyan truffles that were so popular around Murzuq were also gathered by the Bushmen of the Kalahari (Gritzner 1981).

15.8.2 Travelers' Records of Traditional Cooking and Preservation Techniques

Desert truffles are usually cooked before they are consumed, and their specific texture and flavor dictate the best way to prepare them. Since Bedouins were often the caravans' guides and guards (Ross 2011), it is no wonder that the Bedouins' typical truffle cooking techniques influenced the local preparations of desert truffles. According to reports of travelers, these preparations included boiling the truffles alone, in sauces, or with meats. Other popular preparations included roasting, baking, or pounding the boiled truffles and then serving them with *samn* (fermented butter) or *leben* (yogurt-like fermented milk). The truffles were also sliced and fried in fat (Blunt and WSB 2007; Burckhardt 1831; Feeney 2002; Gritzner 1981; Mandaville 2011).

Travelers to the Australian Outback tell of a particular cooking technique used by the Aborigines to prepare a certain kind of desert truffle. This technique involves leaving the already roasted truffles to cool overnight and heating them through the next day prior to eating. Travelers who ate truffles prepared in this manner wrote that they had soft texture and mild flavor reminiscent of delicate ripe cheese. Although the truffles on both continents as well as the two ethnic populations that use them are unrelated, travelers to the African Kalahari report that the Khoisan use the same cooking technique (Trappe et al. 2008a, b; Dixon 2006).

Desert truffles are fast becoming known outside their traditional areas, raising the interest of foodies and food industry professionals alike. Present-day travelers share their observations and describe their experiences in numerous online articles and blogs. In season, desert truffles are available for sale online even outside their growing areas (Feeney 2002; Iddison 2011; Syrian Foodie in London).

15.9 Desert Truffles as a Divine Gift

Perhaps because desert truffles do not produce fruit bodies each year and their origin could not be easily explained, populations who traditionally gather desert truffles have considered them to be a special gift, often a divine gift. As quoted in the Hadith Sahih Muslim, Islam's Prophet Muhammad (seventh century) referred to desert truffles as a gift from God. This feeling is shared by the Khoisan in Namibia who think of desert truffles as *God's given gift from the soil* (Mshigeni 2001). In the Jewish religious literature, there are numerous discussions regarding the type of blessing that should be given to God for the creation of desert truffles (Shavit and Volk 2007; Shemesh 2010).

A first-century BCE story from the Jerusalem area involving the legendary Honi (Onias) the Rain-maker illustrates the belief that truffles were considered a divine gift (Shavit and Volk 2007; Shemesh 2010; Eisenman 1990). The story of Honi made such an impression on the population of the area that it is discussed in the Jewish Mishna and Talmud, and the historian Josephus Flavius (93 CE), who recorded the important historical events of his time, repeated it in his book the *Antiquities of the Jews* (Shavit and Volk 2007; Shemesh 2010; Eisenman 1990). As the story goes, on an early spring day during a period of severe drought that caused crops to fail, the people of the area of Jerusalem commissioned Honi, a well-known yet unconventional man of God, to intervene on their behalf and ask God for immediate rain. Honi was successful, but the rain that followed was torrential and caused devastating flooding, forcing the people to the mountaintop. At the request of the people, Honi intervened with God again, and God stopped the rains, dried the land, and awarded the hungry people a much needed reward:

> *The sun shone bright, and the people went down to the steppe and brought back truffles and mushrooms for everyone* (Talmud (Babylonian), Taanit, 3:23,1).

Desert truffles were an important, traditional food for the ancient Jewish communities of North Africa and the Middle East. The truffles were a staple at the Passover Seder ceremony, which commemorates the exodus of the Israelites from Egypt (Nathan 2010; Kirshenblatt-Gimblett 2003). In the North African Jewish communities, desert truffles were slow cooked with lamb and rice, while in the Iraqi Jewish communities, desert truffles were usually served boiled and salted (Nathan 2010; Oren and Ravid 2012; Ben-Tarbut 2009). Once in Israel, members of these communities replaced their familiar Iraqi, Syrian, Moroccan, or Libyan truffles with the far less widespread Negev desert truffles, preparing them in their customary ways (Oren and Ravid 2012; Ben-Tarbut 2009).

15.9.1 An Aboriginal Legend: Betsy Napangardi Lewis

An Aboriginal legend from Australia demonstrates that the Warlpiri people who gather desert truffles as food consider the truffles a special gift. The legend is depicted in a painting called ***Tjintiparnta*** (bush truffles) (Fig. 15.4) painted by the late Warlpiri artist Betsy Napangardi Lewis (Trappe et al. 2008a). Gathering truffles is primarily the role of women in the Aboriginal society. In her painting, Lewis tells a legend about her people. At one time after the rains, the Napanangka Napangardi women all gathered together at a claypan. Claypans are dry, shallow depressions in the ground with a layer of clay that holds water after a heavy rain (The Free Dictionary).

Fig. 15.4 ***Tjintiparnta*** (Bush truffles): The legend of the truffle-digging sticks, a painting by the late Aboriginal artist Betsy Napangardi Lewis. Courtesy of the Warlukurlangu Artists Aboriginal Corporation in Alice Springs, NT, Australia, who owns the copyright

Desert truffles are often found in claypans. The women performed long ceremonial dances on the pressed claypan (depicted as the concentric circles at the top right side of the painting in Fig. 15.4). Truffle-digging sticks arose from the numerous water holes that remained on the claypan after the rains (seen in blue on the left side of the painting in Fig. 15.4). With the help of these truffle-digging sticks, the Napanangka Napangardi women learned how to gather the many desert truffles that were buried in the sand amidst the water holes (the truffles are the brown, round shapes among the puddles in Fig. 15.4). The bent U-shaped holding sticks on the right side of the painting depict the women digging out the truffles (Trappe et al. 2008a).

15.10 Desert Truffles in Traditional Medicine

Desert truffles are traditionally gathered not just for nourishment, but also for their curative and aphrodisiac properties (Alsheikh and Trappe 1983; Mandeel and Al-Laith, 2007). The juice of desert truffles, mainly of *Tirmania nivea*, *Terfezia claveryi,* and *T. boudieri*, has been used in the Middle East to treat eye and skin diseases (Patel 2012; Mandeel and Al-Laith 2007; see Shavit and Shavit in Chap. 20 of this volume). This is in line with the recommendations of Islam's Prophet Muhammad and of the Persian physician Avicenna (1000 CE) (Hall et al. 2007). These traditional uses have been substantiated by recent scientific studies, which indicate that species of *Tirmania* and *Terfezia* have been found to have antibacterial qualities, particularly against the organisms that cause eye infections (Janakat et al. 2005; Gouzi et al. 2011; see Shavit and Shavit in Chap. 20 of this volume). Recent studies of the bioactive properties of desert truffles show them to have antimicrobial, antiviral, antimutagenic, hepato-protective, and anti-inflammatory properties (Wang and Marcone 2011, see also Shavit and Shavit in Chap. 20 of this volume).

The Australian Aborigines have a similar medicinal use for the desert truffles they gather. A Warlpiri man reported that to heal boils or other skin sores, his people squeeze the juice of a desert truffle they call ***kumpu*** onto them (Trappe et al. 2008a). The same Warlpiri man also reported a cosmetic use for the juice of the *kumpu* desert truffle, claiming that rubbing the underarm with this juice makes the underarm hair fall off (Trappe et al. 2008a).

In the African Kalahari, the dried and powdered fruit body of the *Kalaharituber pfeilii*, called ***Mahupu*** in the Setswana language of Botswana, is used to induce birth in both humans and livestock (Khonga and Mogotsi 2007).

15.10.1 The Belief That Desert Truffles Possess Aphrodisiac Properties

Even though desert truffles have not been found to contain the pheromonal steroid 5α-androstenol, a chemical sexual attractor that is found in species of *Tuber*, many consumers of desert truffles in the Middle East believe that desert truffles possess aphrodisiac properties (Feeney 2002; Mandaville 2011; Hifnawy et al. 2001; Patel 2012). Nearly 95 % of non-Bahraini and 72 % of native Bahraini people who participated in a study conducted among people who use desert truffles in Bahrain said that they would eat desert truffles for their aphrodisiac properties (Mandeel and Al-Laith 2007). This belief must have been particularly strong during the Moorish rule over Spain (711–997 CE) because the *muhtasib* of Seville tried to ban the sale of all truffles in the vicinity of the mosques on the grounds that they were food for debauchery (Wasson and Wasson 1957; Kislinger 1999). The Khoisan people of the African Kalahari also believe that their Kalahari truffle possesses aphrodisiac properties (Trappe et al. 2008b).

15.11 The Place of Desert Truffles in the Culture of the Bedouins

Centuries prior to the solidification of borders between countries, the nomadic, pastoral Bedouins of Arabia would follow their ancient migratory routes in the desert, seeking better pastures for their herds, going deeper into the desert in the rainy season and back toward cultivated lands during the dry months (Mandaville 2011; Jabbur 1995). Mandaville (2011), who documented Bedouin ethnobotany in Saudi Arabia for decades, states that *without doubt, the most important wild food plant for the Bedouins has been the desert truffle.*

Burckhardt (1831) described a typical truffle season for a tribe in the Syrian Desert:

> *The great plain between Damascus, Bagdad and Basra is full of kemeye. If the* [truffle] *is left to attain full maturity, it rises over above the earth to half its volume. The children dig it out with short sticks.* [The truffles] *are sometimes so numerous on the plain that the camels stumble over them. Each family then gathers four or five camel loads; and while this stock lasts, they live exclusively on kemmeye, without tasting either burgul* (bulgur wheat) *or ayesh* (a type of bread).

The knowledge of truffles is passed on orally from generation to generation. To help the children learn the telltale signs of truffles and keep their common names in mind, the Bedouins use riddles, simple limericks, and mnemonic verses, such as:

> *az-zbēidi lil-wlēdi; al-jbēyal lil-bnēyah, al-khlasi hagg rasi, al-hbēri lil-tuwēri* (The *zbedi* for the little boy, the *zbeyah* for the little girl, the *khlasi* for myself [literally my head], the *hberi* for the little bird) (Mandaville 2011).

This last name, the *hberi*, refers to *Picoa lefebvrei,* a type of truffle with small fruit bodies, which the Bedouin children like to eat like treats (Mandaville 2011; Alsheikh and Trappe 1983). *Hberi* fruit bodies are liked by some migratory birds that dig them up and eat them, and the Bedouins use these truffles as bait when hunting the birds (Mandaville 2011; Alsheikh and Trappe 1983).

Also popular are riddles, which are used to pass the time, such as this loosely translated one that teaches how to spot the ripe truffles,

> *Which thing builds a pretty house and drinks water, yet has no throat?* Answer: *al-fag'* (the truffle), which pushes its sand-roof up as it rises to the surface (Mandaville 2011).

Truffles are also used in Bedouin poems and songs, as in the poem that likens the breast of a maiden to a *zbedi*, referring to the truffle's round shape and creamy color (Mandaville 2011).

15.11.1 The Role of Women in Desert Truffle Gathering

In many regions of the world, women are often the main gatherers of desert truffles, mushrooms, and medicinal herbs. This is the case among the Aborigines of Australia and also in many Bedouin tribes. As a result, Bedouin women often possess extensive knowledge of the different fungi they collect and their biology, habitats, time of fruiting, and various uses (Garibay-Orijel et al. 2012). Numerous literary works, even children's books, depict women as the keepers of truffle-gathering traditions. Jabbur (1995) remembers how the Bedouin women in the area of eastern Syria used to sing to the truffles to entice them to emerge. In a book about Basra, Khudayr (2007) reminisces about the places of his childhood, now long gone,

> *The camel kneels chest first at the edge of Wadi al-Nisa (the Women's Valley). The spring rains would cause the earth to explode with white mushrooms and red Khalasi truffles ... The women of Basra used to frequent the valley to gather the truffles* (Khudayr 2007).

The author of a children's book about a Dubai boy describes his own childhood:

> *...We also have the faq'qa, a desert truffle, which blooms after the rains and little girls accompany their mothers to pluck them* (Dias 2011).

Desert truffles have long been an important supplementary source of income for Bedouin women and helped define their role in the family (Lönnqvist 2004; Steinmann 1998). A study to assess the changing gender roles in pastoral lands

Fig. 15.5 Father and Son gathering white desert truffles. Photographer: John Feeney. Courtesy of Saudi Aramco World.

found that while living in the traditional mobile tents, women and men operated with separate sets of resource management tasks. However, as households have settled, these tasks have become increasingly shared (Steinmann 1998). As more Bedouins must change their nomadic way of life, settle down in one place, or move to the cities, the role of women within the family changes as well (Steinmann 1998) (Fig. 15.5).

15.11.2 Desert Truffles as a Survival Food

The relatively high protein and carbohydrate content of desert truffles explains their importance to indigenous populations and their frequent role as a survival food (Wang and Marcone 2011). The composition of desert truffles (in dry matter) is protein, 20–27 % (85 % digestible by humans); fat, 3–7.5 %; crude fiber, 7–13 %; and carbohydrates, 60 % (Patel 2012, see Martínez-Tomé et al. in Chap. 17 and Al-Laith in Chap. 18 of this volume).

Both nomadic tribes and settled populations depend on desert truffles, often using them as meat substitute (Patel 2012; Mandaville 2011). Khalastchy (2003) describes the difficulties of life in Iraq in the 1970s, when Iraq was at war. Food was scarce but truffles were plentiful. Khalastchy writes that for 4 months the population used desert truffles instead of meat in traditional meat dishes, stating that the truffles were *as delicious and nutritious as meat* (Khalastchy 2003).

Although desert truffles do not produce fruit bodies every season, they have a long history as a tie-over food, producing nutritious fruit bodies in a period when stored crops from the previous season have been eaten, but the next season's grains and fruits are not yet ready for harvest (De Roman 2010; Jabbur 1995; Mandaville 2011). In early truffle seasons, the Bedouins would eat only truffles for the whole truffle season. They would also slice truffles and dry the slices on their tent roofs (Addison 1838; Blunt and WSB 2007). These dried slices are called ***shtib***, and the Bedouins say that they can last for years and taste just like fresh truffles when rehydrated (Mandaville 2011). While the traditional preservation methods of salting and pickling are still in use, Bedouins today prefer to preserve their desert truffles by deep freezing (Mandaville 2011; Patel 2012).

15.12 The Decline of Desert Truffles in the Wild

The progressive decline in the production of desert truffles in the wild has been blamed primarily on the global loss of desert truffle habitats and on climate change (Marijcke 2005; Giovannetti et al. 1994; Morte et al. 2009). The accelerated rate of urbanization, changes in the management of grazing lands, settling down of nomadic people, and even the introduction of modern technology are also blamed for the decline. Desert truffles and their host plants play a major role in the maintenance of shrub and grass marginal grazing lands by preventing erosion and desertification, and they are culturally important to indigenous people (Diez et al. 2002; Trappe and Claridge 2010; Morte et al. 2009). Even old and stable desert truffle cultures can disappear within a short period of time when families no longer gather desert truffles, because when the knowledge is not passed on to next generations, it is consequently lost (Trappe et al. 2008a, b; Steinmann 1998; Garibay-Orijel et al. 2012). Efforts to cultivate desert truffles and reintroduce them to areas where they grew only a few decades earlier may contribute to the preservation of indigenous knowledge of desert truffles (see Morte et al. in Chap. 21 and Honrubia et al. in Chap. 22 of this volume).

15.12.1 The Decline of the Egyptian Desert Truffle Culture

For centuries, massive amounts of desert truffles were gathered in the Muqattam Hills near Cairo and in Egypt's Western Desert shores. The truffles were shipped to the market in Cairo, were served to the Fatimid Caliphs (tenth to twelfth centuries), and were so plentiful centuries later that in 1835 the English historian Edward Lane wrote that they had become cheap and common in the Cairo market (Feeney 2002; Trappe 1990). Truffle gathering must have been popular in Egypt in 1878, since an Egyptian calendar included an estimated date for the start of the 1878 desert truffle-gathering season (Michel 1877). However, the prolific truffle fields of the Muqattam Hills were engulfed by urban Cairo. Large areas of Egypt's coastal desert were laced with mines during WWII, and recently other coastal desert areas have undergone development (Morte et al. 2008). Truffles can no longer be gathered in these areas. In 2002, thorough research conducted by John Feeney at the Egyptian Agricultural Museum's library failed to turn up even a single reference to *Terfezia*, let alone to *terfas*, as the truffles are called in Egypt (Feeney 2002). After having lived in Cairo for 40 years, Feeney concluded that few Egyptians have ever heard of desert truffles (Feeney 2002). Kagan-Zur (2001) makes a similar observation, writing that while Egypt is among the

Fig. 15.6 *Terfezia boudieri* Truffles ready to be cooked. Photographer: Elinoar Shavit.

locations where desert truffles grow, truffles are not known by Egyptians. Whatever quantities of desert truffles are still gathered in Egypt's Western Desert and are now brought to the Cairo airport and immediately air-shipped to markets in Abu Dhabi, Doha, Kuwait, and Riyadh, where their popularity and price are high (Hussain and Al-Ruqaie 1999).

15.13 Conclusion

Indigenous traditions are quickly fading away all over the world, and desert truffles and their habitats are dwindling in the wild. Efforts to find solutions to this decline have concentrated on the cultivation of desert truffles in their natural habitats and on better appointed grazing land management. These efforts have the potential to create economic possibilities that will benefit local populations and at the same time preserve the rich desert truffle cultures of these populations.

It is fitting to conclude the way we began with an insight from Alexandre Dumas into the question that has bothered researchers and laymen alike: what is the nature of the truffle? To this Dumas replies,

> *After two thousand years of argument and discussion... the truffles themselves have been interrogated, and their answer is simply: eat us and praise the Lord!* (Dumas and Colman 1958) (Fig. 15.6).

References

Addison CG (1838) Damascus and palmyra: a journey to the East, vol II. Richard Bentley, London

Africanus L, Brown R (eds) (1896) The history and description of Africa. Hakluyt Society, London

Ainsworth GC (1976) Introduction to the history of mycology. Cambridge University Press, New York

Alcock JP (2006) Food in the ancient world. Greenwood, Westport, CT

Alsheikh AM, Trappe JM (1983) Desert truffles: The genus *Tirmania*. Trans Br Mycol Soc 81 (1):83–90

Al-Thani RF (2010) Survey of macrofungi (including truffles) in Qatar. KMB J Biol 1(2):26–29

Athenaeus (2008) The Banquet of the learned. (trans: Yonge CD) The Project Guttenberg EBook, http://www.archive.org/stream/deipnosophistsor01atheuoft/deipnosophistsor01atheuoft_djvu.txt. Accessed 5 Oct 2012

Bajiki A (2011) Arabic Etymological Dictionary—Arabic Hebrew Lexicon, https://sites.google.com/site/arabichebrewlexicon/links/aed-andras-rajiki-s-arabic-etymological-dictionary. Accessed 2 Oct 2012

Ben-Tarbut (2009) Morrocan Culture Forum, a blog, http://www.tapuz.co.il/forums2008/viewmsg.aspx?forumid=928&messageid=128273505 . Accessed 9 May 2013—in Hebrew

The King James Bible (2012) Authorized version of the King James Bible (Pure Cambridge Edition), http://www.kingjamesbibleonline.org/. Accessed 13 Nov 2012

Blunt A, WSB (2007) Bedouin tribes of the Euphrates. Kessinger Publishing's Rare Reprints

Bostock RJ, Reily HT (trans) (1856) The natural history of Pliny, vol IV, Henry G. Bohn, London

Bradlaugh C (2011) The Bible: What it is! by 'Iconoclast'. The project Gutenberg EBook of the Bible, http://www.gutenberg.org/files/36266/36266-h/36266-h.htm. Accessed 3 Oct 2012

Brothwell P, Brothwell DR (1998) Food in antiquity: a survey of the diet of early peoples. The Jones Hopkins University Press, Baltimore

Brueggemann W (1999) The liturgy of abundance, the myth of scarcity. Christian Century, Religion, http://www.religion-online.org/showarticle.asp?title=533. Accessed 1 Oct 2012

Bryant AT (1923) Zulu-English dictionary with notes on pronunciation: a revised orthography and derivations and cognate words from many languages. Reprinted in 2010 by Nabu, Charleston, SC

Burckhardt JL (1831) Notes on the Bedouins and Wahábys: collected during his travels in the East, vol 1. Henry Colburn and Richard Bentley, London, http://archive.org/stream/notesonbedouins01burcgoog/notesonbedouins01burcgoog_djvu.txt. Accessed 26 Sept 2012

Chiera E (1934) Sumerian epics and myths, vol 3. Oriental Institute Publications (OIP 15), Cuneiform Series, Chicago

Clement PA, Hoffleit HA (trans/eds) (1969) Plutarch, *Moralia* VIII 1–6. Harvard Loeb Classical Library, Cambridge

De Roman M (2010) The contribution of wild fungi to diet, income and health: a world review. In: Rai M, Kövics G (eds) Progress in mycology. Springer and Scientific, India

De Varthema L, Jones JW, Badger GP (eds) (1863) The travels of Ludovico de Varthema in Egypt, Syria, Arabia Deserta and Arabia Felix, in Persia, India, and Ethiopia, A.D. 1502–1508, translated from the original De Varthema 1510 edition. Hakluyt Society, London

Dias JC (2011) The magical bone flute. USA, http://www.amazon.com/Magical-Bone-Flute-Joseph-Canisius/dp/1467001821#reader_1467001821. Accessed 3 Oct 2012

Diaz P, Ibáñez E, Reglero G, Señoráns F (2009) Optimization of summer truffles aroma analysis by SPME: comparison of extraction with different polarity fibers. LWT Food science and Technology 42(7):1253–1259

Diez J, Manjón JL, Martín F (2002) Molecular phylogeny of the mycorrhizal desert truffles (Terfezia and Tirmania), host specificity and edaphic tolerance. Mycologia 94(2):247–259

Dixon R (2006) Namibia's four star fungus. The Los Angeles Times, 15 May 2006, http://articles.latimes.com/2006/may/15/world/fg-truffles15. Accessed 3 Oct 12

Dumas A, Colman L (eds) (1958) Alexandre Dumas's dictionary of cuisine. Simon & Schuster, New York

Eisenman R (1990) Eschatological "rain" imagery in the war scroll from Qumran and in the letter of James. Journal of Near Eastern Studies 49(2):173–184

El-Din MB (2005) National report on hunting, country: Arab republic of Egypt: building capacity for sustainable hunting of migratory birds in Mediterranean third countries (SHP). BirdLife International, AAO, SPNL, Lebanon

Faas P (2003) Around the Roman table: food and feasting in ancient Rome. University of Chicago Press, Chicago

Feeney J (2002) Desert truffles galore. Saudi Aramco World, September/October, pp 22–27

Garibay-Orijel, Ramirez-Terrazo A, Ordaz-Valazquez (2012) Women care about local knowledge, experiences from ethnomycology. J Ethnobiol Ethnomed 8:25

Giovannetti G, Roth-Bejerano N, Zanini E, Kagan-Zur V (1994) Truffles and their cultivation. In: Janick J (1994). Horticult Rev 16:71–103
Gottlieb C, Lubetski M, Keller S (1998) Boundaries of the ancient near Eastern world: a tribute to Cyrus H Gordon. Sheffield Academic, Sheffield
Gouzi H, Belyagoubi L, Abdelali KN, Khelifi A (2011) In vitro antibacterial activities of aqueous extracts from Algerian desert truffles (*Terfezia* and *Tirmania*, Ascomycetes) against *Pseudomonas aeruginosa* and *Staphylococcus aureus*. Int J Med Mushrooms 13(6):553–558
Gritzner JA (1981) Staff report: environmental degradation in Mauritania, BOSTID, Commission on International Relations, National Research Council, National Academy Press, Washington, http://*pdf.usaid.gov/pdf_docs/PNAAJ709.pdf*
Gucin F, Dulga B (1997) The researches on the keme truffle – *Terfezia boudieri* Chatin that is edible and showing antimicrobial activities. J Environ 23:27–33
Hadith Sahih Muslim. Book 23, Chapter 26—Excellence of truffles and their use as a medicine for the eyes, number 5086; 23(27) # 5084–5089, http://www.faithinallah.org/sahih-muslim/. Accessed 3 Oct 2012
Hall IR, Brown GT, Zambonelli A (2007) Taming the truffle: the history, lore, and science of the ultimate mushroom. Timber, Portland
Hay JS (1911) The amazing emperor Heliogabalus, Chapter X, Section 254. Macmillan, London
Helttula A (1996) Truffles in ancient Greece and Rome. Arctos: Acta Philologica Fennica 30:33–48
Hifnawy MS, Azzam SM, Soliman SN, Shaheed IB, Abdel Latif SM (2001) Egyptian desert truffles, androgenic natural product, possibility of abusing in sport. In: Schänzer W, Geyer H, Gotzmann A, Mareck-Engelike U (eds) Sport und Buch Straub, Köln
Hussain G, Al-Ruqaie MA (1999) Occurrence, chemical composition, and nutritional value of truffles: an overview. Pakistan J Bio Sci 2(2):510–514
Ibn-Battuta, Gibb HAR (eds) (1929) Travels in Asia and Africa 1325–1354. Broadway House, London
Iddison P (2011) Truffles in Middle Eastern cookery. Tribulus Magazine. Online at: http://enhg.4t.com/iddison/destruf.htm, visited on 29 Oct 2012
Jabbur JS (1995) The Bedouins and the desert: aspects of nomadic life in the Arab East. SUNY, New York
Janakat SM, Al-Fakhiri SM, Sallal AK (2005) Evaluation of antibacterial activity of aqueous and methanolic extracts of the truffle *Terfezia claveryi* against *Pseudomonas aeruginosa*. Saudi Med J 26(6):952–955
Kagan-Zur V (2001) Terfezias, a family of mycorrhizal edible mushrooms for arid zones. In: Pasternak D, Schlissel A (eds) Combating deforestation with plants. Plenum, New York
Khalastchy E (2003) Desert truffles—Eileen Khalastchy writes. The Scribe—J Babylonian Jewry: 76
Khonga EB, Mogotsi KK (2007) Indigenous knowledge on edible, poisonous, and medicinal macrofungi. Botswana Int J Med Mushrooms, Beagle House 9(3–4):262–263
Khudayr M (2007) Basrayata: portrait of a city. American University in Cairo Press, Cairo
Kirshenblatt-Gimblett B (2003) Passover. In: Katz SH, Weaver WW (eds) Encyclopedia of food and culture, vol. 3. Charles Scribner & Sons Publishing, USA, pp 42–46. ISBN 10: 068480567
Kirshenblatt-Gimblett B, Encyclopedia of food and culture, http://www.nyu.edu/classes/bkg/web/Passover.pdf. Accessed on 10 May 2013
Kislinger E (1999) Christians of the East: rules and realities of the Byzantine Diet. In: Flandrin, Montanari (eds) Food: a culinary history from antiquity to the present. Columbia University Press, New York
Klotz F, Oikonomopoulou K (2011) The philosopher's banquet: Plutarch's table talk in the intellectual culture of the Roman empire. Oxford University Press, New York
Latz PK (1995) Bushfires & Bushtucker: aboriginal plant use in Central Australia. IAD Alice Springs, Australia

Lönnqvist MA (2004) Jebel Bishri in Syria and nomadism in the end of the early Bronze Age. In: British Museum Colloquium 2004: The Levant in Transition: The Intermediate Early Bronze Age
Lönnqvist MA (2008) Were nomadic Amorites on the move? Migration, invasion and gradual infiltration as mechanisms for cultural transitions. In: Kuhne H, Czichon M, Kreppner FJ (eds) Offprint from proceedings of the 4th international congress of the archaeology of the ancient Near East (2004), vol 2. Harrassowitz Verlag, Weisbaden
Lönnqvist M, Törmä M (2004) Different Implications of a spatial boundary, Jebel Bishri between the desert and the sown in Syria, section: SYGIS—Jebel Bishri: the Finnish project in Syria. In: Orhan-Altan M (ed) Proceedings from the ISPRS XXth congress, The ISPRS International Archives of the Photogrammetry, Remote Sensing and Spatial Information Sciences, Vol. XXXV
MacFarquhar N (2004) Beneath desert sands, an Eden of truffles. The New York Times, published 14 Apr 2004
Mandaville PM (2011) Bedouin Ethnobotany: plant concepts and uses in a desert pastoral world. University of Arizona Press, Tucson
Mandeel QA, Al-Laith AA (2007) Ethnomycological aspects of the desert truffle among native Bahraini and non-Bahraini people of the Kingdom of Bahrain. J Ethn 110:118–129
Marasas WFO, Trappe JM (1973) Notes on southern African tuberales. Bothalia 11:139–141
Marijcke J (2005) Desert truffles, a disappearing delicacy. Al Shindagah, Issue 63, March–April
Michel RLN (1877) Egyptian calendar for the year 1295 A.H., 1878 A.D. French Printing Office, Alexandria, Egypt
Moreno G, Diez J, Manjon JL (2002) *Terfezia boudieri*, first records from Europe of a rare vernal hypogeous mycorrhizal fungus. Persoonia 17:637–641
Mori L (2007) Land and land use: the Middle Euphrates valley. In: Leick G (ed) The Babylonian World. Routledge, New York
Morse J (2005) Bush resources: opportunities for Aboriginal enterprise in central Australia: Desert Knowledge Cooperative Research Center Report. Alice Springs, Australia
Morte A, Honrubia M, Gutiérrez A (2008) Biotechnology and cultivation of desert truffles. In: Varma A (ed) Mycorrhiza: genetics and molecular biology, eco-function, biotechnology, eco-physiology, structure and systematics, 3rd edn. Springer, Berlin
Morte A, Zamora M, Gutiérrez, Honrubia A (2009) Desert truffle cultivation in semiarid Mediterranean areas. In: Azcon-Aguilar C, Barea JM, Gianinazzi S, Gianinazzi-Pearson (eds) Mycorrhizas: functional processes and ecological impact. Springer, Berlin
Mshigeni KE (2001) The cost of scientific and technological ignorance with special reference to Africa's rich biodiversity. University of Namibia, Windhoek, Namibia
Nathan J (2010) Quiches, Kugels, and Couscous: my search for Jewish cooking in France. Alfred A Knopf, Random House, New York
Oppenheim AL (ed) (1971) Chicago Assyrian Dictionary, Volume 8:K. Oriental Institute, Chicago
Oren S, Ravid L (2012) Ashtidk: The Iraqi cookbook. Lunchbox, Israel
Page TE, Capps DE, Rouse WHD (eds) (1920) Juvenal and Persius, with an introduction by G.G. Ramsay. William Heinemann, London
Patel S (2012) Food, health and agriculture importance of truffles: a review of current scientific literature. Curr Trend Biotechnol Pharm 6(1):15–27
Pegler DN (2002) Useful fungi of the world: The 'poor man's truffles of Arabia' and 'Manna of the Israelites'. Mycologist 16:89
Ramsbottom J (1953) Mushrooms and toadstools. Collins New Naturalist Library, London
Right WB, Patrick B (2002) Lectures on the comparative grammar of the Semitic Languages. Gorgias, London
Ross E (2011) A historical geography of the trans-Saharan trade. In: Kratli G, Lydon G (eds) The trans-Saharan book trade: manuscript culture, Arabic literacy and intellectual history in Muslim Africa. Koninkliijke Brill NV, Leiden, The Netherlands

Saggs HWF (1965) everyday life in Babylonia and Assyria, Assyrian international news agency, Books Online, http://www.aina.org. Accessed on 2 Oct 2012
Sasson J M (2004) The King's table: food and fealty in old Babylonian Mari. In: Grottanelli C, Milano L (eds) Food and identity in the ancient world. History of the ancient Near East, Padua
Shavit E (2008) Truffles roasting in the evening fires: pages from the history of desert truffles. Fungi Magazine 1(3):18–23
Shavit E, Volk TJ (2007) *Terfezia* and *Tirmania*, desert truffles (*terfez, kama, p/faqa*): delicacies in the sand or manna from heaven? Fungus of the month for January 2007, http://botit.botany.wisc.edu/toms_fungi/jan2007.html. Accessed 2 Dec 2012
Shemesh AO (2010) Names of New Agricultural crops in rabbinic literature of the 16th–20th centuries. Mo'ed, 20: 198–216 (in Hebrew) http://www.academia.edu/478319/A.O.Shemesh_Names_of_New_Agricultural_crops_in_Rabbinic_Literature_of_16-20th_Centuries_Moed_20_2010_pp._198-216_in_Hebrew_. Accessed on 3 Oct 2012
Steinberg J (1960) The dictionary of the Bible:Hebrew and Aramaic. Izreel Publishing House Ltd., Tel-Aviv (in Hebrew)
Steinmann SH (1998) Gender, pastoralism, and intensification: changing environmental resource use in Morocco. Yale F & ES Bulletin No 103, pp 81–107
Syrian Foodie in London. A blog, http://syrianfoodie.blogspot.com/2010/03/desert-truffles.html. Accessed on 2 Dec 2012
Talmud (Babylonian)—Ta'anit line 34 and Berahot 40:62; Shabat, 13:30
Talmud (Jerusalem)—Ma'assarot, 71:α1; Ta'anit 83
The Free Dictionary, online, http://www.thefreedictionary.com/claypan. Accessed 28 Apr 2013
The Holy Bible (1989) New and revised standard version. Thomas Nelson, Nashville
Trappe JM (1990) Use of truffles and false truffles around the world. In: Bencivenga M, Granetti B (eds), Proceedings, Atti del Secondo Congresso Internazionale sul Tartufo. Communita Montana dei Monti Martini e del Serano, Spoleto, Italy, pp 19–30
Trappe M, Claridge AW (2010) The hidden life of truffles. Scientific American, 78–84
Trappe JM, Claridge AW, Liddle L (2008a) Desert truffles of the Australian outback: ecology, ethnomycology and taxonomy. Econ Bot 62(3):497–506
Trappe JM, Claridge AW, Arora D, Smit WA (2008b) Desert truffles of the African Kalahari: ecology, ethnomycology and taxonomy. Econ Bot 62(3):521–529
Trufmania (2008) http://www.trufamania.com/Terfezia%20arenaria%20English.htm. Accessed on 8 Oct 2012
Wang S, Marcone MF (2011) The biochemical and biological properties of the world's most expensive underground edible mushroom: truffles. Food Res Int 44:2567–2581
Wasson VP, Wasson RG (1957) Mushrooms, Russia and History. Pantheon, New York

Chapter 16
Enzymes in *Terfezia claveryi* Ascocarps

Manuela Pérez-Gilabert, Francisco García-Carmona, and Asunción Morte

16.1 Introduction

Desert truffles are rich in fiber, proteins, vitamins, and minerals (Murcia et al. 2002, 2003). They are of considerable interest for ecological, agroforestal, and commercial purposes. Their ecological value is derived from their position in arid ecosystems as symbiotic mycorrhizal fungi associated with annual and perennial species of the genera *Cistus* and *Helianthemum* (Cistaceae). The introduction of desert truffle cultivation in dry environments is of strong agroforestry interest since it is a useful way of exploiting lands which have until now been regarded as unproductive (Morte and Honrubia 1995; Morte et al. 2000), thus helping to improve the social and economic level of these dry regions. In addition, the host plants are xerophytic species characteristic of semiarid environments, and their plantation could help to preserve lands from the ravages of erosion (Morte et al. 2000). The most important desert truffles are those included in the genera *Terfezia* (Pezizaceae) and *Balsamia* (Helvellaceae), because of their highly appreciated edible and commercial value.

Terfezia claveryi Chatin is a hypogeous ascomycete, which establishes mycorrhizal symbiosis with several annual and perennial species of the *Helianthemum* genus (Gutiérrez et al. 2003) and *Terfezia claveryi-Helianthemum almeriense* has been taken as a model system to study the process of mycorrhization (Morte et al. 1994, 2008; Gutiérrez et al. 2003). *T. claveryi and H. almeriense* establish

M. Pérez-Gilabert (✉) • F. García-Carmona
Dpto. Bioquímica y Biología Molecular-A, Facultad de Biología, Campus de Excelencia Internacional Regional "Campus Mare Nostrum", Campus de Espinardo, Universidad de Murcia, 30100 Murcia, Spain
e-mail: mpg@um.es

A. Morte
Dpto. Biología Vegetal (Botánica), Facultad de Biología, Campus de Excelencia Internacional Regional "Campus Mare Nostrum", Campus de Espinardo, Universidad de Murcia, 30100 Murcia, Spain

V. Kagan-Zur et al. (eds.), *Desert Truffles*, Soil Biology 38,
DOI 10.1007/978-3-642-40096-4_16,

different mycorrhizal forms, depending, among other factors, on the phosphorus content of the soil or the culture conditions (Gutiérrez et al. 2003; Zaretsky et al. 2006; see also Roth-Bejerano et al. (Chap. 5). Indeed, the importance of phosphorus in the biological cycle of *T. claveryi* led us to characterize and localize the enzymes involved in the metabolism of this macronutrient during the different stages of its life cycle (Navarro-Ródenas et al. 2009, 2012). Another hydrolase (esterase) and two oxidoreductases (lipoxygenase and tyrosinase) were also isolated and characterized from *T. claveryi* ascocarps (Pérez-Gilabert et al. 2001a, b, 2005a, b, c). Although the physiological role of most of these enzymes is not clear, their activity may affect the flavor, color, and texture of their ascocarps. The main methods used to partially purify and characterize these enzymes are summarized in this chapter.

16.2 Tyrosinase and Lipoxygenase: Two Oxidoreductases Isolated from *T. claveryi* Ascocarps

Oxidative and reductive processes may produce adverse physical and biochemical changes in *T. claveryi* ascocarps, which may influence not only the flavor but also the color and texture of the ascocarps, lowering their acceptability and nutritional value. Two oxidases, tyrosinase and lipoxygenase (LOX), which may play an important role in some of these processes have been purified and characterized from *T. claveryi* ascocarps.

Tyrosinase is responsible for the undesired enzymatic browning of mushrooms that takes place during senescence or as a result of damage during postharvest handling. On the other hand, due to the high proportion of polyunsaturated fatty acids present in *T. claveryi* ascocarps, lipid rancidity is one of the main factors limiting its storage life, since lipid peroxidation gives rise to unpleasant odors and tastes which lead to consumer rejection. Enzymes such as LOX can accelerate the spoilage caused by oxidative rancidity. Hydroperoxides produced by this enzyme decompose to form volatile aroma compounds, which are perceived as off-flavors (Gordon 2001). In addition, the free radicals formed during lipid oxidation may also lead to a reduction in nutritional quality by reacting with vitamins, especially vitamin E, which is lost from foods during its action as antioxidant.

16.2.1 Tyrosinase in T. claveryi *Ascocarps Is a Fully Latent Enzyme*

Tyrosinase (EC 1 .14.18.1) is a copper-containing monooxygenase which uses molecular oxygen to catalyze two different reactions: the oxidation of monophenols (monophenolase or cresolase activity) (Pérez-Gilabert et al. 2001a) and their

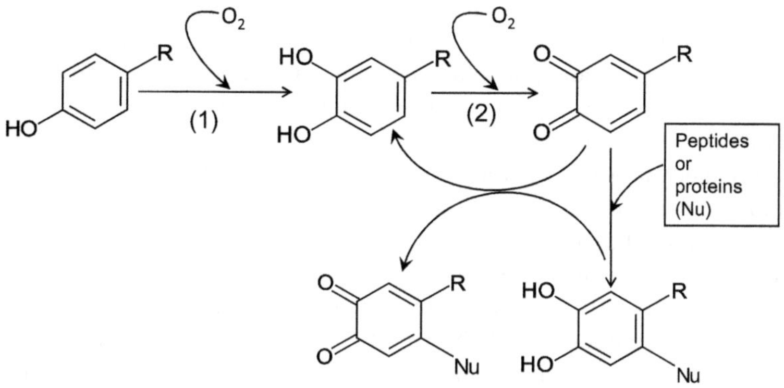

Fig. 16.1 Schematic representation of monophenolase (1) and diphenolase (2) activities catalyzed by tyrosinase. *Nu* nucleophile

subsequent oxidation to *o*-quinones (diphenolase or catecholase activity). The enzymatically generated quinones produce new molecules of *o*-diphenols (Fig. 16.1) (Pérez-Gilabert et al. 2001b). The quinones thus formed are highly reactive substances, which normally react further with other quinones, amino acids, or proteins (nucleophiles) to produce the colored compounds that are responsible for losses in nutrient quality (Sánchez-Ferrer et al. 1993). This is also a severe problem in the isolation of plant enzymes. Enzyme nomenclature differentiates between monophenol oxidase (tyrosinase, EC 1.14.18.1) and catechol oxidase or *o*-diphenol: oxygen oxidoreductase (EC 1.10.3.2). Since both monophenolase and diphenolase cycles overlap (Sánchez-Ferrer et al. 1995), indicating that a true tyrosinase must show both activities, we will use the term tyrosinase when referring to the enzyme present in *T. claveryi* ascocarps which presents both activities. The first step in the biosynthesis of melanin is catalyzed by tyrosinase. In fungi, melanin has been correlated with differentiation of reproductive organs and spore formation, the virulence of pathogenic fungi, and tissue protection after injury (see references in Pérez-Gilabert et al. 2001b). However, although tyrosinase seems to be of almost universal distribution in animals, plants, fungi, and bacteria (see the large number of articles devoted to tyrosinase), much is still unknown about the biological function of this enzyme (Mayer 2006).

One unusual and intriguing characteristic of tyrosinase is its ability to exist in an inactive or latent state, and in many sources, it can be found in either latent or active form as well as in both forms simultaneously (Whitaker 1995). Latent mushroom (*Agaricus bisporus*) tyrosinase represents around 98–99 % of the total activity (van Leeuwen and Wichers 1999). Latent tyrosinase has been activated by a variety of treatments using agents such as proteases, urea, fatty acids, alcohols, polyamines, divalent cations, polyglucan elicitors, acid or basic shock, and detergents such as SDS (see references in Pérez-Gilabert et al. 2004). Its activation may also result from pathogen attack. The use of SDS as an activating agent is particularly interesting because most enzymes are inactivated by this compound. *T. claveryi* tyrosinase is one of the few fully latent tyrosinases which have been characterized to date. The enzyme obtained both from mature and immature ascocarps was fully latent, so that neither monophenolase nor diphenolase activity could be detected unless SDS or trypsin is added to the reaction medium. The effect of different concentrations of SDS on the activity of *T. claveryi* tyrosinase was analyzed using 4-*tert*-butylcatechol or L-DOPA as substrates (Fig. 16.2). For both substrates, little activation occurred until a concentration above 0.5 mM SDS was present in the reaction medium. The enzyme was activated in a sigmoidal manner (Fig. 16.2), with increasing SDS concentrations below the *critical micellar concentration* for SDS (3.5 mM at pH 6.0). Gandía-Herrero et al. (2005) proposed the existence of a regulatory peptide, unrelated with the affinity of the enzyme, whose presence would block the access of phenolic compounds to the active center. The binding of discrete molecules of SDS to tyrosinase could modify the 3D structure of this enzyme, improving the accessibility of the substrates to the active site, without directly affecting its integrity. The presence of this peptide is assumed to be related with "shield region" described in hemocyanins to block the access of phenolic compounds, and also proposed in the structural model of a latent catechol oxidase (Gerdemann et al. 2002a, b). With respect to the physiologically relevant counterpart of the detergent, it was suggested that lipids might fulfill this role (van Gelder et al. 1997). The reversibility of SDS activation was clearly established using sections of *T. claveryi* ascocarps, since, when sections were preincubated with SDS and then rinsed and incubated with L-DOPA, no color was developed (Pérez-Gilabert et al. 2004). No reaction was observed when the sections of both mature and young ascocarps were incubated with tyrosine or L-DOPA without SDS, which confirms that *T. claveryi* tyrosinase, both in young and mature ascocarps, is a fully latent enzyme.

16.2.2 Lipoxygenase

Oxylipins constitute a large family of oxidized fatty acids and metabolites derived from the same. These bioactive lipids, widely distributed on the phylogenetic scale, are involved in regulating developmental processes as well as environmental responses. Compared with the amount of information accumulated during recent years on mammalian and plant oxylipins, knowledge about these bioactive lipids in fungi is still scarce. In a recent review, Brodhun and Feussner (2011) presented a synopsis of the oxylipins identified so far in fungi and the enzymes involved in their biosynthesis. Oxylipins are enzymatically formed by an initial peroxidation reaction of a polyunsaturated fatty acid that is catalyzed by lipoxygenase (LOX), thus starting the so-called LOX pathways. LOXs (linoleate: oxygen oxidoreductase, EC 1.13.11.12) are nonheme iron-containing enzymes that use molecular oxygen in the stereospecific and regiospecific oxidation of a polyunsaturated fatty acid containing a 1Z,4Z-pentadiene system, giving rise to their corresponding hydroperoxides. The hydroperoxy fatty acid formed is converted by other enzymes (allene oxide

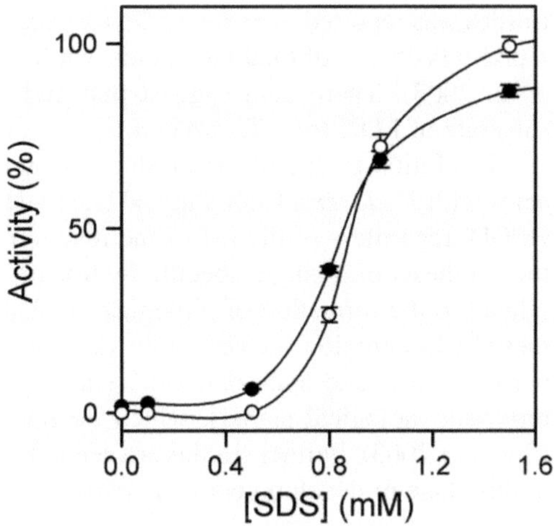

Fig. 16.2 Percentage of activation of *Terfezia claveryi* tyrosinase with different concentrations of SDS and 4-*tert*-butylcatechol (*open circle*) or L-DOPA (*filled circle*) as substrates

synthase, divinyl ether synthase, hydroperoxide lyase, and peroxygenase), yielding a large variety of structurally different products (Brodhun and Feussner 2011). Some plant LOXs are constitutive, whereas others are expressed by wounding and by fungal pathogens (Oliw 2002). Although there is little information on fungal LOX and on its physiological role, it has been proposed that oxylipins act as host-fungus communication signals (Tsitsigiannis and Keller 2007; Christensen and Kolomiets 2011).

Several LOXs have been identified in different fungal species such as *Saprolegnia parasitica* (Hamberg 1986; Herman and Hamberg 1987), *Fusarium* (Bisakowski and Kermasha 1998), *Thermomyces lanuginosus* (Li et al. 2001), *Geotrichum candidum* (Perraud et al. 1999), *Penicillium camemberti, Penicillium roqueforti* (Perraud and Kermasha 2000), and various *Mortierella strains* (Filippovich et al. 2001). With respect to LOX from edible fungi, there is rather little information and only LOXs from *Pleurotus ostreatus* (Kuribayashi et al. 2002) and from *Terfezia claveryi* (Pérez-Gilabert et al. 2005a, b) have been partially purified and characterized.

T. claveryi lipoxygenase was extracted from ascocarps and partially purified using phase partitioning with TX-114 (see Sect. 16.5.1.1 and Fig. 16.4). Then, the enzyme was purified to apparent homogeneity by two steps of cation-exchange chromatography (Pérez-Gilabert et al. 2005a). The enzyme recovered after the first chromatographic step was very unstable, probably because most of the components of the initial homogenate had been eliminated. LOX inactivation was avoided by adding TX-100 (final concentration 0.04 % v/v) to the enzyme. The progress of the purification was followed by SDS-PAGE (Laemmlli 1970). The apparent molecular mass of the purified *T. claveryi* LOX, also determined by SDS-PAGE, was 66 kDa, identical to the molecular weight of *P. ostreatus* LOX (Kuribayashi et al. 2002).

The highest relative enzymatic activity (100 %) was obtained using linoleic acid as the substrate, followed by linolenic acid (91 %). The lowest relative LOX activity (32 %) was exhibited using γ-linolenic acid. The lipid content of raw truffles was reported to be 69.5 g/kg (dry matter) (Murcia et al. 2003). Linoleic acid represents 45.4 % of total fatty acids, while linolenic acid represents 5.8 % (Murcia et al. 2003). These data suggest that endogenous linoleic acid is the preferred substrate of LOX from *T. claveryi*.

The elution profile of the products obtained by incubating linoleic or linolenic acids with *T. claveryi* LOX showed one major peak which corresponded to 13-*Z*,*E*-HPOD, regardless of the pH of the reaction medium. According to Brash (1999), the synthesis of a single specific hydroperoxide from free fatty acid substrates is related to the formation of biological mediators of signaling molecules. In plants, these hydroperoxides serve as substrate for enzymes, such as hydroperoxide lyase, peroxygenase, and hydroperoxide reductase. Some of these fatty acid derivatives represent biological signals, which do not require the prior activation of genes (Spiteller 2003). Further studies are needed to clarify the physiological role of LOX at this stage of development of *T. claveryi*.

16.3 Hydrolases from *T. claveryi* Ascocarps: Esterases and Phosphatases

Hydrolases catalyze the hydrolysis of various bonds. Some of these enzymes have a very wide specificity, and sometimes it is difficult to decide whether two preparations described by different authors correspond to the same enzyme or not. Esterases, hydrolases acting on ester bonds, are subdivided into different groups, depending on the chemical structure of their substrate. The esterases described in this chapter correspond to a carboxylic-ester hydrolase (EC 3.1.1) and phosphoric monoester hydrolases or phosphatases (EC 3.1.3).

16.3.1 Characterization of an Esterase from T. claveryi *Ascocarps*

Esterases (EC 3.1.1) represent a diverse group of enzymes catalyzing the hydrolysis and formation of ester bonds and are widely distributed in animals, plants, and microorganisms (Ollis et al. 1992; Bornscheuer 2002). Esterases are arbitrarily classified as enzymes that hydrolyze substrates in solution, whereas lipases hydrolyze substrate in emulsion (Chahinian et al. 2002). Among esterases, "true esterases" (EC 3.1.1.1) hydrolyze esters of short-chain carboxylic acids ($\leq$12), and lipases (EC 3.1.1.3) display maximum activity toward insoluble long-chain ($\geq$12) acylglycerides (Eggert et al. 2002). Many of them show wide substrate tolerance, which has led to the assumption that they have evolved to enable access to carbon sources or are involved in catabolic pathways.

To date, the characterization of esterases from some mushroom species such as *Sparassis crispa, Amanita vaginata* var. *vaginata, Tricholoma terreum*, and *Lycoperdon perlatum* have been reported (Colak et al. 2009; Ertunga et al. 2009; Chandrasekaran et al. 2011; Akatin et al. 2011). In addition, esterase has been detected in pathogenic fungi, such as *Pestalotia malicola* (Sugui et al. 1998) and *Botrytis elliptica* (Hsiesh et al. 2001), where this enzyme could be involved in the infection processes (Deising et al. 1992; Jansson and Akesson 2003). The esterase isozyme pattern has been used as a tool for the characterization of arbuscular mycorrhizal fungi (Hepper et al. 1986) and in taxonomic studies of edible mushrooms (Yokono et al. 2000; Bonefant Magne et al. 1997; Matsumoto et al. 1995). Several authors have observed that mycorrhizal colonization causes an increase in the esterase activity in roots (Fries et al. 1996; Timonen and Sen 1998; Vázquez et al. 2000), which has been taken as an indication of the enhanced metabolic activity produced during the establishment of mycorrhizal associations (Fries et al. 1996).

The products of the esterase reaction, both short-chain fatty acids and esters, constitute a well-known class of aromatic molecules in foods (Cristiani and Monnet 2001; Nardi et al. 2002; McSweeney and Sousa 2000). For example, esterase from brewer's yeast *Saccharomyces carlsbergensis* was found to contribute to the aroma and flavor of beer (Horsted et al. 1998), whereas esterases and lipases produced by fungi such as *Penicillium* species (McSweeney and Sousa 2000) play an important role in the flavor of cheese.

Our group has reported the partial purification of an esterase from ascocarps of *T. claveryi,* and characterized it to determine the optimum reaction conditions (pH and temperature), the kinetic parameters involved and the effect of various chemicals on the activity (Pérez-Gilabert et al. 2005c). Preliminary results indicated that ascocarps of *T. claveryi* contained an enzymatic activity that was responsible for the hydrolysis of *p*-nitrophenyl esters. In principle, this reaction may have been carried out by a lipase or by an esterase. Lipases differ from esterases in that their natural substrates are insoluble in water and their activity is maximal only when the enzyme is adsorbed to the oil/water interface. Thus, a series of *p*-nitrophenyl esters, prepared in two different reaction media, were assayed in order to distinguish between both enzymes. The substrate was prepared by emulsifying it with TX-100 or by dilution with ethanol. The results obtained with the two reaction media were similar, and no interfacial activation was observed. Together with the fact that maximum activity was displayed with short-chain *p*-nitrophenyl esters such as *p*-NPA (*p*-nitrophenyl acetate) and *p*-NPB (*p*-nitrophenyl butyrate) and that no reaction was observed when *p*-NPL (*p*-nitrophenyl laurate) or *p*-NPP (*p*-nitrophenyl palmitate) was used as substrate, these results indicate that the enzyme responsible for the hydrolysis of *p*-nitrophenyl esters was an esterase and not a lipase. Most lipases and esterases contain the consensus sequence motif Gly-X-Ser-X-Gly (where x represents an arbitrary amino acid residue) around the active site serine (Bornscheuer 2002). Phenylboric acid is a reversible inhibitor of most serine esterases (Gjellesvik et al. 1992), which binds at or near the serine of the active site (Garner 1980).

The inhibition of *T. claveryi* esterase by this compound suggests that serine residues are involved in the enzyme activity. The nature of this inhibition was studied and the results obtained indicate that phenylboric acid is a competitive inhibitor of *T. claveryi* esterase.

16.3.2 Phosphatase Activity in Terfezia claveryi *Ascocarps*

The importance of phosphorus in the biological cycle of *T. claveryi* led us to characterize and localize the enzymes involved in the metabolism of this macronutrient during the different stages of the fungal life cycle (Navarro-Ródenas et al. 2009, 2011).

Phosphatases have traditionally been classified as alkaline (ALP) or acid (ACP), according to whether their optimal pH for catalysis is above or below pH 7.0 (Duff et al. 1994). The measurements of phosphatase activity in the crude extract of *Terfezia claveryi* ascocarps revealed two peaks of activity, one with maximum activity between pH 9.5 and 10.0 and another, of lower activity, with a maximum at pH 5.0. These results suggest that *T. claveryi* ascocarps contain both ACP and ALP, the latter being the main one since the activity measured at pH 10.0 was 2.8 times higher than that measured at pH 5.0 (Navarro-Ródenas et al. 2009). These results contrast with those reported by Reyes et al. (1990) for *Neurospora crassa* (Ascomycota), *Coriolus versicolor,* and *Schizophyllum commune* (Basidiomycota). In these fungi, ACP activity was higher than ALP at the maximum dry weight of their mycelium.

The phosphatase present in many soil organisms has been used as a biochemical indicator in the measurement of biological activity since phosphatase activity responds to adverse conditions such as pollution, soil degradation, drought, and phosphorus limitation (Kuperman and Carreiro 1997; Trasar-Cepeda et al. 1998; Ming and Hui 1999; van Aarle and Plassard 2010). Phosphatase activity, as detected by vital staining of intraradical (e.g., Tisserant et al. 1993; Larsen et al. 1996) and extraradical (Zhao et al. 1997; Vosatka and Dodd 1998) hyphae, has been extensively used as an indicator of metabolic activity in arbuscular mycorrhizal fungi. In a recent paper, our group reported the presence of ALP (but not ACP) in *T. claveryi* mycelium (Navarro-Ródenas et al. 2011). This enzyme responded to water stress and could be used as an indicator of the metabolic activity of this fungus (Navarro-Ródenas et al. 2011). The presence of ALP both in mycelia and ascocarps indicates that this enzyme must play an important role during the life cycle of *T. claveryi*, while ACP might be involved in a process that takes place during the ascocarp stage.

16.4 Localization of Enzymes in the Ascocarp

Localization of fungal enzymes at the whole tissue level using histochemical methods is a useful tool for the characterization and elucidation of the function of many enzymes (Miranda et al. 1997; Pérez-Gilabert et al. 2001a, b). However, the application of this method has been hampered by endogenous substrates, inhibitors, and phenolics that impair the enzymatic activity (Spruce et al. 1987). Nevertheless, the clear color of *T. claveryi* ascocarps permitted the direct localization of several enzymatic activities (esterase, ALP, and mono- and diphenolase) in 10 μm sections, after incubating them with the appropriate substrate. The sections were obtained by cutting ascocarps, previously frozen in liquid nitrogen, with a cryostat.

The histochemical localization of esterase activity was carried out using *p*-nitrophenyl butyrate (*p*-NPB) as a substrate (Pérez-Gilabert et al. 2005c). In the case of tyrosinase and ALP, the substrates used were those described in Sect. 16.5.2 to reveal the enzymatic activity in NativePAGE (Pérez-Gilabert et al. 2001a, b; Navarro-Ródenas et al. 2009). In the case of ALP, besides using NBT (nitroblue tetrazolium chloride, see Sect. 16.5.2) and BCIP (5-bromo-4-chloro-3-indoxyl phosphate, see Sect. 16.5.2) colorimetric method, detection was carried out using the *enzyme-labeled fluorescent substrate* contained in the ELF97 Endogenous Phosphatase Detection Kit (Navarro-Ródenas et al. 2009). All ascocarp sections were observed using an epifluorescence microscope fitted with a mercury lamp, under bright-field settings for visible precipitates or, in the assay with ELF, under epifluorescence settings. For each enzyme, control sections were incubated with an inhibitor or only with buffer; conditions under which no staining was observed.

While esterase activity was localized only in the hypothecium, both monophenolase and diphenolase activities of tyrosinase (Fig. 16.3) and ALP have been localized in the peridium, hypothecium, and the ascogenic hyphae of the gleba, which seem to be the most metabolically active tissues in the truffle ascocarp. The co-localization of monophenolase and diphenolase activities of *T. claveryi* tyrosinase has also been reported by Miranda et al. (1992, 1997) and Ragnelli et al. (1992) for different species of *Tuber,* confirming the bifunctional character of this enzyme. The localization of tyrosinase in *T. claveryi* ascocarps agrees with the localization observed in *Tuber* species (Miranda et al. 1997), and diphenolase localization in *T. claveryi* is more similar to that observed in the white truffles *Tuber magnatum, Tuber puberulum, and Tuber sphaerospermum,* which do not show evident ramification of the basal cavity, than to its localization in *Tuber melanosporum* and *Tuber excavatum*, which possess a different ascocarp morphology. The presence of diphenolase activity in mature ascocarps of *T. claveryi* (Fig. 16.3) at even higher levels than those detected in young ascocarps contrasts with the results reported by Ragnelli et al. (1992) in subadult ascocarps of *T. melanosporum* and *Tuber aestivum*, where no diphenolase activity was observed, even though a weak Schmorl reaction for melanin occurred. These discrepancies, together with the lack of evidence of tyrosinase activity in other desert truffles (Miranda et al. 1992; Ragnelli et al. 1992), might be due to differences in the

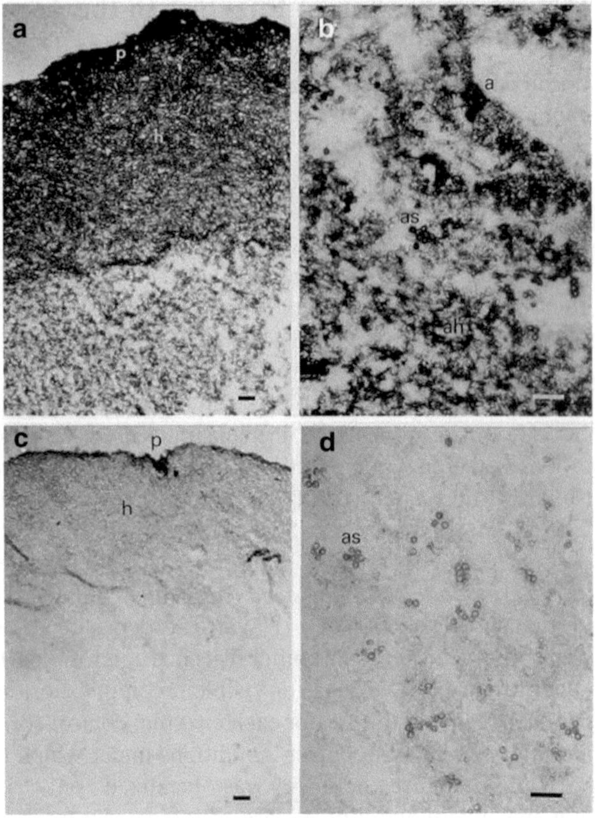

Fig. 16.3 Localization of tyrosinase activity in sections of *Terfezia claveryi* mature ascocarps. The sections were incubated in 10 mM L-DOPA containing 2.5 mM SDS in 0.1 M phosphate buffer pH 5.5 (**a** and **b**) or in 0.1 M phosphate buffer pH 5.5 (**c** and **d**). As, ascospores; a, asci; ah, ascogenic hyphae; h, hypothecium; p, peridium. Bars=40 μm

latency state of this enzyme, which was not analyzed by these authors. If SDS had not been added to the incubation medium of *T. claveryi* sections, the histochemical localization of tyrosinase would not have been possible (Fig. 16.3).

The localization of ALP activity in the peridium, hypothecium, and in the ascogenic hyphae of the gleba carried out by BCIP/NBT staining was confirmed with fluorescent staining (Navarro-Ródenas et al. 2009). The specificity of both staining methods was demonstrated by adding orthovanadate, an inhibitor of ALP, to the incubation medium. No ALP reaction was observed with BCIP/NBT staining, and only very weak fluorescence was detected in the hypothecium and gleba compared to that observed during incubation without inhibitor. This weak fluorescence of the ascocarp section incubated with ELF and orthovanadate may reflect the higher sensitivity of the fluorescent staining method as compared to BCIP/NBT staining. The capacity of the tufts of hyphae that sprout from the peridium of the ascocarps of *Tuber aestivum* and *Tuber melanosporum* to absorb phosphate has previously been confirmed (Barry et al. 1995) using labeled phosphate. These authors demonstrated that the conduction pathways of the absorbed phosphate were the fertile veins or ascogenic hyphae of the gleba and, in particular, the sterile hyphae adjacent to the paraphysis. This phosphate was metabolized rapidly by the ascocarp (Barry et al. 1995). The high levels of ALP detected in the unripe fruiting body of *T. claveryi* suggest that this enzyme may play an important role in transferring phosphate from the peridium to the inner part of the gleba, particularly into the fertile hyphae of the gleba, since the phosphate transport is not determined by water flux (Barry et al. 1994). These results provide additional evidence that the ascocarp of *T. claveryi*, during some stages of its development, may become nutritionally autonomous and independent from the host plant.

16.5 Methods Used to Partially Purify and Characterize the Above-Described Enzymes

16.5.1 Extraction and Partial Purification of Enzymes from T. claveryi *Ascocarps*

16.5.1.1 Phase Partitioning with Triton X-114

Enzyme purification in fungal extracts is difficult because of the large variety and quantity of secondary products that may bind tightly to the enzymes and change their characteristics. To overcome this, different methods have been developed, including acetone powders, ammonium sulfate fractionation, hydrophilic and insoluble polymers, and detergents. Among the last, the use of Triton X-114 (TX-114), a polyoxyethylene-type nonionic detergent, presents many advantages for the purification of proteins from plants and fungi (Sánchez-Ferrer et al. 1994). One of the most important advantages is its ability to form clear solutions in buffers at 4 °C, whereas it separates into two phases at 25 °C due to the formation of large micellar aggregates (Bordier 1981). This characteristic has been used not only to separate integral proteins from hydrophilic proteins, because the former remain in the detergent rich phase (Pryde and Philips 1986), but also to extract plant and fungal proteins while eliminating chlorophylls, phenols, etc. (Pérez-Gilabert and García-Carmona 2000; Pérez-Gilabert et al. 2001b, 2005a, c). In addition to phenolic compounds, truffles present a high amount of lipids that strongly interfere with the spectrophotometric characterization of different enzymes since they increase the turbidity of the reaction medium and may form micellar aggregates with certain hydrophobic substrates. We used phase partitioning with TX-114 to partially purify several enzymes from *T. claveryi* ascocarps: tyrosinase (Pérez-Gilabert et al. 2001b), lipoxygenase (Pérez-Gilabert et al. 2005a), and esterase (Pérez-Gilabert et al. 2005c). In this way, a reduction of 87 % in the triglyceride content was achieved along with removal of 62 % of the phenols (Pérez-Gilabert et al. 2005b, c). This reduction in the phenolic content was sufficient to avoid browning of the enzyme solutions even after several months of storage at −20 °C. The degree of enzyme purification after phase partitioning was similar for the three enzymes and ranged from 1.5-fold for esterase to 2.3-fold for lipoxygenase

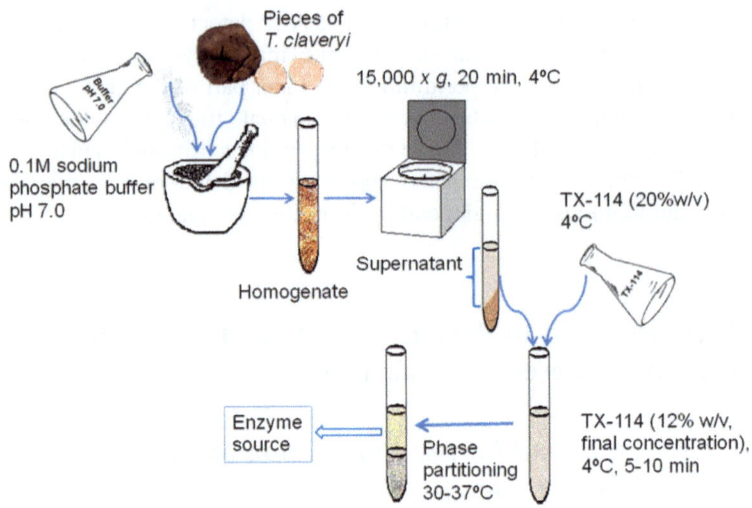

Fig. 16.4 General method used to extract esterase, tyrosinase, and lipoxygenase from ascocarps of *Terfezia claveryi* using phase partitioning with TX-114

(Pérez-Gilabert et al. 2001b, 2005a, c); moreover, little protein loss was observed, the enzyme recovery being around 65 % in the three cases. The method used to extract esterase, tyrosinase, and lipoxygenase is briefly described in the scheme presented in Fig. 16.4. The fact that neither sonication nor the addition of detergent to the extraction buffer was needed, together with the partitioning of these enzymes in the aqueous phase, suggests that tyrosinase, esterase, and LOX from *T. claveryi* ascocarps are soluble enzymes.

16.5.1.2 Partial Purification of ALP Using Precipitation by Polyethylene Glycol

The precipitation by PEG, a non-denaturing hydrophilic nonionic polymer, is a common method to concentrate proteins as the first step of a purification process. PEG acts as an inert solvent sponge, reducing solvent availability. The effective protein concentration increases with increasing concentrations of PEG until solubility is exceeded and precipitation occurs (Atha and Ingham 1981). Precipitation by polyethylene glycol was used to partially purify ALP (Navarro-Ródenas et al. 2009), as an alternative to phase partitioning with TX-114, which did not give satisfactory results with this enzyme. The partial purification of ALP was necessary to avoid interferences from the ACP activity present in the crude extract and to remove most of the turbidity produced by lipids and other substances, which strongly interfere with the spectrophotometric characterization of this enzyme. In the final step of this partial purification, most of the lipids (98 %) and phenolic compounds (88 %) were removed. ALP was purified fivefold, with a recovery of

53 % (Navarro-Ródenas et al. 2009). No ACP activity was detected in this fraction. The enzyme remained stable at −80 °C for >3 months. ALP was extracted without the addition of detergent or sonicating with a buffer of high ionic strength, which indicates that ALP from *T. claveryi* ascocarps is also soluble enzyme. This finding contrasts with the fact that many nonbacterial ALPs are membrane-bound proteins (Asgeirsson et al. 1995). Different approaches were examined to purify ACP; however, this enzyme was very sensitive and its activity was easily lost (data not shown).

16.5.2 Non-denaturing Electrophoresis

To preserve the enzymatic activity of a protein, non-denaturing or native electrophoresis systems must be used. The gels used in NativePAGE are prepared without SDS and the protein sample is not boiled or treated with SDS plus 2-mercaptoethanol as in an SDS-PAGE. The electrophoresis is run at 4 °C to avoid the irreversible inactivation of the enzyme due to an excessive warming. Enzyme activity staining after NativePAGE is a widely used technique that allows the isoforms of a specific enzyme present in a sample to be detected. Several detection methods were used to detect the presence of ALP and tyrosinase (monophenolase and diphenolase activities) in extracts from *T. claveryi* ascocarps to reveal ALP activity: the gel was incubated with BCIP (5-bromo-4-chloro-3-indoxyl phosphate) and NBT (nitroblue tetrazolium chloride) in the presence of NaCl and $MgCl_2$ (Navarro-Ródenas et al. 2009). A single band of ALP activity was observed, which indicates that under these conditions a single ALP enzyme is extracted.

In the case of tyrosinase, after the electrophoresis run, the gel was incubated with L-DOPA (diphenolase activity) or tyrosine (monophenolase activity) plus SDS. Native electrophoresis revealed that both the monophenolase and diphenolase activities of tyrosinase may be attributed to a single protein. This is in contrast to the multiplicity found in PAGE for tyrosinases of different truffles of the genus *Tuber* (Miranda et al. 1992), which may be attributed to covalent bonds of this enzyme with its generated *o*-quinones (Matheis 1987).

Acknowledgements This work has been supported by projects BIO2010-22225-C02-01 and CGL2011-29816 (**MINECO-FEDER**) and Project 04541/GERM/06 (Fundación Séneca, Región de Murcia, Spain).

References

Akatin MY, Colak A, Ertunga NS (2011) Characterization of an esterase activity in *Lycoperdon pyriforme*, an edible mushroom. J Food Biochem. doi:10.1111/j.1745-4514.2011.00621.x

Asgeirsson B, Hartemink R, Chlebowski JF (1995) Alkaline phosphatase from Atlantic cod (*Gadus morhua*). Kinetic and structural properties which indicate adaptation to low temperatures. Comparative biochemistry and physiology. Biochem Mol Biol 110:315–329

Atha DH, Ingham KC (1981) Mechanism of precipitation of proteins by polyethylene glycols. J Biol Chem 256:12108–12117

Barry D, Staunton S, Callot G (1994) Mode of the absorption of water and nutrients by ascocarps of *Tuber melanosporum* and *Tuber aestivum*. A radioactive tracer technique. Can J Bot 72:317–322

Barry D, Jaillard B, Staunton S, Callot G (1995) Translocation and metabolism of phosphate following absorption by the ascocarps of *Tuber melanosporum* and *T. aestivum*. Mycol Res 99:167–172

Bisakowski B, Kermasha S (1998) Characterization of purified lipoxygenase extracts from *Fusarium proliferatum*. J Agric Food Chem 46:2382–2388

Bonefant Magne M, Magne C, Esnault MA, Lemoine C (1997) Characterization of cultivated strains of a new edible mushroom: *Stropharia rugoso-annulata*. 1. Protein variability. Cryptogam Mycol 18:255–265

Bordier C (1981) Phase separation of integral membrane proteins in TX-114 solutions. J Biol Chem 256:1604–1607

Bornscheuer UT (2002) Microbial carboxyl esterases: classification, properties and application in biocatalysis. FEMS Microbiol Rev 26:73–81

Brash AR (1999) Lipoxygenases: occurrence, functions, catalysis, and acquisition of substrate. J Biol Chem 274:23679–23682

Brodhun F, Feussner I (2011) Oxylipins in fungi. FEBS J 278:1047–1063

Chahinian H, Nini L, Boitard E, Dubes JP, Comeau LC, Sarda L (2002) Distinction between esterases and lipases: kinetic study with vinyl esters and TAG. Lipids 37:653–662

Chandrasekaran G, Kim GJ, Shin HJ (2011) Purification and characterization of an alkaliphilic esterase from a culinary medicinal mushroom. Sparassis crispa Food Chemistry 124:1376–1381

Christensen SA, Kolomiets MV (2011) The lipid language of plant-fungal interactions. Fungal Genet Biol 48:4–14

Colak A, Camedan Y, Faiz O, Sesli E, Kolcuoglu Y (2009) An esterolytic activity from a wild edible mushroom, *Lycoperdon perlatum*. J Food Biochem 33:482–499

Cristiani G, Monnet V (2001) Food micro-organisms and aromatic ester synthesis. Sci Aliment 21:211–230

Deising H, Nicholson RL, Haug H, Howard RJ, Mendgen K (1992) Adhesion pad formation and the involvement of cutinase and esterases in the attachment of uredospores to the host cuticle. Plant Cell 4:1101–1111

Duff SMG, Sarath G, Plaxton WC (1994) The role of acid phosphatases in plant phosphorus metabolism. Physiol Plant 90:791–800

Eggert T, Pouderoyen GV, Pencreach G, Douchet I, Verger R, Dijstra BW, Jeager K (2002) Biochemical properties and three dimensional structures of two extracellular lipolytic enzymes from *Bacillus subtilis*. Colloid Surf B 26:37–46

Ertunga NS, Cakmak U, Colak A, Faiz O, Sesli E (2009) Characterisation of esterolytic activity from two wild mushroom species, *Amanita vaginata* var. vaginata and *Tricholoma terreum*. Food Chem 115:1486–1490. doi:10.1016/j.foodchem.2009.01.090

Filippovich SY, Rybakov YA, Afanasieva TP, Bachurina GP, Lukina GP, Ezhova IE, Nosova AV, Arjushkina TV, Sineokii SP, Kritskii MS (2001) Characterization of lipoxygenase from fungi of the genus *Mortierella*. Appl Biochem Microbiol 37:473–479

Fries LLM, Pacovsky RS, Safir GR (1996) Expression of isoenzymes altered by both *Glomus intraradices* colonization and formononetin application in corn (*Zea mays* L.) roots. Soil Biol Biochem 28:981–988

Gandía-Herrero F, Jiménez-Atiénzar M, Cabanes J, García-Carmona F, Escibano J (2005) Differential activation of a latent polyphenol oxidase mediated by sodium dodecyl sulfate. J Agric Food Chem 53:6825–6830. doi:10.1021/jf050505e

Garner CW (1980) Boronic acid inhibitors of porcine pancreatic lipase. J Biol Chem 255:5064–5068

Gerdemann C, Eicken C, Galla HJ, Krebs B (2002a) Comparative modeling of the latent form of a plant catechol oxidase using a molluscan hemocyanin structure. J Inorg Biochem 89:155–158

Gerdemann C, Eicken C, Krebs B (2002b) The crystal structure of catechol oxidase: new insight into the function of type-3 copper proteins. Acc Chem Res 35:183–191

Gjellesvik DR, Lombardo D, Walter BT (1992) Pancreatic bile salt dependent lipase from cod (*Gadus morhua*): purification and properties. Biochim Biophys Acta 1124:123–124

Gordon MH (2001) The development of oxidative rancidity in foods. In: Pokorny J, Yanishlieva N, Gordon M (eds) Antioxidants in food. Practical applications. CRC, Washington DC, USA, pp 7–20

Gutiérrez A, Morte A, Honrubia M (2003) Morphological characterization of the mycorrhiza formed by *Helianthemum almeriense* Pau with *Terfezia claveryi* Chatin and *Picoa lefebvrei* (Pat.) Maire. Mycorrhiza 13:299–307

Hamberg M (1986) Isolation and structure of lipoxygenase from *Saprolegnia parasitica*. Biochem Biophys Acta 876:688–692

Hepper CM, Sen R, Maskall CS (1986) Identification of vesicular arbuscular mycorrhizal fungi in roots of leek (*Allium porrum* L.) and maize (*Zea mays* L.) on the basis of enzyme mobility during polyacrylamide gel electrophoresis. New Phytol 102:529–539

Herman RP, Hamberg M (1987) Properties of the soluble arachidonic acid 15-lipoxygenase 15-hydroperoxide isomerase from the oomycete *Saprolegnia parasitica*. Prostaglandins 34:129–139

Horsted MW, Dey ES, Holmberg S, Kielland-Brandt MC (1998) A novel esterase from *Saccharomyces carlsbergensis*, a possible function for the yeast TIP1 gene. Yeast 14:793–803

Hsiesh TF, Huang JW, Hsiang T (2001) Light and scanning electron microscopy studies on the infection of oriental lily leaves by *Botrytis elliptica*. Eur J Plant Pathol 107:571–581

Jansson HB, Akesson H (2003) Extracellular matrix, esterase and the phytotoxin prehelminthosporol in infection of barley leaves by *Bipolaris sorokiniana*. Eur J Plant Pathol 109:599–605

Kuperman R, Carreiro M (1997) Soil heavy metal concentrations, microbial biomass and enzyme activities in a contaminated grassland ecosystem. Soil Biol Biochem 29:179–190

Kuribayashi T, Kaise H, Uno C, Hara T, Hayakawa T, John T (2002) Purification and characterization of lipoxygenase from *Pleurotus ostreatus*. J Agric Food Chem 50:1247–1253

Laemmlli UK (1970) Cleavage of structural proteins during assembly of head of bacteriophage-T4. Nature 227:680–685

Larsen J, Thingstrup I, Jakobsen I, Rosendahl S (1996) Benomyl inhibits phosphorus transport but not fungal alkaline phosphatase activity in a *Glomus*-cucumber symbiosis. New Phytol 132:127–133

Li DC, Lui ZW, Lu J (2001) Purification and characterization of lipoxygenase from the thermophilic fungus *Thermomyces lanuginosus*. Mycol Res 105:190–194

Matheis G (1987) Polyphenol oxidase and enzymatic browning of potatoes (*Solanum tuberosum*) I. Properties of potato polyphenol oxidase. Chem Mikrobiol Technol Lebensm 11:5–12

Matsumoto T, Mimura K, Fukumasa Nakai Y (1995) Isozyme variation and genetic relatedness among natural populations of *Pleurotus ostreatus*. J Gen Appl Microbiol 41:487–497

Mayer AM (2006) Polyphenol oxidases in plants and fungi: going places? A review. Phytochemistry 67:2318–2331

McSweeney PLH, Sousa MJ (2000) Biochemical pathways for the production of flavor compounds in cheeses during ripening: a review. Lait 80:293–324

Ming T, Hui C (1999) Effects of arbuscular mycorrhizal fungi alkaline phosphatase activities on *Hippophae rhamnoides* drought resistance under water stress conditions. Trees 14:113–115

Miranda M, Bonfigli A, Zarivi O, Ragnelli AM, Pacioni G, Botti D (1992) Truffle tyrosinase: properties and activity. Plant Sci 81:175–182

Miranda M, Zarivi O, Bonfigli A, Amicarelli F, Aimola P, Ragnelli AM, Pacioni G (1997) Melanogenesis tyrosinase expression, and reproductive differentiation in black and white truffles (ascomycotina). Pigm Cell Res 10:46–53

Morte A, Honrubia M (1995) Improvement of mycorrhizal synthesis between micropropagated *Helianthemum almeriense* plantlets with *Terfezia claveryi* (desert truffle). In: Elliott TJ (ed) Science and cultivation of edible fungi. Balkema, Rotterdam, The Netherlands

Morte A, Cano A, Honrubia M, Torres P (1994) *In vitro* mycorrhization of micropropagated Helianthemum almeriense plantlets with *Terfezia claveryi* (desert truffle). Agric Sci Finland 3:309–314

Morte A, Lovisolo C, Schubert A (2000) Effect of drought stress on growth and water relations of the mycorrhizal association *Helianthemum almeriense-Terfezia claveryi*. Mycorrhiza 10:115–119

Morte A, Honrubia M, Gutiérrez M (2008) Biotechnology and cultivation of desert truffles. In: Varma A (ed) Mycorrhiza: biology, genetics, novel endophytes and biotechnology. Springer, Berlin, Heidelberg

Murcia MA, Martínez-Tomé M, Jiménez AM, Vera AM, Honrubia M, Parras P (2002) Antioxidant activity of edible fungi (truffles and mushrooms): losses during industrial processing. J Food Prot 65:1614–1622

Murcia MA, Martínez-Tomé M, Vera A, Morte A, Gutiérrez A, Honrubia M, Jiménez AM (2003) Effect of industrial processing on desert truffles Terfezia claveryi Chatin and Picoa juniperi Vittadini: proximate composition and fatty acids. J Sci Food Agric 83:535–541

Nardi M, Fiez-Vandal C, Tailliez P, Monnet V (2002) The EstA esterase is responsible for the main capacity of *Lactococcus lactis* to synthesize short chain fatty acid esters *in vitro*. J Appl Microbiol 93:994–1002

Navarro-Ródenas A, Morte A, Pérez-Gilabert M (2009) Partial purification, characterization and histochemical localization of alkaline phosphatase from ascocarps of the edible desert truffle *Terfezia claveryi* Chatin. Plant Biol 11:678–685. doi:10.1111/j.1438-8677.2008.00172.x

Navarro-Ródenas A, Lozano-Carrillo MC, Pérez-Gilabert M, Morte A (2011) Effect of water stress on *in vitro* mycelium cultures of two mycorrhizal desert truffles. Mycorrhiza 21:247–253

Navarro-Ródenas A, Pérez-Gilabert M, Torrente P, Morte A (2012) The role of phosphorus in the ectendomycorrhiza continuum of desert truffle mycorrhizal plants. Mycorrhiza 22:565–575. doi:10.1007/s00572-012-0434-2

Oliw EH (2002) Plant and fungal lipoxygenases. Prostaglandins Other Lipid Mediat 68–69:313–323

Ollis DL, Shea E, Cygler MB, Dijstra B, Frolow F (1992) The α/β hydrolase fold. Protein Eng 5:197–211

Pérez-Gilabert M, García-Carmona F (2000) Characterization of catecholase and cresolase activities of eggplant polyphenol oxidase. J Agric Food Chem 48:695–700

Pérez-Gilabert M, Morte A, Honrubia M, García-Carmona F (2001a) Monophenolase activity of latent *Terfezia claveryi* tyrosinase: characterization and histochemical localization. Physiol Plant 113:203–209

Pérez-Gilabert M, Morte A, Honrubia M, García-Carmona F (2001b) Partial purification, characterization, and histochemical localization of fully latent desert truffle (*Terfezia claveryi* Chatin) polyphenol oxidase. J Agric Food Chem 49:1922–1927

Pérez-Gilabert M, Morte A, García-Carmona F (2004) Histochemical and biochemical evidences of the reversibility of tyrosinase activation by SDS. Plant Sci 166:365–370

Pérez-Gilabert M, Sánchez-Felipe I, García-Carmona F (2005a) Purification and partial characterization of lipoxygenase from desert truffle (*Terfezia claveryi* Chatin) ascocarps. J Agric Food Chem 53:3666–3671

Pérez-Gilabert M, Sánchez-Felipe I, Morte A, García-Carmona F (2005b) Kinetic properties of lipoxygenase from desert truffle (*Terfezia claveryi* Chatin) ascocarps: effect of inhibitors and activators. J Agric Food Chem 53:6140–6145

Pérez-Gilabert M, Morte A, Ávila-González R, García-Carmona F (2005c) Characterization and histochemical localization of nonspecific esterase from ascocarps of desert truffle (*Terfezia claveryi* Chatin). J Agric Food Chem 53:5754–5759

Perraud X, Kermasha S, Bisakowski B (1999) Characterization of a lipoxygenase extract from Geotrichum candidum. Process Biochem 34:819–827

Perraud X, Kermasha S (2000) Characterization of lipoxygenase extracts from *Penicillium* sp. J Am Oil Chem Soc 77:335–342

Pryde JG, Philips JH (1986) Fractionation of membrane proteins by temperature-induced phase separation in TX-114 application to subcellular fractions of the adrenal medulla. Biochem J 233:525–533

Ragnelli AM, Pacioni G, Aimola P, Lanza B, Miranda M (1992) Truffle melanogenesis: correlation with reproductive differentiation and ascocarp ripening. Pigm Cell Res 5:205–212

Reyes F, Villanueva P, Alfonso C (1990) Comparative study of acid and alkaline phosphatase during the autolysis of filamentous fungi. Lett Appl Microbiol 10:175–177

Sánchez-Ferrer A, Laveda F, García-Carmona F (1993) Partial purification of soluble potato polyphenol oxidase by partitioning in an aqueous two-phase system. J Agric Food Chem 41:1583–1586

Sánchez-Ferrer A, Pérez-Gilabert M, Núñez E, Bru R, García-Carmona F (1994) Triton X-114 phase partitioning in plant protein purification. J Chromatogr A 668:75–83

Sánchez-Ferrer A, Rodríguez-López JN, García-Cánovas F, García-Carmona F (1995) Tyrosinase: a comprehensive review of its mechanism. Biochim Biophys Acta 1247:1–11

Spiteller G (2003) The relationship between changes in the cell wall, lipid peroxidation, proliferation, senescence and cell death. Physiol Plant 119:5–18

Spruce J, Mayer AM, Osborne DJ (1987) A simple histochemical method for locating enzymes in plant tissues using nitrocellulose blotting. Phytochemistry 26:2901–2903

Sugui JA, Pascholati SF, Kunoh H, Howard RJ, Nicholson RL (1998) Association of *Pestalotia malicola* with the plant cuticle: visualization of the pathogen and detection of cutinase and non-specific esterase. Physiol Mol Plant Pathol 52:213–221

Timonen S, Sen R (1998) Heterogeneity of fungal and plant enzyme expression in intact Scots pine *Suillus bovinus* and *Paxillus involutus* mycorrhizospheres developed in natural forest humus. New Phytol 138:355–366

Tisserant B, Gianinazzi-Pearson V, Gianinazzi S, Gollote A (1993) In plant histochemical staining of fungal alkaline phosphatase activity for analysis of efficient arbuscular mycorrhizal infection. Mycol Res 97:245–250

Trasar-Cepeda C, Leiros C, Gil-Sotres F, Seoane S (1998) Towards a biochemical index for soils: An expression relating several biological and biochemical properties. Biol Fertil Soils 26:100–106

Tsitsigiannis DI, Keller NP (2007) Oxylipins as developmental and host-fungal communication signals. Trends Microbiol 15:109–118

Van Aarle IM, Plassard C (2010) Spatial distribution of phosphatase activity associated with ectomycorrhizal plants is related to soil type. Soil Biol Biochem 42:324–33

Van Gelder CWG, Flurkey WH, Wichers HJ (1997) Sequence and structural features of plant and fungal tyrosinases. Phytochemistry 45:1309–1323

Van Leeuwen J, Wichers H (1999) Tyrosinase activity and isoform composition in separate tissues during development of *Agaricus bisporus* fruit bodies. Mycol Res 103:413–418

Vázquez MM, César S, Azcón R, Barea JM (2000) Interactions between arbuscular mycorrhizal fungi and other microbial inoculants (*Azospirillum, Pseudomonas, Trichoderma*) and their effects on microbial populations and enzyme activities in the rhizosphere of maize plants. Appl Soil Ecol 15:261–272

Vosatka M, Dodd JC (1998) The role of different arbuscular mycorrhizal fungi in the growth of *Calamagrostis villosa* and *Deschampsia exuosa*, in experiments with simulated acid rain. Plant Soil 200:251–263

Whitaker JR (1995) Polyphenol oxidae. In: Wong DWS (ed) Food enzymes structure and mechanism. Chapman and Hall, New York

Yokono T, Tamai Y, Azuma T, Sakuma Y, Miura K, Kojima Y, Sunagawa M, Ohmasa M (2000) Karyotype analysis of interspecific fusants of basidiomycetes by pulsed-field gel electrophoresis. J Wood Sci 46:480–484

Zaretsky M, Kagan-Zur V, Mills D, Roth-Bejerano N (2006) Analysis of mycorrhizal associations formed by *Cistus incanus* transformed root clones with *Terfezia boudieri* isolates. Plant Cell Rep 25:62–70

Zhao B, Trouvelot A, Gianinazzi S, Gianinazzi-Pearson V (1997) Influence of two legume species on hyphal production and activity of two arbuscular mycorrhizal fungi. Mycorrhiza 7:179–185

Chapter 17
Nutritional and Antioxidant Properties of *Terfezia* and *Picoa*

Magdalena Martínez-Tomé, Luana Maggi, Antonia María Jiménez-Monreal, María Antonia Murcia, and Josep Antoni Tur Marí

17.1 Introduction

Desert truffles are a complex family of mycorrhizal hypogeous fungi, consisting mainly of species of the genera *Balsamia*, *Picoa*, *Terfezia* and *Tirmania* (Ackerman et al. 1975; Honrubia et al. 2002). *Terfezia* and *Picoa* are two of the most common edible desert truffles in the world. These fungi constitute a popular food and are particularly appreciated for their flavour and texture profiles in Arab countries and in countries of the Mediterranean Basin as a whole (Murcia et al. 2003). In countries of the Middle East and North Africa, desert truffles are usually used in cooked dishes and have long been used by Arabs of the desert as substitutes for meat in their diet as their preparation and cooking are similar to those of meat. Although *Terfezia* provides large quantities of a rich and much-appreciated vegetable in the cuisine of some countries, particularly during early spring when other vegetables are still in the fields, little information is available regarding their composition, nutritional status and biological activities (Murcia et al. 2003). Various studies indicate that truffles are rich in proteins, amino acids, minerals, fibre and carbohydrates (Al-Delaimy and Ali 1970;

M. Martínez-Tomé • A.M. Jiménez-Monreal • M.A. Murcia (✉)
Department of Food Science, Faculty of Veterinary, Murcia University, Campus of Espinardo, 30100 Murcia, Spain

CIBER: CB12/03/30038, Fisiopatología de la Obesidad y la Nutrición, CIBERobn, Instituto de Salud Carlos III (ISCIII), Madrid, Spain
e-mail: mamurcia@um.es

L. Maggi
Institute of Oenology and Agro-Food Engineering, Universita Cattolica del Sacro Cuore, Via Emilia Parmense 84, Piacenza 29122, Italy

J.A. Tur Marí
Dpto Biología Fundamental y Ciencias de la Salud, Facultad de Ciencias de Mallorca, Islas Baleares, Spain

CIBER: CB12/03/30038, Fisiopatología de la Obesidad y la Nutrición, CIBERobn, Instituto de Salud Carlos III (ISCIII), Madrid, Spain

V. Kagan-Zur et al. (eds.), *Desert Truffles*, Soil Biology 38,
DOI 10.1007/978-3-642-40096-4_17, © Springer-Verlag Berlin Heidelberg 2014

Ahmed et al. 1981; Bokhary et al. 1989; Bokhary and Parvez 1993; Hussain and Al-Ruqaie 1999; Murcia et al. 2003; Kagan-Zur and Roth-Bejerano 2008; Slama et al. 2010). The edible portion of desert truffles (the fruit body) has a size ranging from that of a walnut to a large apple, with a chocolate colour (*Approved Methods of the American Association of Cereal Chemists* 1983). It could be an important and alternative source of proteins for a substantial part of the population in developing countries that suffer from protein malnutrition (Buckingham 1985).

Since ancient times, wild mushrooms have been used as food and food-flavouring materials in soups and sauces, due to their unique flavour and texture, and numerous investigations have been carried out into their sensory (Gray 1970; Cuppett et al. 1998; Díaz et al. 2003; Cullere et al. 2009; Díaz et al. 2009; Hiraide et al. 2010) and nutritional characteristics (Sawaya et al. 1985; Bokhary and Parvez 1993; Hussain and Al-Ruqaie 1999; Díez and Alvarez 2001; Mattila et al. 2002; Murcia et al. 2003; Barros et al. 2007; Barros et al. 2008; Kagan-Zur and Roth-Bejerano 2008; Ouzouni et al. 2009; Beluhan and Ranogajec 2011; Liu et al. 2012; Pereira et al. 2012; Reis et al. 2012). More recently food scientists have focused special attention on the bioactivities of truffles, including their antioxidant, anticancer, antimicrobial, hepatoprotective, anti-mutagenic and anti-inflammatory activities (Breene 1990; Sagakami et al. 1991; Chang and Buswell 1996; Hussain and Al-Ruqaie 1999; Isabelle et al. 2010; Murcia et al. 2002; Janakat et al. 2004; Janakat et al. 2005; Saleh 2006; Fratianni et al. 2007; Stanikunaite et al. 2007; Heleno et al. 2010; Al-Laith 2010; Chen et al. 2010; Janakat & Nassar 2010; Wang and Marcone 2011.

The current paper provides an overview of *Terfezia* and *Picoa* investigations on their chemical and biological qualities, including antioxidant properties, during recent decades.

17.2 Nutritional Properties of *Terfezia* and *Picoa*

17.2.1 General Nutritional Values

Various studies have shown that the nutritional characteristics of truffles vary from species to species (Ahmed et al. 1981; Sawaya et al. 1985; Bokhary and Parvez 1993; Hussain and Al-Ruqaie 1999; Dabbour and Takruri 2002b; Murcia et al. 2003; Barros et al. 2007; Barros et al. 2008; These authors did not discuss variability between species Liu et al. 2012; Pereira et al. 2012; Reis et al. 2012).

Kagan-Zur and Roth-Bejerano (2008) reviewed the chemical composition of desert truffles, finding that the dry matter (DM) consisted of 20–27 % protein, 85 % of which is digestible by humans; 3–7.5 % fat (unsaturated and saturated fatty acids); and approximately 60 % carbohydrates with 7–13 % crude fibre, highlighting its 2–5 mg/100 g of ascorbic acid. Similar results were obtained by Sawaya et al. (1985) who compared the compositional difference among three popular Saudi Arabian truffles including two black desert truffles Gibbah and Kholeissi (both are *Terfezia claveryi* according to Hussain and Al-Ruqaie 1999) and one white truffle *Tirmania nivea* (Zubaidi). In this study, the truffle protein content ranged from 19.59 to 27.18 %, fat 2.81 to 7.42 %, crude fibre 7.02 to 13.02 %, ash 4.64 to 6.39 % and ascorbic acid 1.56 to 5.1 mg/100 g. White truffle had the highest protein, fat and crude fibre content, whereas *Terfezia claveryi* had the highest ascorbic acid and ash contents.

In *T. claveryi* Hussain and Al-Ruqaie (1999) observed a 26.23 % protein, 54.48 % carbohydrate and 19.29 % ash content. Moreover, they found that there were differences in some chemical constituents such as carbohydrates, protein, ash and phosphorus between three Iraqi truffle species (*T. claveryi*, *Tirmania nivea* and *Tirmania pinoyi*). However, Bokhary and Parvez (1993) found that the nutritional composition of the desert truffle *T. claveryi* contained 17 % protein, 2 % total crude fat, 28 % carbohydrate and 4 % total crude fibre. The values of protein (15.95 %), fat (6.95 %), carbohydrate (64.55 %), crude fibre (8.32 %) and ash (4.25 %) presented by Murcia et al. (2003) for *T. claveryi* were in agreement with data indicated for *Terfezia boudieri* (Libya) by Ahmed et al. (1981), who obtained 17.19 % crude protein, 6.4 % crude fat, 59.63 % carbohydrate, 3.80 % crude fibre and 12.88 % ash. The following amino acids were also present: alanine 1.11 g, arginine 0.44 g, aspartic acid 1.56 g, cystine 0.28 g, glutamatic acid 2.25 g, glycine 0.77 g, histidine 0.34 g, isoleucine 0.75 g, leucine 1.11 g, lysine 0.56 g, methionine 0.33 g, phenyl alanine 0.67 g, proline 0.98 g, serine 0.94 g, threonine 1.10 g, tyrosine 0.50 g and valine 0.84 g (expressed/100 g dry weight). Recently Slama et al. (2010), also studying *Terfezia boudieri* (Tunisia), observed 10.15 % protein, 15.4 % total sugar and 1.9 % soluble sugar. Dundar et al. (2012) obtained for *Terfezia boudieri* Chatin the following results: protein 14 %, fat 8 % and carbohydrates 54 %.

With regard to the nutritional composition of the genus *Picoa*, only one reference was found (Murcia et al. 2003), reporting the following nutritional profile: 22.54 % protein, 19.94 % fat, 36.66 % carbohydrate, 13.04 % fibre and 8.21 % ash.

17.2.2 Protein and Amino Acid Nutritional Value

Note that the above protein levels (see Sect. 17.2.1) exceed those of any food of animal origin. For the first time, Sawaya et al. (1985) demonstrated that all essential amino acids were present in all three truffle species of Saudi Arabia including the sulphur-containing amino acids (methionine, cystine), besides good levels of tryptophan, and lysine, which are usually the limiting amino acids in many foods of plant origin. These sulphur-containing amino acids were present in amounts of 1.2–5.9 g per 100 g of protein, well within respectable limits. Later, Dabbour and Takruri (2002b) applied protein digestibility-corrected amino acid score (PDCAAS) to evaluate the protein quality of four Jordanian truffle/mushroom species (*Agaricus macrosporus* (cultivated), *Tricholoma terreum*, *Pleurotus ostreatus* and *Terfezia claveryi*). PDCAAS is a value that takes into account the chemical score of the limiting amino acid (for humans at various developmental stages) multiplied by the true digestibility of the protein. According to their evaluation, the true protein digestibility values of *A. macrosporus*, *T. terreum*, *P. ostreatus* and *T. claveryi* are

80.5, 52.6, 73.4 and 61.4, respectively. These authors further calculated PDCAAS based on the essential amino acid pattern requirement for children, the PDCAAS values of *A. macrosporus*, *T. terreum*, *P. ostreatus* and *T. claveryi* being 0.40, 0.35, 0.45 and 0.43, respectively. Based on the essential amino acid pattern requirements for laboratory rats, the PDCAAS values of *A. macrosporus*, *T. terreum*, *P. ostreatus* and *T. claveryi* were 0.20, 0.17, 0.22 and 0.34, respectively.

17.2.3 Mineral Contents

Singer (1961) found great variance in both the minerals present (Si, K, Na, Ca, Mg, Mn, Fe, Al, P, S, Cu and Zn) and their amounts in fresh truffles. Ahmed et al. (1981) obtained for *Terfezia boudieri* the following mineral compositions: Ca 68.0 mg, Na 29.0 mg, K 996.0 mg, Fe 17.0 mg, Mn 2.2 mg, Cu 8.3 mg and Zn 13.0 mg (expressed/100 g DM). On the other hand, Sawaya et al. (1985) observed for *T. claveryi* and *Tirmania nivea* high amounts of K 1,730 mg and P 756 mg and fair levels of Ca 129 mg, Na 199 mg, Mg 104 mg, Fe 10.68 mg, Cu 1.69 mg, Zn 5.10 mg and Mn 0.48 mg (expressed/100 g DM). Also Bokhary et al. (1987) analysed different species of *Tirmania* and *Terfezia,* obtaining various amounts of Mn, Cu, Na, Ca, K, Fe, Mg, P and Co. Beuchat et al. (1993) detected as major elements in *Tuber texense* (a forest truffle) K (2.49 %), P (1.29 %) and S (0.68 %). Hashem (1997) found in Saudi truffles (*T. claveryi* and *Phaeangium lefebveri*) high amounts of Ca, Cu, Fe, K, Mg, Mn, Na, Pb and Zn. Dabbour and Takruri (2002b) concluded in their study that *T. claveryi* is relatively rich in Zn, Mg and Zn and very rich in Fe. Recently, Saltarelli et al. (2008) found K, P, Fe and Ca to be particularly abundant in European truffles, while Slama et al. (2010) also observed that *Terfezia boudieri* (Tunisia) are rich in Ca (1,423), K (1,346), P (346), Mg (154) and Na (77 mg/100 g dry weight).

Dabbour and Takruri (2002b) studied the vitamin content and concluded that *T. claveryi* is relatively rich in thiamin and very rich in riboflavin, niacin, pantothenic acids and pyridoxine.

17.2.4 Fatty Acids

The fatty acid profile and composition obtained by several authors is compared in Table 17.1, which shows the different levels of fatty acids. Linoleic (C18:2), oleic (C18:1), palmitic (16:0) and stearic (C18:0) acids were the four main fatty acids detected for *Terfezia* and *Picoa*. The data obtained by Murcia et al. (2003) are similar to the findings indicated by Bokhary et al. (1989), although they detected an unknown peak (5.3 %) related to a longer chain fatty acid. Murcia et al. (2003) observed a signal representing 3.9 % of the total as 24:0 fatty acid and this was also verified by GC–mass spectrometry.

Table 17.1 Fatty acid composition of *Terfezia* and *Picoa* obtained by other authors

		Fatty acid composition[a]															
Specie	Origin	C14:0	C15:0	C16:0	C16:1	C17:0	C18:0	C18:1	C18:2	C18:3	C19:0	C20:0	C21:0	C22:0	C22:1	C24:0	Authors
Terfezia	Saudi Arabia	ND	0.64	12.28	1.05	0.35	3.00	16.38	22.14	ND	ND	1.32	ND	0.33	0.20	–	Bokhary et al. (1989)
Terfezia claveryi	Saudi Arabia	4	2	16	7	18	7	5	3	–	11	5	1	1	8	–	Bokhary and Parvez (1993)
Terfezia claveryi	Spain	2.1	1.3	17.0	1.4	1.4	4.5	6.9	45.4	5.8	–	3.7	0.4	4.0	1.9	3.9	Murcia et al. (2003)
Picoa juniperi	Spain	1.2	0.4	8.0	2.5	0.9	2.2	23.5	53.3	2.1	–	2.1	ND	0.8	2.0	1.5	Murcia et al. (2003)

ND not detected
[a]Relative peak area (methyl esters)
C14:0 myristic acid
C15:0 pentadecanoic acid
C16:0 palmitic acid
C16:1 palmitoleic acid
C17:0 margaric acid
C18:0 stearic acid
C18:1 oleic acid
C18:2 linoleic acid
C18:3 linolenic acid
C19:0 nonadecanoic acid
C20:0 arachidic acid
C21:0 heneicosanoic acid
C22:0 docosanoic acid
C22:1 erucic acid
C24:0 lignoceric acid

Volatile compound profiles have been investigated and numerous papers have been published regarding species belonging to the genus *Tuber*, known for its unique and subtle aroma (Pelusio et al. 1995; Díaz et al. 2003; Mauriello et al. 2004; March et al. 2006; Cullere et al. 2009; Díaz et al. 2009). However nothing is known, at present, as far as desert truffle aroma components are concerned.

17.3 Antioxidant Activities of *Terfezia* and *Picoa*

Besides research into the sensory and nutritional importance of desert truffles, several investigations have been carried out into their biological activities. Some of their bioactivities include antiviral and antimicrobial activities (Janakat et al. 2004; Janakat et al. 2005), antioxidant capacities (Murcia et al. 2002) and hepatoprotective activity (Janakat and Nassar 2010).

The biological activities of desert truffles that have traditionally been investigated are antimicrobial activities.

Special attention has been paid to the antioxidant capacity of *Terfezia* and *Picoa* truffles in order to assess their suitability for the nutritional enrichment of food. Murcia et al. (2002) evaluated the antioxidant effects of *Terfezia claveryi* Chatin and *Picoa juniperi* Vittadini. The data obtained are summarised in Table 17.2. Dundar et al. (2012) concluded in their study that *Terfezia boudieri* Chatin has excellent antioxidant activity.

17.3.1 Scavenging of Peroxyl Radical

One way to test the antioxidant ability of a substance directly is to examine whether it inhibits the peroxidation of artificial lipid systems (ox-brain phospholipid liposomes) by scavenging peroxyl radicals as described by Aruoma et al. (1996). Both the desert truffles studied exhibited higher inhibition percentages (95 % inhibition) than did the common food antioxidants analysed (α-tocopherol E307, BHA E320, BHT E321 and propyl gallate E310).

17.3.2 Scavenging of Hydroxyl Radical

The deoxyribose assay, described by Aruoma et al. (1993), evaluates whether a compound is a scavenger of the hydroxyl radicals (OH•) generated in the human body under physiological conditions, although the compound can also be produced from peroxyl radicals, in which case it will compete with deoxyribose for the OH• and inhibit deoxyribose degradation. Hydroxyl radicals are extremely reactive and

Table 17.2 Antioxidant activity of raw *Terfezia* and *Picoa* compared with the activity of common food antioxidants (Murcia et al. 2002)

Substance added to reaction mixtures	Lipid peroxidation % inhibition	Deoxyribose assay[a] For RM + DR (absorbance units)	% inhibition	Without ASC (absorbance units)	Peroxidase assay ($a_{436\ nm}$)	Rancimat test[b] IP(h)	PF	Linoleic acid[c] % inhibition	TEAC assay[d]
None (control)		1.226 ± 0.01	–	0.257	0.622 ± 0.02	7.58			–
Trolox (0.05 mM)									1.00 ± 0.0
Trolox (0.5 mM)									10.00 ± 0.0
Terfezia (raw)	95.7 ± 1	0.149 ± 0.03	87.8	0.110	0.162 ± 0.03	9.26	1.22 ± 0.1	77.0 ± 2	4.77 ± 0.1
Picoa (raw)	95.3 ± 1	0.070 ± 0.02	94.3	0.067	0.168 ± 0.04	8.75	1.15 ± 0.1	85.3 ± 3	3.91 ± 0.1
α-Tocopherol	15.3 ± 1	1.186 ± 0.03	3.2	0.240	0.711 ± 0.02	20.10	2.65 ± 0.2	27.0 ± 1	1.16 ± 0.1
BHA	71.4 ± 1	0.914 ± 0.02	25.4	0.201	0.770 ± 0.03	8.68	1.14 ± 0.1	84.0 ± 2	0.44 ± 0.1
BHT	22.3 ± 2	1.116 ± 0.05	8.9	0.559	0.700 ± 0.04	7.18	0.95 ± 0.2	96.0 ± 1	0.26 ± 0.1
Propyl gallate	52.5 ± 1	2.070 ± 0.01	–	1.537	0.400 ± 0.02	14.21	1.87 ± 0.1	95.0 ± 2	3.47 ± 0.1

[a] *RM* reaction mixtures, *DR* deoxyribose, *ASC* ascorbate
[b] Rancimat tested at 120 °C; *IP* induction period, *PF* protection factor. PF = IP (olive oil + samples)/IP (olive oil)
[c] Percentage of inhibition on the 30th day of storage
[d] TEAC value is the millimolar concentration of a Trolox solution showing the antioxidant capacity equivalent to that of the dilution of the substance under investigation. Propyl gallate dilution was selected to reduce the measurement within the appropriate part of the Trolox standard curve
(–) No inhibition detected

can react with non-selective compounds such as proteins, DNA, unsaturated fatty acids and almost every biological membrane. The data of deoxyribose damage caused by the OH• radical in the presence of raw truffles along with the activity of common food antioxidants (Table 17.2) showed a high percentage of inhibition for both truffles (88 and 94 % inhibition). When ascorbate was omitted, the attack on deoxyribose was less intense, because the absence of ascorbate decreased the concentration of OH• in the reaction mixture (Martínez-Tomé et al. 2001a). The samples lacking ascorbate exhibited lower absorbance levels than the control sample, since *Terfezia* and *Picoa* scavenged OH• radicals and could be considered as primary antioxidants. All samples showed better performance than the food antioxidants considered (α-tocopherol, BHA, BHT and propyl gallate).

17.3.3 Scavenging of Hydrogen Peroxide

Hydrogen peroxide is generated in vivo by several oxidase enzymes and by activated phagocytes, and it is known to play an important role in the killing of several bacterial and fungal strains (Halliwell et al. 1995). There is increasing evidence that H_2O_2, either directly or indirectly via its reduction product, OH•, can act as a messenger molecule in the synthesis and activation of several inflammatory mediators. When a scavenger is incubated with H_2O_2 using a peroxidase assay system, the loss of H_2O_2 can be measured (Martínez-Tomé et al. 2001b). Table 17.2 shows the reaction of hydrogen peroxide with raw truffles investigated and with common food antioxidants (α-tocopherol, BHA, BHT and propyl gallate). The inhibition percentage for *Terfezia* and *Picoa* was 74 % and 73 %, respectively.

17.3.4 Rancimat Test for Oxidative Stability

The Rancimat test is used to obtain information on whether the antioxidant activity resists heating at high temperature. This assay is also applied during storage to evaluate the protection that a sample supplies to a food (rich in oils or fats) and to study whether manipulation or industrial processing generates free radicals (Murcia et al. 2001). To assess oxidative stability, the food industry uses the Rancimat test, in which the scavenger to be tested is added to a lipidic food and the degree of protection is evaluated (Schwarz and Ernst 1996). The induction period and the protection factor obtained by the Rancimat method for refined olive oil with truffles and for refined olive oil with common food antioxidants were determined (Table 17.2). Refined olive oil alone (control) started the radical chain reactions of the propagation phase of autoxidation after 7.58 h. The time required for the formation of a sufficient concentration of initiating radicals (initiation phase) was slightly longer when food antioxidants or truffles were added, delaying the onset of the propagation phase of the radical chain reaction and showing the protection factors of these products. *Terfezia* provided higher protection than *Picoa*, with induction period of 9.26 h and 8.75 h, respectively. This technique has been questioned by some authors (Frankel 1993), but Martínez-Tomé et al. (2001b) and Murcia et al. (2001) decided to use it because it is a commonly used procedure in the food industry and in governmental analytic laboratories.

17.3.5 Total Antioxidant Activity Evaluation During Storage

The linoleic acid system assay which is used to determine antioxidant activity during storage at unfavourable temperatures (40 °C), measures the inhibition of linoleic acid autoxidation during 30 days of storage (Murcia et al. 2002). The results for *Terfezia* (77 % inhibition) and *Picoa* (85 %), showed a very high level of antioxidant activity similar to that of propyl gallate, BHA and BHT during all the days of storage. The results are expressed as inhibition percentages on the 30th day of storage (Table 17.2).

17.3.6 Measurement of Total Antioxidant Activity by TEAC Assay

Finally, TEAC values can be assigned to all compound able to scavenge ABTS by comparing the scavenging capacities of these compounds with that of Trolox, a water-soluble vitamin E. The quantitative evaluation of antioxidant capacity based on TEAC can be used to provide an order of antioxidants (van den Berg et al. 1999). The TEAC values of the two truffles and of common food antioxidants are presented in Table 17.2. Of the food antioxidants analysed, *Terfezia* and *Picoa* samples exhibited the highest TEAC value, compared with propyl gallate, α-tocopherol, BHA and BHT.

The biological and protective properties of mushrooms and truffles can be attributed to several compounds, such as water-soluble polysaccharides, which are effective in protection against hydroxyl and superoxide radical-scavenging activities (Liu et al. 1997). In fact, polysaccharides (lentinan and schizophyllan are pure b-glucans, whereas polysaccharide (PSK, Krestin) is a protein-bound b-glucan) isolated from *L. edodes* and others exhibit free radical-scavenging activity (Ooi and Liu 2000). Other structures, including vitamins, pigments and phenolic compounds (detected in *Terfezia boudieri* Chatin, Dundar et al. 2012), may also be related to the capacity to reduce and chelate the ferric iron that catalyses lipid peroxidation. The molecular structures of these compounds include an aromatic ring with hydroxyl groups containing mobile hydrogens (Aruoma et al. 1996) that lies within the phytin and phytin-P antinutritional constituents of the mushrooms. The ability to chelate certain mineral elements, especially Ca, Mg, Fe and Zn, has also been described (Aletor 1995). Other compounds isolated from different mushrooms, such as inoscavin A, curtisians A through D (*p*-terphenyl) compounds (Yun et al. 2000), hydrazine and the indole derivates, are also associated with antioxidant capacity and the inhibition of lipid peroxidation.

17.4 Conclusions

The enrichment of stews and casseroles with desert truffles and their consumption would represent an increase in the antioxidant levels of our diet. Dessert truffles can be considered promising candidates for industrial processing to replace synthetic antioxidants with natural truffle extracts. The several studies carried out on sensory, nutritional and biological activities of desert truffles during recent years point to their potential for use as functional food and therapeutic agent, but further investigations are recommended on how to incorporate their chemical and biological properties into value-added truffle or truffle-related products.

References

AACC (1983) Approved methods of the American Association of Cereal Chemists, 8th edn. AACC, St Paul, MN

Ackerman LGJ, Van Wyk PJ, du Plassis LM (1975) Some aspects of the composition of the Kalahari truffle or N'abba. S Afr Food Rev 2:145–147

Ahmed AA, Mohamed MA, Hami MA (1981) Libyan truffles "Terfezia boudieri Chatin": chemical composition and toxicity. J Food Sci 11:927–929

Al-Delaimy KS, Ali SH (1970) Storage, spoiled and proximate food composition of Iraqi truffles. Berj trop subtrop ianstwirt: tropen med 88:77–81

Aletor VA (1995) Compositional studies on edible tropical species of mushrooms. Food Chem 54:265–268

Al-Laith AAA (2010) Antioxidant components and antioxidant/antiradical activities of desert truffle (*Tirmania nivea*) from various Middle Eastern origins. J Food Compos Anal 23:15–22

Aruoma OI, Murcia A, Butler J, Halliwell B (1993) Evaluation of the antioxidant and prooxidant actions of gallic acid and its derivatives. J Am Oil Chem Soc 73:1617–1625

Aruoma OI, Spencer JPE, Rossi R, Aeschbach R, Khan A, Mahmood MA, Murcia A, Butler J, Halliwell B (1996) An evaluation of the antioxidant and antiviral action of extracts of rosemary and provençal herbs. Food Chem Toxicol 34:449–456

Barros L, Baptista P, Correia DM, Casal S, Oliveira B, Ferreira ICFR (2007) Fatty acid and sugar compositions, and nutritional value of five wild edible mushrooms from Northeast Portugal. Food Chem 105:140–145

Barros L, Venturini BA, Baptista P, Estevinho LM, Ferreira ICFR (2008) Chemical composition and biological properties of Portuguese wild mushrooms: a comprehensive study. J Agric Food Chem 56:3856–3862

Beluhan S, Ranogajec A (2011) Chemical composition and non-volatile components of Croatian wild edible mushrooms. Food Chem 124:1076–1082

Beuchat LR, Brenneman TB, Dove CR (1993) Composition of the pecan truffle (*Tuber texense*). Food Chem 46:189–192

Bokhary HA, Parvez S (1993) Chemical composition of desert truffles *Terfezia claveryi*. J Food Compos Anal 6:285–293

Bokhary HA, Suleiman AAA, Basalah MO, Parvez S (1987) Chemical composition of desert truffles from Saudi Arabia. Calif Inst Sci Tech 20:336–341

Bokhary HA, Suleiman AAA, Basalah MO (1989) The fatty acid components of the desert truffle "Al Kamah" of Saudi Arabia. J Food Prot 52:668–669

Breene WM (1990) Nutritional and medicinal value of specially mushrooms. J Food Prot 53:883–894

Buckingham KW (1985) Effect of dietary polyunsaturated/saturated fatty acid ratio and dietary vitamin E on lipid peroxidation in the rat. J Nutr 115:1425–1435

Chang ST, Buswell JA (1996) Mushroom nutriceuticals. World J Microbiol Biotechnol 12:473–476

Chen X-H, Xia L-X, Zhou H-B, Qiu G-Z (2010) Chemical composition and antioxidant activities of *Russula griseocarnosa* sp. nov. J Agric Food Chem 58:6966–6971

Cullere L, Ferreira V, Chevret B, Venturini ME, Sanchez-Gimeno AC, Blanco D (2009) Characterisation of aroma active compounds in black truffles (*Tuber melanosporum*) and summer truffles (*Tuber aestivum*) by gas chromatography– olfactometry. Food Chem 122:300–306

Cuppett SL, Parkhurst AM, Chung W, Weyer M, Bullerman LB (1998) Factors affecting sensory attributes of oyster mushrooms. J Food Qual 21:383–395

Dabbour JR, Takyuri HR (2002a) Protein quality of four types of edibles moushroom found in Jordan. Plant Foods Human Nutr 57:1–11

Dabbour IR, Takruri HR (2002b) Protein digestibility using corrected amino acid score method (PDCAAS) of four types of mushrooms grown in Jordan. Plant Foods Hum Nutr 57:13–24

Díaz P, Ibáñez E, Señoráns FJ, Reglero G (2003) Truffle aroma characterization by headspace solid-phase microextraction. J Chromatogr A 1017:207–214

Díaz P, Ibáñez E, Reglero G, Señoráns FJ (2009) Optimization of summer truffle aroma analysis by SPME: comparison of extraction with different polarity fibres. LWT- Food Sci Technol 42:1253–1259

Díez VA, Alvarez A (2001) Compositional and nutritional studies on two wild edible mushrooms from northwest Spain. Food Chem 75:417–422

Dundar A, Faruk O, Acay H, Okumus V, Ozdemir S, Yildiz A (2012) Antioxidant properties, chemical composition and nutritional value of *Terfezia boudieri* (Chatin) from Turkey. Food Sci Technol Int 18:317–328

Frankel EN (1993) In search of better methods to evaluate natural antioxidants and oxidative stability in food lipids. Trends Food Sci Tech 4:220–225

Fratianni F, Di Luccia A, Coppola R, Nazzaro F (2007) Mutagenic and antimutagenic properties of aqueous and ethanolic extracts from fresh and irradiated *Tuber aestivum* black truffle: a preliminary study. Food Chem 102:471–474

Gray WD (1970) The use of fungi as foods and in food processing. Butterworths and Co, London

Halliwell B, Murcia MA, Chirico S, Aruoma OI (1995) Free radicals and antioxidants in food and in vivo: what they do how they work? Crit Rev Food Sci Nutr 35:7–20

Hashem AR (1997) Effect of heavy metal ions on the mycelial growth of some fungi isolated from soil of industrial Al-Jubail city, Saudi Arabia. JKS Univ Sci 9:119–124

Heleno SA, Barros L, Sousa MJ, Martins A, Ferreira ICFR (2010) Tocopherols composition of Portuguese wild mushrooms with antioxidant capacity. Food Chem 119:1443–1450

Hiraide M, Kato A, Nakashima T (2010) The smell and odorous components of dried shiitake mushroom, *Lentinula edodes* V: changes in lenthionine and lentinic acid contents during the drying process. Jpn Wood Res Soc 56:477–482

Honrubia M, Gutiérrez A, Morte A (2002) Desert truffle plantation from south-east Spain. Edible mycorrhizal mushrooms and their cultivation. In: Hall IR, Wang Y, Danell E, Zambonelli A (eds) Proceedings of the 2nd international conference on edible mycorrhizal mushrooms,

Christchurch, New Zealand, 3–6 July 2001. New Zealand Institute for Crop & Food Research Limited, Christchurch

Hussain G, Al-Ruqaie IM (1999) Occurrence in chemical composition, and nutritional value of truffles: overview. Pakistan J Biol Sci 2:510–514

Isabelle M, Lee BL, Lim MT, Koh W-P, Huang D, Ong CN (2010) Antioxidant activity and profiles of common vegetables in Singapore. Food Chem 120:993–1003

Janakat S, Nassar M (2010) Hepatoprotective activity of desert truffle (*Terfezia claveryi*) in comparison with the effect of *Nigella sativa* in the rat. Pakistan J Nutr 9:52–56

Janakat S, Al-Fakhiri S, Sallal AK (2004) A promising peptide antibiotic from *Terfezia claveryi* aqueous extract against *Staphylococcus aureus* in vitro. Physiother Res 18:810–813

Janakat S, Al-Fakhiri S, Sallal AK (2005) Evaluation of antibacterial activity of aqueous and methanolic extracts of the truffle *Terfezia claveryi* against *Pseudomonas aeruginosa*. Saudi Med J 26:952–955

Kagan-Zur V, Roth-Bejerano N (2008) Dessert truffles. Fungi 1 (Special issue—truffles): 32–37

Liu F, Ooi VE, Chang ST (1997) Free radical scavenging activities of mushroom polysaccharide extracts. Life Sci 60:763–771

Liu Y-T, Sun J, Luo Z-Y, Rao S-Q, Su Y-J, Xu R-R, Yang Y-J (2012) Chemical composition of five wild edible mushrooms collected from Southwest China and their antihyperglycemic and antioxidant activity. Food Chem Toxicol 50:1238–1244

March RE, Richards DS, Ryan RW (2006) Volatile compounds from six species of truffle—head-space analysis and vapor analysis at high mass resolution. Int J Mass Spectrom 249:60–67

Martínez-Tomé M, García-Carmona F, Murcia MA (2001a) Comparison of antioxidant and pro-oxidant properties of broccoli amino acids with those of common food additives. J Sci Food Agric 81:1019–1026

Martínez-Tomé M, Jiménez AM, Ruggieri S, Frega N, Strabbioli R, Murcia MA (2001b) Antioxidant properties of Mediterranean spices compared with common food additives. J Food Prot 64:1412–1419

Mattila P, Salo-Vaananen P, Konko K, Aro H, Jalava T (2002) Basic composition and amino acid contents of mushrooms cultivated in Finland. J Agric Food Chem 50:6419–6422

Mauriello G, Marino R, D'Auria M, Cerone G, Rana GL (2004) Determination of volatile organic compounds from truffles via SPME–GC–MS. J Chromatogr Sci 42:299–305

Murcia MA, Jiménez A, Martinez-Tomé M (2001) Evaluation of the antioxidant properties of Mediterranean and tropical fruits compared with common food additives. J Food Prot 64:2037–2046

Murcia MA, Martinez-Tomé M, Jiménez A, Vera A, Honrubia M, Parras P (2002) Antioxidant activity of edible fungi (truffles and mushrooms): losses during industrial processing. J Food Prot 65:1614–1622

Murcia MA, Martínez-Tomé M, Vera A, Morte A, Gutiérrez A, Honrubia M, Jiménez AM (2003) Effect of industrial processing on desert truffles *Terfezia claveryi* Chatin and *Picoa juniperi* Vittadini): proximate composition and fatty acids. J Sci Food Agric 83:535–541

Ooi VEC, Liu F (2000) Immunomodulation anti-cancer activity of polysaccharide-protein complexes. Curr Med Chem 7:715–729

Ouzouni PK, Petridis D, Koller W-D, Riganakos KA (2009) Nutritional value and metal content of wild edible mushrooms collected from West Macedonia and Epirus, Greece. Food Chem 115:1575–1580

Pelusio F, Nilsson T, Montanarella L, Tilio R, Larsen B, Facchetti S, Madsen J (1995) Headspace solid-phase microextraction analysis of volatile organic sulphur compounds in black and white truffle aroma. J Agric Food Chem 43:2138–2143

Pereira E, Barros L, Martins A, Ferriera ICFR (2012) Towards chemical and nutritional inventory of Portuguese wild edible mushrooms in different habitats. Food Chem 130:394–403

Reis FS, Barros L, Martins A, Ferriera ICFR (2012) Chemical composition and nutritional value of the most widely appreciated cultivated mushrooms: an inter-species comparative study. Food Chem Toxicol 50:191–197

Sagakami H, Aohi T, Simpson A, Tanuma S (1991) Induction of immunopotentiation activity by a protein-bound polysaccharide, PSK. Anticancer Res 11:993–1000

Saleh A (2006) Antimicrobial and antioxidant activity of two desert truffles. Tirmania and terfezia. Masters Abstracts Int 46:109

Saltarelli R, Ceccaroli P, Cesari P, Barbieri E, Stocchi V (2008) Effect of storage on biochemical and microbiological parameters of edible truffle species. Food Chem 109:8–16

Sawaya WN, Al-Shalhat A, Al-Sogair A, Mohammad M (1985) Chemical composition and nutritive value of truffles of Saudi Arabia. J Food Sci 50:450–453

Schwarz K, Ernst H (1996) Evaluation of antioxidative constituents from thyme. J Sci Food Agric 70:217–223

Singer R (1961) Mushroom and truffles XXIII: 272. Leonard Hill Publisher Ldt, London, p 189

Slama A, Neffati M, Boudabous A (2010) Biochemical composition of desert truffle *Terfezia boudieri* Chatin. Acta Horticulturae 853:285–289, Proceedings of the International Symposium on Medicinal and Aromatic Plants, 2009

Stanikunaite R, Trappe JM, Khan SI, Ross SA (2007) Evaluation of therapeutic activity of *Hypogeous Ascomycetes* and *Basidiomycetes* from North America. Int J Med Mushrooms 9:7–14

van den Berg R, Haenen GRMM, van den Berg H, Bast A (1999) Applicability o fan improved Trolox equivalent antioxidant capacity (TEAC) assay for evaluation of antioxidant capacity measurements of mixtures. Food Chem 66:511–517

Wang S, Marcone MF (2011) The biochemistry and biological properties of the world's most expensive underground edible mushroom: truffles. Food Res Int 44:2567–2581

Yun B-S, Lee L-K, Kim J-P, Yoo I-D (2000) Curtisians A D, new free radical scavengers from the mushroom Paxillus curtisii. J Antibiot 53:114–122

Chapter 18
Nutritional and Antioxidant Properties of the White Desert Truffle *Tirmania nivea* (Zubaidi)

Abdul Ameer A. Al-Laith

Abbreviations

ABTS	2,2′-Azino-bis(3-ethyl-benzothiazoline-6-sulphonic acid)
DNPH	2,4-Dinitrophenylhydrazine
DP	Degree of polymerization
DPPH	Diphenylpicrylhydrazyl
EC_{50}	Effective concentration which gives 50 % radical inhibition activity
EDTA	Ethylenediamine tetra acetic acid
ET	Electron transfer
FC	Folin–Ciocalteu reagent (or method)
FRAP	Ferric reducing/antioxidant power
GAE	Gallic acid equivalent
HAT	Hydrogen atom transfer
MDA	Malondialdehyde
NO	Nitric oxide
PVPP	Polyvinyl polypyrrolidone
TBARS	Thiobarbituric acid reactive substances
TEAC	Trolox equivalent antioxidant capacity
TPTZ	2,4,6-Tripyridyl-*s*-triazine
VCEAC	Vitamin C equivalent antioxidant capacity

A.A.A. Al-Laith (✉)
Department of Biology, College of Science, University of Bahrain, P.O. Box 32038, Sakhir Campus, Building S41, Sakhir, Kingdom of Bahrain
e-mail: biolaith@gmail.com; allaith@sci.uob.bh

V. Kagan-Zur et al. (eds.), *Desert Truffles*, Soil Biology 38,
DOI 10.1007/978-3-642-40096-4_18, © Springer-Verlag Berlin Heidelberg 2014

18.1 Introduction

Currently, searching the Internet for scientific publications addressing the topic "*Tirmania nivea*" yields rather meager results (700 hits), stressing the scarcity of relevant information. The number of hits pertinent to antioxidant properties of *T. nivea* were less than 10, referring, mainly, to two published works, as of October 2012. This evident paucity in the aforementioned topics obliged the author to resort extensively to his own published work (Al-Laith 2010).

As a food commodity, truffles resemble wild mushrooms in many chemical, nutritional, and ecological aspects which make the latter a good and justifiable candidate for comparison purposes to compensate for the scant available information on truffles. The antioxidant properties of wild and commercially available mushrooms have been reported by many researchers (Mau et al. 2002; Murcia et al. 2002; Yang et al. 2002; Badalyan 2003; Cheung et al. 2003; Mau et al. 2004; Valentão et al. 2005; Puttaraju et al. 2006; Gezer et al. 2006; Barros et al. 2007; Lee et al. 2008; Gaafar et al. 2010; Vidović et al. 2010; Pal et al. 2010; Abdulla et al. 2011; Akata et al. 2011) to name a few. Recently, the antioxidant properties of wild mushrooms and the biochemistry and biological properties of truffles have been reviewed by Ferreira et al. (2009) and Wang and Marcone (2011), respectively. Murcia et al. (2013) review the antioxidant activity of other desert truffles. Desert truffles form mycorrhizal relationships with plants. The couple *Terfezia boudieri* Chatin with the host plant *Helianthemum sessiliflorum* was reported to effect many changes concerning plant parameters (Turgeman et al. 2011). Mycorrhiza-derived modifications may also include the production of secondary metabolites such as phenolics, potent plant antioxidants, to enhance the competence of the plant to cope with stressful environmental conditions, hence increase the antioxidant properties of the product of this association, the truffle (see Sect. 18.4.2.9).

Based on their morphological and phylogenic attributes, *Tirmania* and *Terfezia* species are closely related and form a distinct clade (genera) in the family *Pezizaceae*, order *Pezizales* (Diez et al. 2002; Jamali and Banihashemi 2012).

Truffles were regarded by Ancient Greek as "the miracle of nature" presumably due to their (then) claimed medicinal and nutritional values (Patel 2012, see also Shavit as well as Shavit and Shavit this volume). Moreover, and because of the ambiguity surrounding their natural life cycle, Ancient Greek considered truffles as "a natural phenomenon of great complexity . . ." (Patel 2012). The Arab Bedouins (nomadic) as well as sedentary populations living east and south to the Mediterranean basin in the arid and semiarid areas used desert truffles as a meat substituent.

Truffles are generally known as **Kamah** or **Fagaa** in Arabic. Kamah literally means hidden or to hide, while Fagaa literally means to crack, burst, or to expose. Both names are descriptive and informative of the nature of truffles. Among the many but limited types of truffles commonly found in the desert and semidesert areas of the Middle East, *T. nivea* stands prominent (Fig. 18.1). It is one of the highly regarded truffles by many Arabs and is the most preferred and expensive

Fig. 18.1 *Tirmania nivea*. A composite image showing a natural scene of the truffle. The crack on the soil surface at bottom right is a sign for the presence of the fruiting body. (**a**) A close picture of *T. nivea*. (**b**) A typical dish in which *T. nivea* is cooked and served with rice and spices. *Source*: http://ar.wikipedia.org/w/index.php?title=%D9%85%D9%84%D9%81:B.k.Kookherd._25_%D9%83%D9%88%D8%AE%D8%B1%D8%AF.jpg&filetimestamp=20090827061919

truffle in the Gulf Cooperative Countries (GCC) due to its musky smell, delicacy, and soft white tissues (Mandeel and Al-Laith 2007). The Arabic name for *T. nivea* is ***Zubaidi***, which literally means butter-like, a reflection of both the color and the merit of this valuable truffle, since butter used to be a very valuable food commodity. Many truffle collectors, and collectors of *T. nivea* in particular, still believe that desert truffle spawn when lightning and thunder strikes (Mandeel and Al-Laith 2007).

Their collection time is still a fun and ceremonial to some. Truffles, including *T. nivea* are not eaten raw. They are peeled, soaked, cooked, and presented in many ways (Fig. 18.1). They can be stored as frozen or as dried after cutting into cubes or slices. They can also be grounded and added to other dishes as a supplement (Mandeel and Al-Laith 2007).

This chapter is devoted to the antioxidant properties of *T. nivea*. The second section (18.2) covers physical description, chemical composition, and nutritional quality of this truffle. The third (18.3) and fourth (18.4) sections cover antioxidant properties, methodology, constituents, and activities. The final section (18.5) summarizes conclusions and presents future outlooks.

18.2 Chemical Composition and Nutritional Quality

18.2.1 Tirmania nivea *Description*

Ascocarps of *T. nivea* grow and mature below ground (hypogeous). They develop in basic soil. When they reach maximum size, they begin to crack the ground surface and burst (Fig. 18.1). It is not uncommon to find fully exposed *T. nivea* lying on the ground (Jabbur 1995). Ascocarps of *T. nivea* have a roughly spherical shape (subglobose or pyriform, glabrous). Their skin is creamy white or light brown, and the pith (gleba) is white or yellowish white becoming yellowish brown with age (Jamali and Banihashemi 2012). The pith is usually soft. The smooth outer layer of the fruit body (peridium) superficially resembles plant parenchyma (pseudoparenchymatous) and is made of an interwoven large mass hyphae and inflated cells (asci). These large cells have ellipsoid to obovoid shape and commonly contain eight amyloid spores (Trappe 1979; Al-Sheikh and Trappe 1983; Bokhary 1987; El-Kholy 1989).

18.2.2 Chemical Composition and Nutritional Quality

Chemical composition and nutritional quality of *T. nivea* and other related truffles have been reported by several groups from different countries (Ahmed et al. 1981; Sawaya et al. 1985; Al-Naama et al. 1988; Abdallah et al. 1997; Hussain and Al-Ruqaie 1999; Mattila et al. 2000; Al-Alawi 2009; Al-Rawi and Taha 2010). More recently, Wang and Marcone (2011) reviewed the chemical properties of truffles. There is a general agreement among these studies that truffles, including *T. nivea,* have a high content of protein, carbohydrate, and fiber. In particular *T. nivea* has a high water content with an overall average of 76.4 % which is similar to values reported for other desert truffle (Al-Laith 2010). A greater moisture content (82 %) was observed using the freeze dry method. This noticeable variation must be taken into consideration when reporting antioxidant values based on dried matter as this may affect the reported value. Dry matter constitutes 18–25 %, and consists of 40–60 % carbohydrate, 20–27 % protein, 2.4–7.5 % lipid, 7–13 % crude fiber, and 7.4–9.6 % ash.

T. nivea was found to be superior to several other desert truffles with regard to protein, fat, and crude fiber (Sawaya et al. 1985). Depending on the geographical location, values of these parameters may vary considerably. Whether carbohydrate and protein contribute towards the antioxidant properties of *T. nivea* has not been, as yet, adequately examined. Nutritionally, digestible and nondigestible carbohydrates make *T. nivea* a very healthy food providing both energy and fibers which are essential for the function of the intestinal track. *T. nivea* contains almost all the amino acids commonly found in proteins, and they represent 16.4 % of the dry matter (Sawaya et al. 1985). Both essential and nonessential amino acids are present in appreciable amounts. Besides, nitrogen content accounts for about 16–17 % (DM). These findings qualify *T. nivea* as a rich source of good quality protein and justify the practice of the Arab Bedouins of using truffle as a meat replacer. *T. nivea* contains high levels of potassium, phosphorus, and magnesium. Among their ecological roles, desert truffles are known to help their associated plants in mineral acquisition by taking up many important minerals such as P, N, Zn, K, Cu, Sr, and S and thus becoming a good source of these minerals (Awameh and Al-Sheikh 1979; Al-Rahmah 2001).

18.3 Assessment of Antioxidant Properties of *T. nivea*

18.3.1 Sample Preparation

Fresh samples of *T. nivea* of different origins were collected or purchased from the local markets. Origins of samples tested include Bahrain, Iran, Morocco, Kuwait, and Saudi Arabia. Identification of the truffle was confirmed by Professor Qaher Maneel of the Department of Biology at the University of Bahrain. Fresh samples were either air dried or lyophilized. Except for results comparing antioxidant activity of both fresh and dried samples, the majority of the findings were generated using dried powdered *T. nivea*.

18.3.2 Methods Used for Measuring Antioxidant and Antiradical Activities

Total antioxidant capacity of food and plant materials are currently evaluated using different methods. These methods differ in their chemistry and in what they measure. There is no single method that can provide a full assessment of the TAC of a food sample. It is well known that different compounds with antioxidant properties act differently in these assays. However, assays used to evaluate the antioxidant capacity fall into one of the following two mechanisms: hydrogen atom transfer (HAT) or electron transfer (ET) (Prior et al. 2005).

The truffle, like any other organic substance originated from living organisms, is a complex food system. Measuring the antioxidant capacity of complex systems necessitates the utilization of several assays with different reaction mechanisms.

Samples of *T. nivea* were evaluated for their antioxidant properties in our laboratory. Ten different methods were employed (Benzie and Strain 1996; Re et al. 1999; Brand-Williams et al. 1995; Sreejayan and Rao 1997; Halliwell et al. 1987; Waterhouse 2002; Prieto et al. 1999; Oyaizu 1986; Puntel et al 2005; Yagi 1998). Details of most methods may be found elsewhere (Al-Laith 2010).

18.3.3 Calculation and Expression of the Antioxidant Activity

Antioxidant capacity is defined as the millimolar concentration of a known antioxidant solution having the antioxidant capacity equivalent to 1.0 mM solution of the substance under investigation (Rice-Evans et al. 1995) and expressed as TEAC and VCEAC in the case of Trolox and vitamin C, respectively. Whereas, the Trolox equivalent (TE) or gallic acid equivalent (GAE) is defined as the μmol of standard antioxidant necessary to provide the same antioxidant capacity as a gram (or 100 g) of the sample. TEAE (VCEAC) is the concentration of antioxidant giving the same % of change in absorbance or same % of the inhibition as that of 1 mM.

$$\text{Antioxidant capacity (mmol TE/g)} = (A/\varepsilon) \times (V_{\text{ass}}/V_{\text{aliq}}) \times (V_{\text{ext}}/m) \times (\text{DF})$$

Whereas: A is the absorbance at a given wavelength specific for the assay; ε is the molar absorptivity of the chromophore measured; V_{ass} is the total volume of the reaction mixture; V_{aliq} is the sample aliquots; V_{ext} is the total volume of the extract; m is the mass of the extracted sample in gram; and DF is the dilution factor. The following molar absorptivities (l mol^{-1} cm^{-1}) were used for (Apak et al. 2007) ABTS-Trolox (1.6×10^4), FRAP-Trolox (4.625×10^3), and FC-GA (3.25×10^3). AC values are sometimes reported as mmol per 100 g for convenience. Percent inhibition of a radical (% scavenging activity) is calculated as $=[(1 - A_{sample}/A_{control})] \times 100$. Metal-chelating activity is similarly calculated. The antiradical activity is also reported as EC_{50}, the effective or the relative concentration of an antioxidant (mg/ml or μg/ml) to reduce the initial radical concentration (i.e., ABTS, DPPH) by 50 %.

18.3.4 Methods Used to Estimate Antioxidant Constituents

Total, free and bound phenolics, total flavonoids, ascorbic acid, and tannins were estimated as described previously (Al-Laith 2010). Tannin contents were estimated using the FC method after precipitation with polyvinyl polypyrrolidone (PVPP) (FAO/IAEA 2000). The vanillin–HCl method (Chang et al. 1994) was used to estimate condensed tannins (proanthocyanidins). Flavanone content was evaluated using 2,4-DNPH (Nagy and Grancai 1996).

18.4 Antioxidant Properties

18.4.1 Antioxidant Constituents

18.4.1.1 Phenolics

Phenolic compounds are the major constituents contributing towards the antioxidant activity of *T. nivea*. Available data indicate the presence of free and bound phenolics including flavonoids and non-flavonoids constituency. Table 18.1 summarizes the means, standard deviations, and percents of these phenolics and other non-phenolic antioxidants. Free phenolics represent the majority of total phenolics (~73 %). Percent of total free phenolics to total hydrolysable phenolics varies widely and ranged from as low as 24 % to as high as 83 % and averaged 55.5 % (Fig. 18.2). This observation warrants the usage of acidified methanolic solution for extraction of truffles. However, works reported here were conducted using 80 % methanolic solution, unless otherwise stated. The usage of non-acidified methanolic solution may tend to underestimate the phenolic content, hence the potential antioxidant activity by up to 40–75 %. Mild acid hydrolysis of truffle samples liberates bound phenolics. Levels of flavonoids and non-flavonoid phenolics represent 22–25 % and 75–78 % of the total phenolics, respectively, based on FC, $AlCl_3$, and 2,4-DNPH methods. Statistical analysis using coefficient of variance indicates that the level of flavonoids is the least variable (9 %) as compared to the level of non-flavonoids (17 %) (data not shown).

Levels of total phenolics vary widely in plant foods. When expressed as mg GAE/g DM, the total phenolic contents in cereals, vegetables, and fruits was (0.2–1.3), (0.4–6.6), and (10–51), respectively (Kahkonenn et al. 1999). Significant amounts of total phenolics are also found in herbs (9–23 mg GAE/g) and some medicinal plants (32–40 mg GAE/g).

When compared with other foods of plant origin, the total phenolic contents of *T. nivea* were very high. In fact, they are higher than phenolic-rich foods such as strawberry (330 mg/100 g FW) or red plum (320 mg/100 g FW) (Apak et al. 2007) on fresh-weight basis (*T. nivea* = 445 mg/100 g FW). They are comparable or higher than the total phenolics levels of 12 medicinal species of mushrooms (Song and Van Griensven 2008), Turkish wild edible mushrooms (Elmastas et al. 2007), cultivated mushrooms (Dubost et al. 2007), and Portuguese wild edible mushrooms (Barros et al. 2007). Phenolic compounds of wild mushrooms have been recently reviewed by Ferreira et al. (2009). Simple phenolic acids are the major type of phenolics found in mushrooms. This is consistent with our findings using solid extraction and fractionation using C_{18} Sep-pak described below.

Higher tannins content was recorded with the Folin–Ciocalteu method as compared to the vanillin–HCl method. It is known that Folin–Ciocalteu method is prone to interference by many potential metabolites such as sugars and organic acids and tend to overestimate the total phenolics content. It is also known that different phenolics exhibit different reactivity with FC reagent.

Table 18.1 Means, standard deviations and percents of various antioxidant constituents found in the truffle *Tirmania nivea* (*Zubaidi*) from six different sources in the Middle East

Antioxidant constituents		mg/100 g (DM) (±SD)	%	Remarks
Phenolics	Total[a]	1,811.5 ± 330.15	100	
	Bound[a]	494.67 ± 207.77	27.3	
	Free[a]	1,316.83 ± 195.81	72.7	
	Non-flavonoids phenolics[a]	1,027.0 ± 176.16	78.8	
	Flavonoids phenolics[a]	289.83 ± 26.99	22	Estimated by $AlCl_3$ method
	Flavanones	48.67 ± 29.63	3.6	Estimated by DNPH method
	Total flavonoids	338.50 ± 28.31	24.8	
	Tannins (TAE)[b,c]	335.08 ± 22.08	33.90	Estimated by FC method after PVPP precipitation
	Tannins (CE)[b,d]	225.31 ± 82.40	22.8	Estimated by vanillin–HCl method after PVPP precipitation
	Anthocyanins[e]	11.98 ± 9.20	0.012	
Others	Ascorbic acid[e]	9.63 ± 2.52	0.01	
	Total carotenoids[e]	0.681 ± 0.27	0.0007	

[a]Average of triplicate samples from six different sources
[b]Average of triplicate samples from three different sources
[c]Calculated as tannic acid equivalent (TAE) (mg/100 g)
[d]Calculated as catechin equivalent (CE) (mg/100 g)
[e]Average of triplicate samples from four different sources

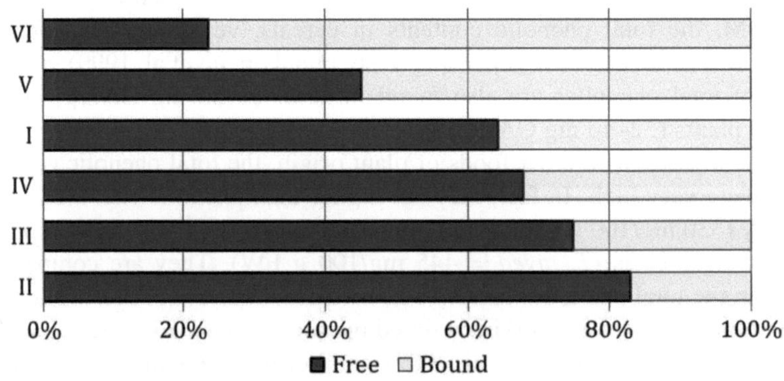

Fig. 18.2 Percent of free and bound phenolics in *T. nivea* samples from six different sources arranged in increasing order of free phenolics

Usually, high phenolic content is correlated with antioxidant activity. Phenolic compounds react with many active oxygen species such as hydroxyl radical ($^{\bullet}OH$), superoxide anion radical ($O2^{\bullet-}$), and lipid peroxyl radicals ($ROO^{\bullet}$) (Eberhardt 2001).

Flavonoids content was only 22–25 % of the total phenolics. Contrary to our findings, Barros et al. (2011) recently reported that Portuguese wild mushrooms contain no flavonoids. This group used HPLC/DAD/MS to profile the phenolics, as compared to the $AlCl_3$ and 2,4-dinitrophenylhydrazine (DNPH) methods used to grossly quantify total flavonoids in our study. Furthermore, the 2,4-DNPH method was used to specifically estimate flavanone content in *T. nivea*. According to Chang et al. (2002), the $AlCl_3$ method can be used to estimate flavones, flavonols, and isoflavones, whereas 2,4-DNPH specifically reacts with flavanones. Results of both methods may be compiled for the estimation of total flavonoids.

Flavonoids play many diverse biological roles in plants (Gould and Lister 2005; Samanta et al. 2011). Among these are effects on seed and fungal spore germinators. They have been implicated in cellular processes affected by temperature (Kaplan et al. 2004), and the levels of some flavonoids gradually increase during drought (Hernández et al. 2009).

The degree of polymerization (DP), estimated as the ratio between total phenolics and total flavonoids, was calculated for *T. nivea* samples obtained from six sources. They had an average of 6.25 and ranged between 4.9 and 7.6, which is a moderate value. DP is commonly used as a marker for ecological adaptation. They are also associated with some bioactivity such the antioxidant activity (see below).

Based on analysis of *T. nivea* samples from two different sources, the average tannin content was 335 $\pm$ 22.53 mg/100 g DM. No significant difference was found between the two sources. This level of tannins constitutes around 34 % of the total phenolics. Tannins have been correlated with antioxidant activity in general and with metal-chelating capacity in particular. In fact, tannic acid forms an inactive complex with iron in Fenton-type reactions (Lopes et al. 1999).

As a group, tannins are mixed polyhydroxy-flavan-3-ol oligomers and polymers (Schofield et al. 2001), and their active sites increase in number with increased degree of polymerization. Recently, Tharayil et al. (2011) showed that warmer and drier climatic conditions, typical of deserts, led to higher concentrations of flavonoids in Acer plants, to doubling the concentration of total tannins, and to the production of condensed tannins with lower polymerization and a greater proportion of procyanidin units, thus providing more protective compounds.

Tannins are involved in antioxidant activity, radical scavenging, and metal and protein binding. The antioxidant activity of tannins correlates with the increase of the molecular weight and possesses greater antioxidant potential than simple phenolics and flavonoids (Hagerman et al. 1998). Tian et al. (2012) reported that free radical scavenging ability of persimmon tannin increased with the mean degree of polymerization. With this capacity, tannins in *T. nivea* may act both as primary and secondary antioxidants in many in vitro systems.

18.4.1.2 Anthocyanins

Generally, mushrooms and other higher fungi do not contain anthocyanins, a pigment widely associated with plant colors, and only a limited number of higher fungi contain carotenoids (Velisek and Cejpek 2011). *T. nivea* was reported to contain a low level of anthocyanins (12 mg/100; or 0.012 %) using pH differential assay (Al-laith 2010). A possible further evidence of the presence of anthocyanins in *T. nivea* comes from the solid-phase extraction experiment described below. Anthocyanins are very potent antioxidants, and their performance exceeds that of both ascorbic acid and α-tocopherols in many systems (Kahkonen and Heinonen 2003). The role of anthocyanins as a protector against oxidative stress in plants has been recently reviewed (Hatier and Gould 2009). Anthocyanins reduce chlorophyll bleaching and lipid peroxidation which are symptoms of oxidative stresses. They act as transition metal chelators, optical masking agents leading to decreased production of ROS due to photooxidation (photoabatement), and as direct scavengers for ROS and NOS.

18.4.1.3 Ascorbic Acid

Average content of ascorbic acid in *T. nivea* from four different sources was 9.63 ± 1.52 mg/100 g, representing ~0.01 % of the total dried matter (Al-Laith 2010). Significant difference was observed between sources, with twofold difference between the lowest and the highest. Values of ascorbic acid of *T. nivea* are comparable to values reported for Libyan (Shamekh et al. 1985) and Saudi (Sawaya et al. 1985) truffles. Somewhat higher values were reported in wild mushrooms from Portugal (13–35 mg/100 g) (Barros et al. 2007) and Turkey (24.9 mg/g) (Keleş et al. 2011). All these studies demonstrated a low concentration of ascorbic acid in truffle and mushrooms. In fact, Keleş et al. (2011) showed that ascorbic acid was not detected in many wild mushrooms. At such low levels, it is reasonable to assume that ascorbic acid, though a potent antioxidant, has only a minor role to play in the antioxidant activity of *T. nivea* and other truffles as compared to the phenolic compounds. It should be noted, however, that higher levels of ascorbic acid (~5 %) was reported in some truffles (Hussain and Al-Ruqaie 1999).

18.4.1.4 Total Carotenoids

Total carotenoids content of *T. nivea* varied significantly from source to source and averaged at 0.68 mg/100 g dw. Generally, higher fungi such as wild mushrooms and truffles contain low concentrations of carotenoids (Barros et al. 2007; Keleş et al. 2011). It is not expected to have high content of total carotenoids in organisms that spawn and grow underground unexposed to sunlight, like truffles. Carotenoids play various and important roles in photosynthetic organisms (Murthy et al. 2005). They have many other functions in higher organisms including mammals, but mainly, they act as antioxidants and radical scavengers. It is reasonably safe to speculate that some carotenoids may act as antioxidant–antiradical agents in truffles knowing that they are fat-soluble and may protect against possible damages to the cell membrane triggered by abiotic stresses.

18.4.2 Antioxidant and Antiradical Activities

Experimental results from our laboratory, both published (Al-Laith 2010) and unpublished, indicate that samples of *T. nivea* from various regions of the Middle East possess antioxidant and antiradical activities in all analytical methods used and in a dose-dependent manner. These samples differ in the content and the activity of the analyzed antioxidant(s). *T. nivea* from one source may show high antioxidant activity in one assay but low activity in another assay. Preliminary work showed that a methanolic extraction of fresh and dried *T. nivea* was more efficient than aqueous phosphate buffer. It is well documented that phenolic compounds, a diverse group, are better extracted with such an extraction system. For this reason, a methanolic solution was used to prepare extracts for the majority of this work.

18.4.2.1 Antioxidant Properties of Fresh Versus Dried *T. nivea*

Fresh and dried *T. nivea* samples were both evaluated using several assays. In the FRAP assay, higher FRAP values were reported for dried samples than for fresh samples on equal solid weight basis (assuming 75 % of moisture) (Fig. 18.3a). A higher antioxidant activity was also found for dried samples using DPPH assay; again, on an equivalent weight basis (Fig. 18.3b), deoxyribose assay for hydroxyl radical (Fig. 18.3c), as well as for sodium nitroprusside–Griess reagent assay for measuring nitric oxide (NO) radical activity (Fig. 18.3d). These findings obtained using assays measuring different antioxidant properties indicate that *T. nivea* can be dried and preserved without significant loss of antioxidant activity. Drying is still a common practice to preserve truffles. This indicates that the major constituents contributing to the antioxidant activity are apparently not adversely affected by the drying process.

18.4.2.2 Antioxidant Activities of Dried *T. nivea*

The average values of antioxidant–antiradical activities of dried *T. nivea* measured using various methods are given in Table 18.2, whereas the EC_{50} values are given in Table 18.3. Since studies covering the antioxidant activity of truffles are very limited, comparison will be made to mushrooms, which have been more frequently studied. This is justifiable because of the phylogenetic relationship and resemblance of many ecological and chemical aspects. For example, both types are rich in carbohydrate, protein, and phenolics.

The range of antioxidant activity values of dried samples of *T. nivea*, expressed as mmol TE/100 g dw, were 10–20, 2.5–4.9, and 12.2–34.1 for FRAP, ABTS, and DPPH assays, respectively. The VCEAE values measured by FRAP assay ranged between 10.4 and 18.6 mmol VCAE/100 g (Table 18.2) and were similar in rank

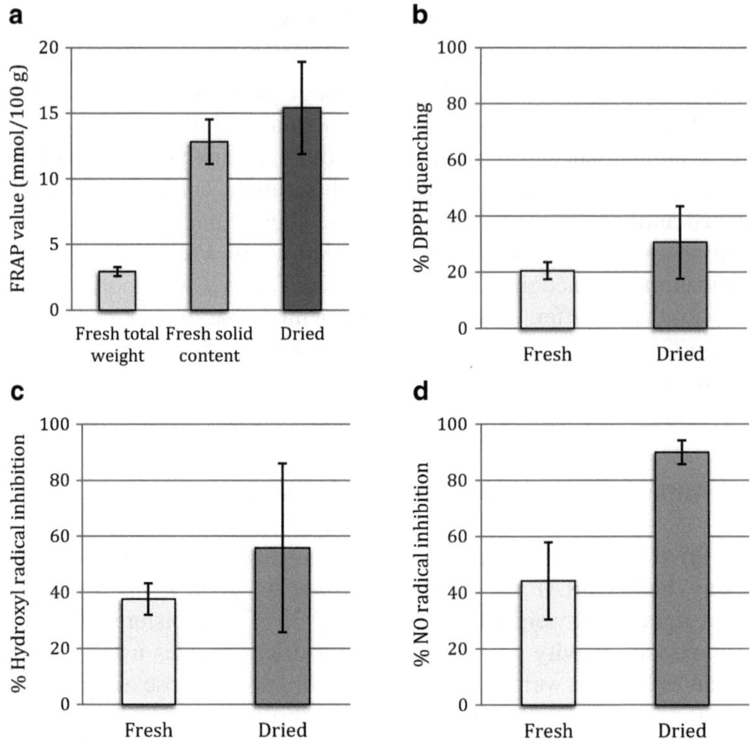

Fig. 18.3 Antioxidant–antiradical activities of fresh and dried *T. nivea*. Calculation of equivalent weight was based on 75 % moisture content. Values are averages of samples from four different sources. Error bars represent ± SD. (**a**) FRAP values (expressed as mmol ascorbic acid/100 g) and calculated based on fresh total weight, fresh solid content, and dried weight. (**b**) % inhibition of DPP quenching. Based on 2.0 mg and 0.5 mg per assay for fresh and dried samples, respectively. (**c**) % inhibition of hydroxyl radical. (**d**) % inhibition nitric oxide (NO) radical. For (**c**) and (**d**), the amount of fresh truffle used was 10 mg/ml, and the equivalent weight of dried truffle was 2.5 mg/ml

Table 18.2 Antioxidant–antiradical activities of *T. nivea* determined by FRAP, ABTS, DPPH, and FC assays

	FRAP	FRAP	ABTS	DPPH	FC
	VCEAC (mmol AAE/100 g)	mmol TE/100 g	mmol TE/100 g	mmol TE/100 g	mmol GAE/100 g
Average	15.77 ± 3.65	14.81 ± 3.23	3.84 ± 1.25	23.23 ± 8.37	12.44 ± 2.88

Values are average ± SD of triplicate samples from six different sources. All values are on a dry weight basis

Table 18.3 EC_{50} values for scavenging/inhibition activity of *T. nivea* extracts from different sources

Assay/radical	EC_{50}
DPPH ($n = 4$)[a]	0.60 ± 0.35 mg/ml
Hydroxyl radical scavenging activity ($n = 3$)	1.2 ± 0.50 mg/ml
Nitric oxide ($n = 4$)	159.25 ± 60.29 μg/ml
Iron chelating ($n = 2$)	6.12 ± 6.25 mg/ml
Lipid peroxidation ($n = 1$)[b]	10 ± 0.63 mg/ml

EC_{50} is the amount of dried truffle required to cause 50 % inhibition specific for the assay system
[a]*n* refers to the number of different sources tested, each in triplicate
[b]Two sources tested, one showed pro-oxidant activity at 2.5 mg/ml and not included

and magnitude to FRAP values calculated as TE. The antioxidant activity based on GAE ranged between 9.8 and 16.2 mmol GAE/100 g.

FRAP values reported here are comparable to those reported for truffles and mushrooms. Keleş et al. (2011) studied 21 wild edible mushrooms from Turkey and reported FRAP values as high as 62,771 μmol Fe^{2+}/g (dw) for some mushrooms. They are also as high as values reported for other plant edible foods considered to be at of the top list (Halvorsen et al. 2006). Few, uncommon, berries contain exceptionally high FRAP values, but other, common, fruits have comparable or lower values (Carlsen et al. 2010).

Results from the DPPH assay indicate that *T. nivea* possesses high antiradical activity. At a level of about 1 mg/ml, 90 % DPPH scavenging activity can be achieved (Al-Laith 2010). The average EC_{50} is 0.60 ± 0.35 mg/ml. Huang et al. (1999) reported that methanolic extract of several mushrooms showed high DPPH scavenging activity at low concentration. *Agaricus blazei* showed, at 2.5 ml/ml, more than 90 % DPPH inhibition, a value similar to that of *T. nivea* from some tested sources. Many wild and commercially available mushrooms possess high antiradical scavenging activity in DPPH assay amounting to 70–100 % at levels of 1–5 mg/ml (Gezer et al. 2006; Gaafar et al. 2010). Methanolic extracts of some *Boletus edulis* and *Suillus luteus* at 25 mg/ml exhibited more than 90 % DPPH scavenging activity (Keleş et al. 2011). However, other wild mushrooms possessed much lower activities in this study. *T. nivea* has higher antiradical activity than several common fruits (average 7.53 mg TE/100 g) and vegetables (average of 1.32 mg TE/100 g) as reported by Apak et al. (2007).

When the antioxidant activity of the *T. nivea* samples from different sources was ranked using the four methods (FRAP, DPPH, ABTS, FC), some samples showed clear patterns (Fig. 18.4), being ranked high (source I) or low (source VI) by the four methods. Other samples showed contrasting ranking orders. For example, sample from source V was ranked high by FRAP assay but low by FC method. Although the FRAP value is generally correlated to the total phenolics in many food systems, some reports have shown otherwise. For example, Keleş et al. (2011) reported that the wild mushroom *Suillus luteus* possessed a relatively high FRAP value of 58,529 μmol Fe^{2+}/g with a relatively low total phenolic (5,064 mg/kg). The antioxidant activity of the truffles, like any other complex system, cannot be

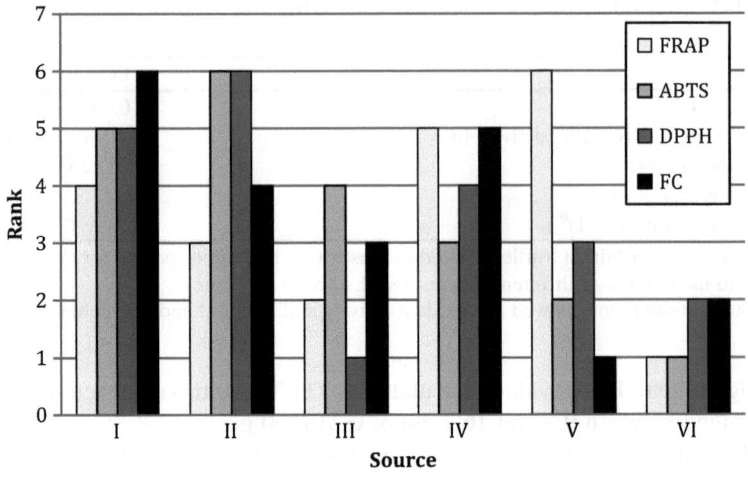

Fig. 18.4 Ranking of sources of *T. nivea* according to their antioxidant activity using four analytical methods from 1 (being the lowest) to 6 (being the highest) in each method

attributed to a single compound. Many phenolic compounds in plant materials exist as monomers, dimmers, trimers, and polymers, with different molecular weights and degrees of polymerization. These constituents possess different free phenolic hydroxyl groups per molecules, thus they contribute differently to the total antioxidant activity (Aaby et al. 2004).

18.4.2.3 Nitric Oxide ($NO^•$) Radical Scavenging Activity

Dried samples of *T. nivea* from several sources showed high antiradical activity against nitric oxide radical ($NO^•$) (Al-Laith 2010). A significant variation range of this activity was evident and manifested as large standard deviations on both the VCEAC (2.8-fold) and EC_{50} (2.5-fold) values. At low concentrations (up to 50 μg/ml DM), a linear relationship was demonstrated between *T. nivea* extract concentration and $NO^•$-inhibition and this was used for calculating VCEAC values. At higher concentrations (50–500 μg/ml), a hyperbolic scavenging activity was observed. Maximum % $NO^•$ inhibition was 70–84 % at *T. nivea* concentration of 500 μg/ml. Green tea (*Camellia sinensis*) and rosemary (*Rosmarinus officinalis*) are known as very potent $NO^•$-scavengers (Tsai et al. 2007). The reported EC_{50} values of these two herbs were 144 μg/ml and 200 μg/ml, respectively. The average EC_{50} value of *T. nivea* is 159 μg/ml indicating that *T. nivea* is as potent as green tea and perform better than rosemary as $NO^•$-scavenger. Ascorbic acid is a very effective $NO^•$ scavenging agent in the reaction system employed, whereas Trolox is not. The EC_{50} of ascorbic acid is 15.8 μg/ml, representing about one-tenth (1/10) of the average value of *T. nivea* scavenging capacity. In addition, *T. nivea* possesses comparable total antioxidants (mmol TE/g, ABTS assay) and total phenolics (mg GAE/g) as green tea.

18.4.2.4 Hydroxyl (•OH) Radical Scavenging Activity

The deoxyribose method is widely used to measure the hydroxyl radical scavenging activity. The sugar is attacked by hydroxyl radical generated by Fenton-type reaction and the MDA is formed as a product. The formation of MDA is either prevented or decreased in the presence of an antiradical. Extracts of *T. nivea* from all sources tested showed variable anti-hydroxyl activity ranging between 11 and 91 % •OH scavenging activity at 2.5 mg/ml (Al-Laith 2010). Very recently, Boda et al. (2012) showed a very high variability of •OH scavenging activity of several edible mushrooms from Kashmir using the same method. In this study, the highest activity (91 %) was reported for *Pleurotus sajor-caju* at 300 μg/ml, eightfold less than our finding. In our study, high correlation was found between the hydroxyl radical scavenging activity and the total flavonoids but not total phenolics.

18.4.2.5 Total Antioxidant Capacity and Total Reducing Power

A *T. nivea* sample extract from one source was examined for both total antioxidant capacity (TAC) using the phosphomolybdenum method (Prieto et al. 1999) and total reducing power (TRP) by the ferricyanide method (Oyaizu 1986). This sample possessed an average TAC of 47.81 ± 0.93 mg GAE/g DM, a value which agrees with those reported for *Pleurotus sajor-caju and Pleurotus florida* (Shinde and Deshmukh 2012). The average value of the TRP was 60.92 mg GAE/g.

The reducing power of a substance is an indicator of its potential antioxidant activity (Meir et al. 1995). Both TAC and TRP are relatively high in *T. nivea.* Compounds capable of donating a hydrogen atom or a single electron can break the free radical chain. Specific hydroxyl groups in phenolic compounds increase both the polarity and reducing power.

18.4.2.6 Iron Chelating Activity of *T. nivea*

Metal-chelating activity of *T. nivea* extracts from two different sources was estimated using 1,10-phenanthroline to probe Fe^{2+} and compared to a strong chelating agent (EDTA) in the concentration range used (100–1,000 μg/ml). Large variation was found between these two samples. The highest iron-chelating activity, expressed as EC_{50}, was 1.76 ± 0.33 mg/ml and the lowest was 10.47 ± 0.74 mg/ml or sixfold difference. Both samples were significantly lower in their iron-chelating activity as compared to EDTA (the positive control), which has an EC_{50} of 118.82 ± 4.34 μg/ml. Recently, Witkowska et al. (2011) reported iron-chelating activity (EC_{50} values) of 16 wild edible mushrooms from Poland which ranged between 8.0 and 12.1 mg/ml, similar to that of the least effective. The highest and the least effective samples of *T. nivea* exhibited iron-chelating activity comparable to 74.80 and 12.47 mg EDTA equivalent per g sample. Iron, an important constituent of many antioxidant enzymes (ascorbate and nonspecific peroxidases, catalases, and Fe-superoxide dismutase), is also known to serve as a pro-oxidant in a Fenton-type reaction (Fraga and Oteiza 2002). This pro-oxidant activity is observed when an antioxidant reduces the Fe^{3+} to Fe^{2+}. This activity was not observed under the experimental conditions employed.

18.4.2.7 Anti-lipid Peroxidation Activity

Measuring lipid peroxidation products using the TBARS method is considered a direct biomarker of oxidative stress (Farmer and Davoine 2007). Some products of lipid peroxidation (i.e., malondialdehyde, MDA) react with TBA to form a pinkish chromophore. *T. nivea* extracts of lyophilized samples showed significant anti-lipid peroxidation activity when tested using egg yolk (as a substrate) and APPH (a radical and inducer of lipid oxidation). At a level of 1 mg/ml, samples of *T. nivea* from two different sources were 31 % and 81 % as effective as tannic acid and 37 % and 94 % as effective as Trolox, respectively (Fig. 18.5). At higher levels (2.5 mg/ml) the sample from the more effective source showed reduced anti-lipid peroxidation activity or pro-oxidation activity, whereas the least effective sample exhibited an increased activity in a dose–response manner with no sign of pro-oxidant activity. This observation was not further investigated. However, since the induction of lipid peroxidation in the assay employed did not involve the use of Fe^{2+} as an inducer, Fe^{2+} may not be the cause of this observation. On the other hand, the capacity of flavonoids to donate an electron and hydrogen play a role in terminating lipid peroxidation chain reaction (Samanta et al. 2011). Depending on the surrounding conditions, this reducing power qualifies some phytochemicals to exhibit both anti- and pro-oxidant. The pro-oxidant activity is observed in the absence of free radicals. Catechin was shown to act as an antioxidant of LDL oxidation at the initiation phase, but as a pro-oxidant during the propagation phase. Similar effect was also reported for ascorbic acid in other systems. Recently, Wie and Van Griensven (2008) showed that extracts of medicinal mushrooms acted as anti- and pro-oxidant in vivo.

18.4.2.8 Fractionation of Antioxidant Activity

Solid-phase extraction and fractionation of phenolics of *T. nivea* further confirms the presence of several constituents which contribute towards the antioxidant activity, including ascorbic acid, free and bound phenolics, and flavonoids. When a C_{18} Sep-pak column was used according to Oszmianski et al. (1988), and fractions were tested using the DPPH assay, 43 % of the total antiradical activity was detected in the eluent and the washing fractions. These fractions usually contain

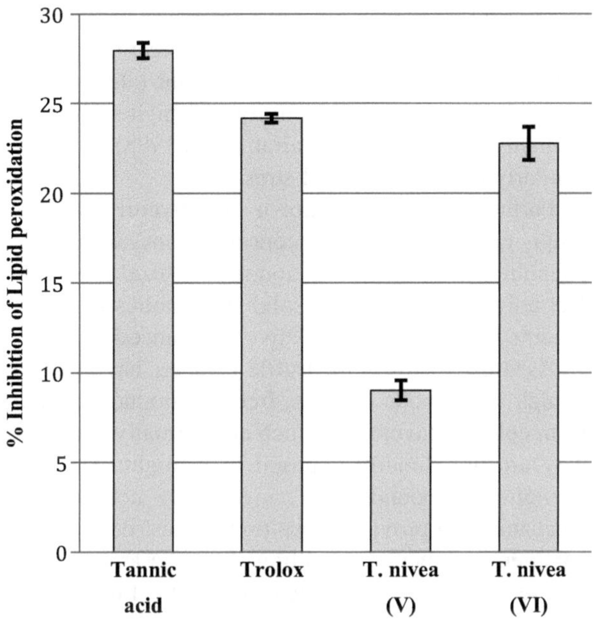

Fig. 18.5 Percent inhibition of lipid peroxidation by standard antioxidants (tannic acid, Trolox). Extracts of *T. nivea* at 1 mg/ml DM were obtained from two different sources (V and VI) and measured as TBARS-MDA in the egg yolk-APHH system. Results are average of duplicate readings with error bars representing SD

unretained compounds including sugars and ascorbic acid, if present. Of the retained materials (57 % of the total activity), fractions I, II, III, and IV possessed 24.3 %, 27.7 %, 17.2 %, and 30.8 % of the activity, respectively. These fractions represent free phenolic acids, the catechins group, the quercetin group, and polymeric anthocyanins, in that order. The last three fractions comprise the total flavonoids and account for the 43 % of the total activity. Besides, *T. nivea* may also contain flavanones as revealed by the method described by Chang et al. (2002) using the 2,4-DHPH reagent.

18.4.2.9 Ecological Importance of Antioxidant Activity

In their natural habitats, plants and higher fungi are subjected to many harsh environmental conditions as recently reviewed by Sharma et al. (2012). In deserts, for example, contrasting conditions may occur on the same day, i.e., very cold nights and very hot days, on top of seasonality. Affected plants have evolved several strategies for survival, which include nonenzymatic and enzymatic defense systems. A nonenzymatic system involves alteration of the level of natural antioxidants; ascorbic acid, glutathione, tocopherols, carotenoids, and phenolic compounds. Synthesis of phenolic compounds as secondary metabolites is known to be affected by environmental conditions (Sharma et al. 2012). The ability of certain plants to withstand stressful conditions is probably due to their ability to neutralize ROS and NOS (Toor et al. 2006) by increasing the level of antioxidants, especially phenolic compounds, which are known for their superior performance as compared to non-phenolic antioxidants (Ksouri et al. 2008). Enzymatic antioxidant systems also play a crucial role in protecting plants against deleterious effects of abiotic stresses. For example, drought-tolerant wheat showed higher ascorbate peroxidase and catalase activity and higher antioxidant capacity as compared to drought-sensitive wheat (Sairam et al. 1998). It is expected that truffle may respond similarly to environmental stress.

Furthermore, it was shown that mycorrhizal associations between plants and fungi contained higher concentrations of secondary metabolites, including phenolics, as compared to non-mycorrhizal roots (Schützendübel and Polle 2002). Not only was the level of soluble phenolics generally higher—the level of certain specific phenolics was selectively enhanced due to mycorrhizal association.

As already shown, the truffle *T. nivea*, having been in a mycorrhizal association, is high in phenolic content, free and bound, including flavonoids, though devoid from colored flavonoids which are normally associated with UV-B radiation, since they are not directly exposed to sunlight. Taken together, the high content of phenolic compounds in *T. nivea* may not only be considered as an adaptive mechanism to provide protection against deleterious oxygen species to cope with harsh environmental conditions but also as manifestation of mycorrhizal association serving both parties, the plant and the fungus. Furthermore, we have shown that *T. nivea* also contains appreciable amount of condensed tannins (proanthocyanidins), and the degree of polymerization is high, which is considered to be a marker of environmental adaptation of plants which enhances their antioxidant capacity to protect against oxidative damages (Falleh et al. 2012).

18.5 Conclusions and Future Perspective

Evidently, samples of *T. nivea* collected from many geographical locations possess variable but high antioxidant activities. These high activities place *T. nivea* among the top antioxidant-rich foods. Available evidence also suggests that *T. nivea* methanolic extracts act as both a primary (i.e., chain-breaking) and secondary (i.e., metal-chelating) antioxidants. Most of the antioxidant activity discussed in this chapter comes from the high content of phenolic compounds and has a role either in helping the plants and/or fungus survival or in reducing the adverse effect of the stress agent(s). Many factors play roles in causing the high variability of the antioxidant properties found for *T. nivea* including geographical origin and soil characteristics. Nonetheless, the general trend of high antioxidant activity shown using ten different assays is still evident. Natural antioxidants are known to exert their protection via several mechanisms. Ample studies have shown that natural antioxidants also provide protection to humans when consumed (Sen and Chakraborty 2011, for a recent review). With such high antioxidant activity, *T. nivea* provides a healthy meat alternative that satisfies the nutritional requirements as well as a very rich source of beneficial phytochemicals, especially for non-meat eaters. The major drawback is its being costly, seasonal, and scarce.

Our current knowledge is deficient. Future work should include, but not limited to, phenolic profiling using HPLC/DAD/MS which is essential for a better understanding of the antioxidant properties; investigating the contribution of other constituents such as peptides/proteins, carbohydrates, and glutathione towards the antioxidant activity of *T. nivea*; studying the enzymatic antioxidants; assessing the pro-oxidant activity; studying the stability of antioxidant activity under various industrial conditions; and investigating the antioxidant activity within the context of the ecology of *T. nivea* and the influence of environmental stresses. With the advancement in relevant research techniques (such as metabolomics and functional genomics), it should be possible and desirable to study the antioxidant properties of truffle and other foods in more detail. One of the anticipated difficulties is the availability of fresh *T. nivea* samples, where the fluctuation in yield is the norm.

Acknowledgment This work was financially supported by the Department of Biology, College of Science, University of Bahrain.

References

Aaby K, Hvattum E, Skrede G (2004) Analysis of flavonoids and other phenolic compounds using high-performance liquid chromatography with coulometric array detection: relationship to antioxidant activity. J Agr Food Chem 52:4595–4603

Abdallah YM, Hassan YM, Mossa JS, Al-Yahya MA (1997) Studies on the nutritive value of Saudi truffles and possibility of their preservation by canning. Proc Arab Conf Food Sci Technol 2:369–376

Abdulla N, Ismail SM, Aminudin N, Shuib AS, Lan BF (2011) Evaluation of selected culinary-medicinal mushrooms for antioxidants and ACE inhibitory activities. Evid base Compl Alternative Med 464238:12 p. doi:10.1155/2012

Ahmed AA, Mohamed MA, Hami MA (1981) Libyan truffles (*Terfezia boudieri* Chatin), chemical composition and toxicity. J Food Sci 46:927–929

Akata I, Ergonul B, Kalyoncu F (2011) Chemical compositions and antioxidant activities of 16 wild edible mushroom species grown in Anatolia. Int J Pharmacol 8:134–138

Al-Alawi NAM (2009) Food habits related with feeding the truffle fangs and effect of some of their treatments on experimental rats. M.Sc. Thesis. Umm Al-Qura University, Education Collage For Home Economy, Department of Nutrition & Food Science, Makkah. Saudi Arabia

Al-Laith AAA (2010) Antioxidant components and antioxidant/antiradical activities of desert truffle (*Tirmania nivea*) from various Middle Eastern origins. J Food Compos Anal 23:15–22

Al-Naama NM, Ewaze JO, Nema JH (1988) Chemical constituents of Iraqi truffles. Iraqi J Agr Sci 6:51–56

Al-Rahmah AN (2001) Truffle of desert and jungles. King Saud University Publications, Riyadh, Saudi Arabia (In Arabic)

Al-Rawi AA, Taha A (2010) Chemical study for three Iraqi truffles types. Al-Anbar J Agr Sci 8:33–41 (In Arabic)

Al-Sheikh AM, Trappe JM (1983) Desert truffles: the genus *Tirmania*. Trans Br Mycol Soc 81:83–90

Apak R, Güçlü K, Demirata B, Özyürek M, Çelik SE, Bektaşoğlu B, Berker KI, Özyurt D (2007) Comparative evaluation of various total antioxidant capacity assays applied to phenolic compounds with the CUPRAC assay. Molecules 12:1496–1547

Awameh MS, Al-Sheikh A (1979) Characteristics and ascospore germination of white kame (*Tirmania nivea* and *T. pinoyi*). Ann Phytopathol 11:223–229
Badalyan SM (2003) Edible and medicinal higher Basidiomycetes mushrooms as a source of natural antioxidants. Int J Med Mushrooms 5:153–162
Barros L, Dueñas M, Ferreira ICFR, Baptista P, Santos-Buelga C (2011) Phenolic compounds profile in wild edible greens from Portugal obtained by HPLC-DADESI/MS. Food Chem 127:169–173
Barros L, Ferreira MJ, Queiros B, Ferreira ICFR, Baptista P (2007) Total phenols, ascorbic acid, β-carotene and lycopene in Portuguese wild edible mushrooms and their antioxidant activities. Food Chem 103:413–419
Benzie IFF, Strain JJ (1996) The Ferric reducing ability of plasma (FRAP) as a measure of antioxidant power: the FRAP assay. Anal Biochem 239:70–76
Boda RH, Wani AH, Zargar MA, Ganie BA, Wani BA, Ganie SA (2012) Nutritional values and antioxidant potential of some edible mushrooms of Kashmir valley. Pak J Pharm Sci 25:623–627
Bokhary HA (1987) Desert truffles "al-kamah" of the Kingdom of Saudi Arabia. I. Occurrence identification and distribution. Arab Gulf J Sci Res 5:245–255
Brand-Williams W, Cuvelier ME, Berset C (1995) Use of a free radical method to evaluate antioxidant activity. Lebensm Wiss Technol 28:25–30
Carlsen MH, Halvorsen BL, Holte K, Bøhn SK, Dragland S, Sampson L, Willey C, Senoo H, Umezono U, Sanada C, Barikmo I, Berhe N, Willett WC, Phillips KM, David R, Jacobs DV Jr, Blomhoff R (2010) The total antioxidant content of more than 3100 foods, beverages, spices, herbs and supplements used worldwide. Nutr J 9:3
Chang CC, Yang MH, Wen HM, Chern JC (2002) Estimation of total flavonoid content in propolis by two complementary colorimetric methods. J Food Drug Anal 10:178–182
Chang MJ, Collins JL, Baily JW, Coffey DL (1994) Tannins related to cultivar, maturity, dehulling, and heating. J Food Sci 59:1034–1036
Cheung LM, Cheung PCK, Ooi VEC (2003) Antioxidant activity and total phenolics of edible mushroom extracts. Food Chem 81:249–255
Diez J, Manjon JL, Martin F (2002) Molecular phylogeny of the mycorrhizal desert truffles (*Terfezia* and *Tirmania*), host specificity and edaphic tolerance. Mycologia 94:247–259
Dubost NJ, Ou B, Beelman RB (2007) Quantification of polyphenols and ergothioneine in cultivated mushrooms and correlation to total antioxidant capacity. Food Chem 105:717–735
Eberhardt MK (2001) Reactive oxygen metabolites: chemistry and medical consequences. CRC Press, Boca Raton, FL
El-Kholy H (1989) Genetical and physiological studies on truffles. Ph.D. Thesis, Faculty of Agric., Cairo University, Egypt
Elmastas M, Isildak O, Turkekul I, Temur N (2007) Determination of antioxidant activity and antioxidant compounds in wild edible mushrooms. J Food Compos Anal 20:337–345
Falleh H, Ksouri R, Boulaaba M, Guyot S, Abdelly C, Christian Magné C (2012) Phenolic nature, occurrence and polymerization degree as marker of environmental adaptation in the edible halophyte *Mesembryanthemum edule*. S Afr J Bot 79:117–124
FAO/IAEA (2000) Quantification of tannins in tree foliage, a laboratory manual, animal production and health sub-programme. Joint FAO/IAEA Division of Nuclear Techniques in Food and Agriculture
Farmer EE, Davoine C (2007) Reactive electrophile species. Curr Opin Plant Biol 10:380–386
Ferreira ICFR, Barros L, Abreu RMV (2009) Antioxidant in wild mushrooms. Curr Med Chem 16:1543–1560
Fraga CG, Oteiza PI (2002) Iron toxicity and antioxidant nutrients. Toxicology 180:23–32
Gaafar AM, Yossef HE, Ibrahim H (2010) Protective effects of mushroom and their ethyl extract on aging compared with L-carnitine. Int J Nutr Metab 2:63–69

Gezer K, Duru ME, Kivrak I, Turkoglu A, Mercan N, Turkoglu H, Gulcan S (2006) Free-radical scavenging capacity and antimicrobial activity of wild edible mushroom from Turkey. Afr J Biotechnol 5:1924–1928

Gould KS, Lister C (2005) Flavonoid functions in plants, Ch. 8. In: Oyvind MA, Markham KR (eds) Flavonoids chemistry, biochemistry and applications. CRC Press, Boca Raton, FL, pp 397–441

Hagerman AE, Riedl KM, Jones GA, Sovik KN, Ritchard NT, Hartzfeld PW, Riechel TL (1998) High molecular weight plant polyphenolics (tannins) as biological antioxidants. J Agric Food Chem 1998(46):1887–1892

Halliwell B, Gutteridge JM, Aruoma OI (1987) The deoxyribose method: a simple "test-tube" assay for determination of rate constants for reactions of hydroxyl radicals. Anal Biochem 165:215–219

Halvorsen BL, Carlsen MH, Phillips KM, Bohn SK, Holte K, Jacobs DR Jr, Blomhoff R (2006) Content of redox-active compounds (i.e., antioxidants) in foods consumed in the United States. Am J Clin Nutr 84:95–135

Hatier J-HB, Gould KS (2009) Anthocyanin function in vegetative organs. In: Gould KS, Davies KM, Winfield C (eds) Anthocyanins: biosynthesis, functions, and applications. Springer, New York, NY, pp 1–20

Hernández I, Alegre L, Breusegem FV, Munné- Bosch S (2009) How relevant are flavonoids as antioxidants in plants? Trends Plant Sci 14(3):125–132

Huang SJ, Huang LC, Chen CC, Mau JL (1999) Antioxidant properties of Agaricus blazei. In: Broderick A, Nair T (eds) Proceedings of the third international conference on mushroom biology and mushroom products, Sydney, Australia, pp. 266–274.

Hussain G, Al-Ruqaie IM (1999) Occurrence, chemical composition, and nutritional value of truffle: an overview. Pak J Biol Sci 2:510–514

Jabbur JS (1995) The Bedouins and the desert: aspects of nomadic life in the Arab East. State University of New York Press, Albany, NY

Jamali S, Banihashemi Z (2012) Fungi associated with ascocarps of desert truffles from different parts of Iran. J Crop Protect 1:41–47

Kahkonenn MP, Hopia AI, Vuorela HJ, Rauha J-P, Pihlaja K, Kujala TS, Heinonen M (1999) Antioxidant activity of plant extracts containing phenolic compounds. J Agric Food Chem 47:3954–3962

Kahkonen MP, Heinonen M (2003) Antioxidant activity of anthocyanins and their aglycones. J Agric Food Chem 51:628–633

Kaplan F, Kopka J, Haskell DW, Zhao W, Schiller KC, Gatzke N, Sung DY, Guy CL (2004) Exploring the temperature-stress metabolome of Arabidopsis. Plant Physiol 136:4159–4168

Keleş A, Koca İ, Gençcelep H (2011) Antioxidant properties of wild edible mushrooms. J Food Process Technol 2:130. doi:10.4172/2157-7110.1000130

Ksouri R, Megdiche W, Falleh H, Trabelsi N, Boulaaba M, Smaoui A, Abdelly C (2008) Influence of biological, environmental and technical factors on phenolic content and antioxidant activities of Tunisian halophytes. C R Biol 331:865–873

Lee Y, Jian S, Lian P, Mau J (2008) Antioxidant properties of extracts from a white mutant of the mushroom *Hypsizygus marmoreus*. J Food Compos Anal 21:116–124

Lopes GKB, Herbert M, Schulman HM, Hermes-Lima M (1999) Polyphenol tannic acid inhibits hydroxyl radical formation from Fenton reaction by complexing ferrous ions. Biochim Biophys Acta 1472:142–152

Mandeel Q, Al-Laith AAA (2007) Ethnomycological aspects of the desert truffle among native Bahraini and non-Bahraini peoples of the Kingdom of Bahrain. J Ethnopharmacol 110:118–129

Mattila P, Sounpa K, Piironen V (2000) Functional properties of edible mushrooms. J Nutr 16:7–8

Mau JL, Chang CN, Huang SJ, Chen CC (2004) Antioxidant properties of methanolic extracts from *Grifola frondosa*. Morchella esculenta and Termitomyces albuminosus mycelia. Food Chem 87:111–118

Mau JL, Lin HC, Chen CC (2002) Antioxidant properties of several medicinal mushrooms. J Agric Food Chem 50:6072–6077

Meir S, Kanner J, Akiri B, Hadas SP (1995) Determination and involvement of aqueous reducing compounds in oxidative defense systems of various senescing leaves. J Agric Food Chem 43:1813–1819

Murcia MA, Matinez M, Jimanez AM, Veera AM, Honrubia M, Parras P (2002) Antioxidant activity of edible fungi (truffles and mushrooms): losses during industrial processing. J Food Protect 65:1614–1622

Murcia MA et al (2013) Antioxidant properties: Terfezia, Picoa. In: Kagan-Zur V, Roth-Bejerano N, Sitrit Y, Morte A (eds) Desert truffles: phylogeny, Physiology, Distribution and Domestication. Springer, Heidelberg

Murthy KNC, Vanitha A, Rajesha R, Swamy MM, Sowmya PR, Ravishankar GA (2005) In vivo antioxidant activity of carotenoids from *Dunaliella salina* - a green microalga. Life Sci 76:1381–1390

Nagy M, Grancai D (1996) Colorimetric determination of flavanones in Propolis. Pharmazie 51:100–101

Oszmianski J, Ramos T, Bourzeix M (1988) Fractionation of phenolic compounds in Red Wine. Am J Enol Vitic 39:259–262

Oyaizu M (1986) Studies on products of browning reactions: antioxidative activities of products of browning reaction prepared from glucosamine. Jpn J Nutr 44:307–315

Pal J, Ganguly S, Tahsin KS, Acharya K (2010) In vitro free radical scavenging activity of wild edible mushroom, Pleurotus squarrosulus (Mont.) Singer. Id. J Exp Biol 47:1210–1218

Patel S (2012) Food, health and agricultural importance of truffles: a review of current scientific literature. Curr Trend Biotechnol Pharm 6:15–27

Prieto P, Pineda M, Aguilar M (1999) Spectrophotometric quantitation of antioxidant capacity through the formation of a phosphomolybdenum complex: specific application to the determination of vitamin E. Anal Biochem 269:337–341

Prior RL, Wu XL, Schaich K (2005) Standardized methods for the determination of antioxidant capacity and phenolics in foods and dietary supplements. J Agric Food Chem 53:4290–4302

Puntel RL, Nogueira CW, Rocha JBT (2005) Krebs cycle intermediates modulate thiobarbituric acid reactive species (TBARS) production in rat brain in vitro. Neurochem Res 30:225–235

Puttaraju NG, Venkateshaiah SU, Dharmesh SM, Urs SMN, Somasundaram R (2006) Antioxidant activity of indigenous edible mushrooms. J Agric Food Chem 54:9764–9772

Re R, Pellegrini N, Proteggente A, Pannala A, Yang M, Rice-Evans C (1999) Antioxidant activity applying an improved ABTS radical cation decolorization assay. Free Radic Biol Med 26:1231–1237

Rice-Evans CA, Miller NJ, Bolwell PG, Bramley PM, Pridham JB (1995) The relative antioxidant activities of plant-derived polyphenolic flavonoids. Free Radic Res 22:375–383

Sairam RK, Deshmukh PS, Saxena DC (1998) Role of antioxidant systems in wheat genotypes tolerance to water Stress. Biol Plantarum 41:387–394

Samanta A, Das G, Sanjoy Kumar Das SK (2011) Roles of flavonoids in plants. Int J Pharm Sci Technol 6:12–35

Sawaya WN, Al-Shalhat A, Al-Sogair A, Al-Muhammad A (1985) Chemical composition and nutritive value of truffles of Saudi Arabia. J Food Sci 50:450–453

Schofield P, Mbugua DM, Pell AN (2001) Analysis of condensed tannins: a review. Anim Feed Sci Technol 91:21–40

Schützendübel A, Polle A (2002) Plant responses to abiotic stresses: heavy metal-induced oxidative stress and protection by mycorrhization. J Exp Bot 53:1351–1365

Sen S, Chakraborty R (2011) The role of antioxidants in human health. In: Silvana A, Hepel M (eds) Oxidative stress: diagnostics, prevention, and therapy (ACS symposium series). American Chemical Society, Washington DC, pp 1–37

Shamekh SS, Ahmed AA, El-Mabsout YE (1985) Effect of various methods of preservation on microbial load, vitamins and fatty acids of Libyan truffles. In: Hamdan, LY, El-Nawawy A,

Mameesh (eds) Advances in food industries development in the Arab World, A Proceedings. Kuwait Institute for Scientific Research and Kuwait Foundation for Advancement of Science, pp. 261–270

Sharma P, Jha AB, Dubey RS, Pessarakli M (2012) Reactive oxygen species, oxidative damage, and antioxidative defense mechanism in plants under stressful conditions. J Bot 2012:Article ID 217037. doi:10.1155/2012/217037

Shinde MN, Deshmukh SR (2012) Comparative study for antioxidant activity of cold and hot water extracts of *Pleurotus sajor-caju* and *Pleurotus florida*. www.biomedscidirect.com/journalfiles/IJBMRF2012737.doc. Accessed 1 Nov 2012

Song W, Van Griensven LJLD (2008) Pro- and antioxidative properties of medicinal mushroom extracts. Int J Med Mushrooms 10:315–324

Sreejayan N, Rao MN (1997) Nitric oxide scavenging by curcuminoids. J Pharm Pharmacol 49:105–107

Tian Y, Zou B, Dong X-Q, Chang Y, Du J, Chen J-Y, Li C-M (2012) Antioxidant activity of Persimmon tannin fractions with different degree of polymerization in vitro and in vivo. Food Sci. http://124.205.222.100/Jwk_spkx/EN/

Tharayil N, Suseela V, Daniella J, Triebwasser DJ, Caroline M, Preston CM, Patrick D, Gerard PD, Dukes JS (2011) Changes in the structural composition and reactivity of Acer rubrum leaf litter tannins exposed to warming and altered precipitation: climatic stress-induced tannins are more reactive. New Phytol 191:132–145

Toor PK, Savage GP, Lister CE (2006) Seasonal variation in antioxidant composition of greenhouse grown tomatoes. J Food Compos Anal 19:1–10

Trappe JM (1979) The order, families and genera of hypogeous *Ascomycotina* Truffles and their relatives. Mycotaxon 9:297–340

Tsai PJ, Tsai TH, Yu CH, Ho SC (2007) Comparison of NO-scavenging and NO-suppressing activities of different herbal teas with those of green tea. Food Chem 103:181–187

Turgeman T, Asher Ben J, Roth-Bejerano N, Kagan-Zur V, Kapulnik Y, Sitrit Y (2011) Mycorrhizal association between the desert truffle Terfezia boudieri and Helianthemum sessiliflorum alters plant physiology and fitness to arid conditions. Mycorrhiza 21:623–630

Valentão P, Lopes G, Valente M, Barbosa P, Andrade PB, Silva BM, Baptista P, Seabra RM (2005) Quantification of nine organic acids in wild mushrooms. J Agric Food Chem 53:3626–3630

Velisek J, Cejpek K (2011) Pigments of higher fungi: a review. Czech J Food Sci 29:87–102

Vidović SS, Mujić IO, Zeković ZP, Lepojević ZD, Tumbas VT, Mujić AI (2010) Antioxidant properties of selected boletus mushrooms. Food Biophys 5:49–58

Wang S, Marcone MF (2011) The biochemistry and biological properties of the world's most expensive underground edible mushroom: truffles. Food Res Int 44:2567–2581

Waterhouse A (2002) Folin-Ciocalteau micro method for total phenol in wine. http://waterhouse.ucdavis.edu/faqs/folin-ciocalteau-micro-method-for-total-phenol-in-wine

Wie S, Van Griensven LJLD (2008) Pro- and antioxidative properties of medicinal mushroom extracts. Int J Med Mushrooms 10:315–324

Witkowska AM, Zujko ME, Mirończuk-Chodakowska I (2011) Comparative study of wild edible mushrooms as sources of antioxidants. Int J Med Mushrooms 13:335–341

Yagi K (1998) Simple assay for the level of total lipid peroxides in serum or plasma. Free Rad Antiox Protoc 108:101–106

Yang JH, Lin HC, Mau JL (2002) Antioxidant properties of several commercial mushrooms. Food Chem 77:229–235

[illegible] Phytochemistry [illegible]

[illegible] Stud [illegible] 2012 [illegible]

[illegible]

[illegible]

Chapter 19
Preservation of Truffles

María Antonia Murcia, Luana Maggi, Ghulam Hussain, Antonia María Jiménez-Monreal, Magdalena Martínez-Tomé, and Ibrahim M. Alruqaie

19.1 Introduction

19.1.1 History

The "Poor Man's Truffle," "Truffles of the Desert" or "Fagaa" were much appreciated in earlier times, particularly by the Ancient Greeks and Romans who imported them largely from Carthage and Libya (see Chap. 15 by Shavit, this volume). All fungal fruit bodies that grow underground are loosely termed as "truffles." The desert truffles, however, are only distantly related to the highly prized true truffles (Tuber species) of southern Europe. The most valuable desert truffles belong to two genera: Terfezia Tul. and C. Tul., which has verrucose spores, and Tirmania, with its smooth spores. These are white to yellowish-brown truffles, rather soft and fleshy, can grow to a size of 3–10 cm across and may weigh 60–150 g. However, fruit bodies weighing up to 1,000 g have been recorded in Tirmania nivea Desfontaines (Pegler 2002). The distribution of these truffles

M.A. Murcia (✉) • A.M. Jiménez-Monreal • M. Martínez-Tomé
Department of Food Science, Faculty of Veterinary, Murcia University, Campus of Espinardo, 30100 Murcia, Spain

CIBER: CB12/03/30038, Fisiopatología de la Obesidad y la Nutrición, CIBERobn, Instituto de Salud Carlos III (ISCIII), Madrid, Spain
e-mail: mamurcia@um.es

L. Maggi
Institute of Oenology and Agro-Food Engineering, Universita Cattolica del Sacro Cuore, Via Emilia Parmense 84, 29122 Piacenza, Italy

G. Hussain • I.M. Alruqaie
National Center for Agriculture Technology (NCAT), King Abdulaziz City for Science and Technology (KACST), P. O. Box 6086, Riyadh 11442, Kingdom of Saudi Arabia
e-mail: iruqaie@kacst.edu.sa

V. Kagan-Zur et al. (eds.), *Desert Truffles*, Soil Biology 38,
DOI 10.1007/978-3-642-40096-4_19, © Springer-Verlag Berlin Heidelberg 2014

extends throughout southern Spain, Algeria, Morocco, Tunisia, Cyprus, Egypt, Syria, Iraq, Iran and Kuwait (see Part III, this volume).

Mandeel and Al-Laith (2007) investigated the general knowledge, attitude and practices of native Bahraini and non-Bahraini peoples of Bahrain concerning ethnomycological aspects and traditional folklore related with the desert truffle. A total of three species were identified from the southern parts of Bahrain. Tirmania nivea "Zubaidi" was found to be the most preferred, expensive and common type of truffle in the region due to its good light smell, delicacy and soft white tissues. This was followed by Terfezia claveryi "Khlassi."

19.1.2 Problems of Fresh Truffles and Their Preservation

Truffles are extremely perishable and have a very short postharvest shelf life. Moreover, truffles are collected at the point of full maturity and so senescence occurs immediately after harvest. Deterioration of their organoleptic quality is caused by the loss of water, which dramatically changes the texture and the mycelium growth that affects the visual appearance, aroma and flavour. Although truffles are sometimes eaten raw, current legislation does not establish limits for microbial counts or pathogen presence. The high initial microbial load and the presence of pathogenic microorganisms (Rivera, et al. 2010) have created a need for early decontamination procedures (physical and/or chemical) that decrease the initial microbial population of truffles. Invasions of fungal pathogens, such as chestnut blight (*Cryphonectria parasitica*) and Dutch elm disease (*Ophiostoma nova-ulmi*), have resulted in devastating effects that are well documented (Desprez-Loustau et al. 2007; Loo 2009). Therefore, to maintain its quality, texture, aroma, shelf life and economic value of truffles, it is important to review the techniques available for enhancing shelf life and quality.

19.1.3 Nutritional Composition

Truffles are healthy foods, low in calories and fat—their lipid composition includes neutral lipids, fatty acids (unsaturated fatty acids such as oleic and linoleic acids, among which linoleic acid is predominant) (Nöel-Suberville et al. 1996; Murcia et al. 2003) (Table 19.1) and phospholipids (such as phosphatidylethanolamine and cardiolipin) (Feofilova et al. 1998; Longvah and Deosthale 1998). Fungi are also rich in non-starch polysaccharides, such as beta-glucans, which are a good source of soluble and insoluble dietary fibre for humans (Bokhary and Parvez 1993; Manzi and Pizzoferrato 2000; Murcia et al. 2003). Moreover, truffles are richer in proteins than most vegetables, whereas their amino acid composition is comparable to that of animal proteins (Danell and Eaker 1992).

Table 19.1 Composition (% dry matter) of *Terfezia claveryi* and *Picoa juniperi*

	Origin	Process	Moisture	Ash	Crude protein	Lipid	Fibre	Carbohydrate[a]
Terfezia claveryi	Lorca, Spain	Raw	73.1 ± 0.3	4.2 ± 0.3	15.9 ± 0.1	6.9 ± 0.2	8.3 ± 0.1	64.5 ± 0.0
		Freezing	74.8 ± 0.1	3.5 ± 0.0	15.1 ± 0.1	6.6 ± 0.3	7.8 ± 0.1	67.0 ± 0.1
		Canning	75.2 ± 0.0	3.8 ± 0.0	14.9 ± 0.1	6.2 ± 0.3	9.2 ± 0.1	66.0 ± 0.0
Picoa juniperi	Lorca, Spain	Raw	63.8 ± 0.0	8.2 ± 0.1	22.6 ± 0.4	19.9 ± 0.0	13.0 ± 0.1	36.7 ± 0.0
		Freezing	64.4 ± 0.0	7.4 ± 0.0	22.2 ± 0.1	19.5 ± 0.1	13.2 ± 0.2	37.7 ± 0.1
		Canning	75.3 ± 0.1	7.7 ± 0.0	22.1 ± 0.3	19.0 ± 0.0	17.1 ± 0.1	34.2 ± 0.1

Source: Murcia et al. (2003)
[a]Carbohydrates calculated by difference

Table 19.2 Proximate analysis of composition (%) of *Tirmania nivea* and *Terfezia claveryi* truffles

Parameters	Truffle varieties	
	T. nivea (Zubaidi)	*T. claveryi* (Khlassi)
Moisture	81.6	79.2
Crude fibre	2.15	1.53
Ash	1.30	1.15
Crude protein	6.58	11.9
Crude fat	1.10	0.89
Carbohydrates	7.25	5.32

Source: Al-Ruqaie (2006)

Al-Ruqaie (2006) in Table 19.2 shows the composition of two types of local truffles, *Tirmania nivea* (Zubaidi) and *Terfezia claveryi* (Khlassi). All the food ingredient parameters were higher in *T. nivea* than in *T. claveryi* truffles, except the protein content, which was higher in *T. claveryi*. Such variability in the protein content could be due to differences in the physiology of the tubers, growth conditions and the environment. Also, nitrogenous compounds tend to accumulate more in *T. claveryi* than in *T. nivea* truffles. This difference between the two types of truffles could be attributed to genetic variation, handling or storage conditions or preservation techniques.

Truffles seem to be a good source of vitamins A, B and C, as well as carotene and minerals (Ahmed et al. 1981; Mattila et al. 2002; Murcia et al. 2002). β-Tocopherol was the vitamin detected in highest amounts in mushrooms, while δ-tocopherol was only detected in some samples (Heleno et al. 2010). All the above constituents have a protective effect due to their antioxidant and antiradical properties. Truffles also contain many phenols, which are efficient scavengers of peroxyl radicals. The action of phenolic compounds is also related to their capacity to reduce and chelate ferric iron, which catalyses lipid peroxidation (Gazzani et al. 1998a; Gazzani et al. 1998b).

Numerous studies have been carried out into the biological activities of mushrooms and truffles (Breene 1990; Sagakami et al. 1991; Chang and Buswell 1996; Hussain and Al-Ruqaie 1999; Murcia et al. 2002; Janakat et al. 2004, 2005; Saleh 2006; Stanikunaite et al. 2007; Fratianni et al. 2007a; Al-Laith 2010; Chen et al 2010; Heleno et al. 2010; Isabelle et al. 2010; Janakat and Nassar 2010; Wang and Marcone 2011), which have been seen to possess medicinal properties, including the ability to inhibit platelet aggregation (Hokama and Hokama 1981) and blood cholesterol (Aletor 1995). Moreover, other research has suggested that mushrooms could form part of an antihypertensive diet (Manzi et al. 1999); contribute to a reduction in blood glucose levels (Manzi and Pizzoferrato 2000); help in the prevention of infections caused by bacterial, viral, fungal and parasitic diseases (Breene 1990; Manzi and Pizzoferrato 2000); and have immunomodulation and anticancer effects (Ooi and Liu 2000) (see also Chap. 20 by Shavit and Shavit, this volume).

19.1.4 Preservation Techniques

The quality of truffles depends on their organoleptic (colour, flavour, aroma) and nutritional properties. These factors, as well as their postharvest conservation, influence their price. After harvest, truffles change in texture, lose weight and become increasingly brown. This change in colour, due to the enzymatic oxidation of polyphenols, indicates a decrease in quality (Heleno et al. 2010).

Truffle quality was recently studied in relation with different postharvest technologies used to maintain them: canning (Coskuner and Özdemir 1997; Murcia et al. 2002), modified atmosphere packaging (Kim et al. 2006; Rivera et al. 2010), freezing (Murcia et al. 2002; Al-Ruqaie 2006; Saltarelli et al. 2008) and irradiation (Akram and Kwon 2010; Nazzaro et al. 2007; Rawi and Aldin 1979; Reale et al. 2009; Rivera et al. 2011).

Truffles are processed using different preservation techniques to protect their typical aroma and flavour profiles and to increase their shelf life after harvest (Nazzaro et al. 2007; Rawi and Aldin 1979; Saltarelli et al. 2008). However, the content of certain nutrients may suffer because of heat degradation or leaching during sterilisation, pasteurisation and dehydration, as well as during storage and home handling and cooking (Kalt 2005). In addition, truffles may be subjected to other chemical changes, during thermal treatments, such as those resulting from the Maillard reaction that lead to the formation of a wide variety of brown melanoidins. The loss of natural antioxidants in heated foods may be minimised or compensated by the formation of non-nutrient antioxidants such as Maillard reaction products (MRPs) (Nicoli et al. 1997).

19.2 Different Industrial Preservation Techniques

Industrial processing (Gebczynski and Kmiecik 2007; Rickman et al. 2007a; Rickman et al. 2007b) allows the rapid processing of foods, which makes the preparation of dishes easier by eliminating not only laborious preliminary cleaning and preparation steps but also, frequently, cooking, which may cause changes in their nutritional values (Murcia et al. 2009).

19.2.1 Refrigeration

Low temperatures inhibit the development of microorganisms and slow the kinetics of chemical and enzymatic reactions. The low-temperature storage of food can basically be divided into refrigeration and freezing. Refrigeration has been used for hundreds of years through the use of natural ice or overwinter storage and is one of the most efficient methods for maintaining the biochemical properties of fresh truffles (Murcia et al. 2002).

Short-term storage of food can be attained by using temperatures close to 0 °C, and foods obtained from live materials will maintain their enzymatic systems for some time.

This method is widely accepted for reducing both the microbial and physiological spoilage of fresh truffles with minimal alterations to their biochemical properties (Saltarelli et al. 2008; Wang and Marcone 2011). However, refrigeration at 4 °C is not suitable for all truffle species, as demonstrated by Saltarelli et al. (2008). For example, white truffles *T. magnatum* and *T. borchii* (white truffles) are more sensitive to biochemical spoilage during refrigeration than the black truffles *T. melanosporum* and *T. aestivum* (black truffles) kept in similar conditions. No specific study of refrigeration effects was performed on desert truffles.

19.2.2 Drying

Drying removes water from the food, which inhibits the growth of microorganisms. Water is usually removed by evaporation (air drying, sun drying, smoking or wind drying) but, in the case of freeze drying, food is first frozen and then the water is removed by sublimation. Because bacteria, yeasts and moulds need the water in foodstuffs to grow, drying effectively prevents them from surviving in the food. Different methods for drying food include dryers, drum drying, freeze drying, spray drying and sunlight drying, while equipment ranges from shelf dryers to household ovens. Preserved dried truffles are sometimes blanched with salt and vinegar treatments (Al-Ruqaie 2006).

To evaluate drying methods, sensory analyses of truffles (Stone and Siedel (1983)) used a panel of 50 scientists from various disciplines within KACST institutes and used a descriptive 5-point hedonic scale—highly desirable, mildly desirable (MD), neutral (N) (no taste [desirable or undesirable]), mildly undesirable (MUN) and highly undesirable—and three characteristics of colour (40 points), texture (20 points) and flavour (40 points).

Frozen truffles (Table 19.3) treated with vinegar became very dark after defrosting, whereas the colour of dried truffles was brown or less dark than the frozen ones. The vinegar treatment adversely also affected the flavour/taste. Therefore, it seems that drying could be better if truffles are blanched in NaCl solution alone without a vinegar treatment (Al-Ruqaie 2006).

Table 19.4 shows the scores obtained for the hedonic testing without vinegar. It is clear that the truffles can easily be preserved by blanching in 2 or 4 % boiling NaCl solution for 2 or 4 min without vinegar. Table 19.5 (with vinegar) shows the combined effect of vinegar and NaCl treatment on the desirability level of *T. nivea* and *T. claveryi* truffles. As can be observed, truffles can be preserved by treating with NaCl solutions of different concentrations without vinegar.

Table 19.3 Organoleptic scoring of blanched dried truffles

	Total score					
Blanched sample treatment	NaCl conc.	Boiling time (min)	Colour 40 mean	Texture 20 mean	Flavour 40 mean	Total 100 mean
Without vinegar						
T. nivea	0	2.0	20.05ab	19.64a	5.05e	44.74
	2.0	2.0	24.34ab	19.90a	30.04c	74.28
	4.0	2.0	29.50c	15.03c	29.56c	74.09
	0	4.0	25.33c	19.74a	25.19d	70.26
	2.0	4.0	34.67b	20.00a	40.00a	94.67
	4.0	4.0	39.34a	20.00a	39.00a	98.34
T. claveryi	0	2.0	38.33a	20.00a	38.00a	96.33
	2.0	2.0	39.01a	20.00a	35.00b	94.01
	4.0	2.0	35.67b	20.00a	40.00a	95.67
	0	4.0	34.90b	19.55a	9.85bc	64.30
	2.0	4.0	38.66a	20.00a	38.00a	96.66
	4.0	4.0	40.00a	20.00a	40.00a	100.00
With vinegar						
T. nivea	0	2.0	39.76a	19.74a	30.25c	89.75
	2.0	2.0	35.05b	19.03b	20.15d	74.23
	4.0	2.0	5.33e	9.87e	5.33e	20.53
	0	4.0	35.66b	20.00a	35.00b	90.66
	2.0	4.0	25.20c	19.55a	19.98d	64.73
	4.0	4.0	4.67e	10.05d	5.11e	19.83
T. claveryi	0	2.0	38.91a	20.00a	38.00a	96.91
	2.0	2.0	30.07c	19.64a	20.31d	70.02
	4.0	2.0	20.51ab	19.75a	15.11ab	55.37
	0	4.0	39.94a	20.00a	40.00a	99.94
	2.0	4.0	25.33c	19.91a	35.21b	80.45
	4.0	4.0	15.05bc	14.96c	30.31c	60.32

Figures followed by the same letters are not significantly different by least significant difference $(LSD)_{0.05}$
Source: Al-Ruqaie (2006)

19.2.3 Freezing

This is a process for preserving food for long period of time and involves transforming the liquid water in food into ice at −18 °C (0 °F) or lower. In this way the water molecules are not available to participate in chemical reactions or in microbial metabolism, thus increasing the shelf life of products (Ghazalas 1998).

One important aspect related to freezing in the food industry is the formation of ice crystals during the freezing of the water present in the food material (Chemat et al. 2011). The problems include non-uniform crystal development, the destruction of the food material structure and losses in sensory quality. This state of affairs

Table 19.4 Hedonic testing of blanched dried truffles

	% of judgments						
Blanched sample treatment	NaCl conc.	Boiling time (min)	Very desirable	MD	N	MUN	Very undesirable
Without vinegar							
T. nivea	0	2	0	10	80	10	0
	2	2	5	80	15	0	0
	4	2	0	0	0	4	96
	0	4	5	5	4	34	52
	2	4	5	80	15	0	0
	4	4	95	5	0	0	0
T. claveryi	0	2	5	5	3	32	55
	2	2	0	0	0	10	90
	4	2	95	5	0	0	0
	0	4	0	0	0	5	95
	2	4	0	0	0	5	95
	4	4	100	0	0	0	0
With vinegar							
T. nivea	0	2	0	5	80	15	0
	2	2	0	0	0	90	10
	4	2	5	80	15	0	0
	0	4	0	5	80	15	0
	2	4	0	0	0	90	10
	4	4	5	80	15	0	0
T. claveryi	0	2	5	80	15	0	0
	2	2	0	0	0	90	10
	4	2	0	5	80	15	0
	0	4	5	80	15	0	0
	2	4	80	20	0	0	0
	4	4	0	0	0	90	10

Source: Al-Ruqaie (2006)
Conc. concentration, *MD* mildly desirable, *N* neutral, *MUN* mildly undesirable

has given rise to some innovative technologies such as air blast, plate contact, fluidised-bed freezing, immersion freezing, cryogenic freezing, high-pressure freezing and combinations of the same (Lakshmisha et al. 2008; Norton et al. 2009).

In desert truffles, statistical analysis of truffles points to no significant variations in the moisture content ($p < 0.05$), although compared with raw truffles, the value increases by 2 % and 1 % in *T. claveryi* and *P. juniperi*, respectively, as it does in popular Saudi Arabian truffles (Sawaya et al. 1985; Murcia et al. 2003). The ash content (mineral content) decreases during the freezing process by 19 % and 11 % in *T. claveryi* and *P. juniperi*, respectively (Murcia et al. 2003), which agrees with data obtained by Collins et al. (1994), who mention a decrease of up to 50 % as a result of the freezing process.

Table 19.5 Organoleptic scoring of blanched and unblanched dried truffles

	Total score			
Fresh sample treatment	Colour 40 mean	Texture 20 mean	Flavour 40 mean	Total
Tirmania nivea				
Control	39.67a	19.33a	39.33a	98.33
Sprinkled NaCl granules	39.33a	19.67a	40.00a	99.00
Vinegar + sprinkled NaCl granules	39.33a	19.33a	39.00a	97.66
Vinegar	19.00b	17.67ab	28.00b	64.67
2 % NaCl + vinegar spray[a]	20.33b	12.00c	12.33d	44.66
4 % NaCl + vinegar spray[a]	18.67b	18.33ab	19.67c	46.67
2 % NaCl[a]	19.33b	15.67b	05.00e	40.00
4 % NaCl[a]	19.67b	20.33a	23.00c	63.00
Terfezia claveryi				
Control	39.00a	19.67a	39.67a	98.34
Sprinkled NaCl granules	36.67ab	19.33a	39.00a	95.00
Vinegar + sprinkled NaCl granules	35.00b	19.67a	39.33a	94.00
Vinegar	08.33c	08.33c	07.00c	23.66
2 % NaCl + vinegar spray[a]	07.00c	17.67a	08.67bc	33.34
4 % NaCl + vinegar spray[a]	10.00c	14.67b	11.00b	35.67
2 % NaCl[a]	36.67ab	18.67a	38.00a	93.34
4 % NaCl[a]	35.00b	19.67a	39.00a	93.67

Figures followed by the same letters are not significantly different by least significant difference $(LSD)_{0.05}$

Source: Al-Ruqaie (2006)

[a]Blanching treatment of truffles with NaCl Salt with and without vinegar

However, the freezing process also reduces the levels of protein content by 5.6 % and 1.3 % in *T. claveryi* and *P juniperi*, respectively, with no significant differences ($p < 0.05$). Freezing does not affect the dietary fibre content. With regard to the lipid content, the freezing process produced no significant differences ($p < 0.05$) in losses between truffles studied (5 % in *T. claveryi* and 2 % in *P. juniperi*). The physical actions of processing, such as washing, blanching in boiling water or freezing, cause injuries in the hyphae that are held together in a mat, and this causes cellular fluids to leak out. Thus, the surfaces are less matted, resulting in increased permeability, and therefore, compounds exude from the fractured hyphae (Murcia et al. 1999).

Also, researchers have attempted to employ synergistic combinations of different treatments to extend the shelf life of truffles; for example, a combination of blanching in 4 % boiling NaCl solution for 4 min and immediate flash freezing at -18 °C was seen to be the best combined treatment for preserving desert truffles (Al-Ruqaie 2006).

19.2.4 Canning

Canning is another postharvest preservation technology commonly used to maintain food quality. In the case of truffles, they are introduced into glass jars filled with hot (85 °C) filling medium (20 g NaCl per litre of water), and then the jars are closed and heated at 121 °C for 30 min before being cooled in water (Murcia et al. 1999).

During thermal treatment, truffles are exposed to high temperatures, which inactivate microorganisms, natural toxins or other detrimental constituents, leading to prolongation of the shelf life and improved digestibility and bioavailability of nutrients. However, some unintentional consequences include the loss of certain nutrients, the formation of compounds with negative effects on flavour perception and texture or colour alterations (Murcia et al. 2009).

Significant variations may be observed in the water content of truffles after the canning process (3 % and 18 % increase for *T. claveryi* and *P. juniperi*, respectively). One explanation for the increased moisture content may be mechanical damage during handling, which causes the hyphae to fracture and parts of the surface to be less extensively matted, thus encouraging water retention (Honrubia et al. 1992; Murcia et al. 2003).

However, only a slight decrease in the ash content was observed in the canned product (11 % and 5.8 % decrease in *T. claveryi* and *P. juniperi*, respectively) due to the addition of salt (NaCl) during canning, as has been reported for other vegetables (Murcia et al. 1999). The canning process led to a protein decrease of 6.9 % in *T. claveryi* and 1.7 % in *P. juniperi*. This effect of industrial processing on the protein levels of truffles agrees with the findings for other vegetables, which, furthermore, pointed to no significant differences in the industrial processing methods studied. These losses with respect to raw proteins are probably due to denaturation and solubility during the industrial process (Martinez-Tome et al. 2001).

In contrast, the dietary fibre content of the canned products increases by 10 % and 30 % in *T. claveryi* and *P. juniperi*, respectively, which agrees with the results obtained by Murcia et al. (1992) for other canned vegetables and seems to be due to the effect of overheating on the fibre residues, in which the amount of resistant starch is increased (Murcia et al. 2003).

19.2.5 Ultrasounds

Ultrasounds are well known to have a significant effect on the speed with which various procedures can be carried out in the food industry. Using this technique, food processing can be completed in seconds or minutes with high reproducibility, reducing costs, simplifying manipulation, reducing the processing time and improving the shelf life. Many processes, such as freezing, cutting, drying, or sterilisation can be used together with ultrasounds (Chemat et al. 2011).

Several studies have been carried out on the influence of ultrasounds during freezing (Li and Sun 2002a; Li and Sun 2002b, Delgado et al. 2008; Song et al. 2009) especially as regards the preservation of high-value foods and pharmaceutical products (Chemat et al. 2011). The use of ultrasounds technology along with conventional cooling is rapid and leads to a shorter cooling time (Acton and Morris 1992). It improves heat transfer and the cooling is accelerated (Li and Sun 2002a, b).

In addition, the final size of the ice crystals is smaller and as a consequence cell damage is reduced (Sun and Li 2003): Acoustic cavitation occurs and acts as nuclei for crystal growth or disrupts the nuclei already present. In freezing, this phenomenon would lead to fine ice crystals being formed and shortening of the time between the onset of crystallisation and the complete formation of ice (Roberts 1993; Chow et al. 2005), thus reducing damage to cellular structure.

19.2.6 Modified Packaging Atmospheres (MAP)

MAP involves altering the atmosphere surrounding the product by reducing the respiration rate, thus retarding the compositional changes associated with maturation and senescence, reducing microorganism growth (Irtwange 2006; Kim et al. 2006) and retaining all the attributes that consumers consider to be freshness markers (Murcia et al. 2003).

The normal packing atmosphere is replaced by an appropriate mixture of gases at various concentrations, which protects the product against alterations caused by oxidation, microbiological growth, colour and aroma variations. A mixture of 15 % CO_2, 7 % O_2 and microperforated films have been found to prolong the shelf life of fresh Périgord black truffles up to 28 days by slowing down respiration (Rivera et al. 2010). However, the cutting operations involved in their preparation means that a great number of cells are disrupted, which causes the release of enzymes and their substrates, and increases the possibility of oxidative enzyme-catalysed processes. Reduced O_2 and high CO_2 levels have been seen to effectively control enzymatic browning, softening and decay in fresh-cut vegetables (Oms-Oliu et al. 2008).

All these operations must be carried out in cold and the product, once in MAP, should be stored at 3–4 °C. However, MAP is quite costly and may result in water accumulation on the surface of the product, promoting microbial growth and sliminess, which impairs the very objective of MAP (Wang and Marcone 2011).

The full beneficial effect of MAP is obtained by correctly choosing the packaging materials (Allende et al. 2006). Taking into account that the sales of MAP ready-to-use or ready-to-eat fresh vegetables have grown rapidly in recent decades as a result of changes in consumer attitudes (Rico et al. 2007), more innovative ways are expected to be found to preserve truffles.

MAP, combined with surface disinfection, prolonged the shelf life of Tuber aestivum and Tuber melanosporum truffles. This method used washes with sodium hypochlorite (500 ppm chlorine), hydrogen peroxide (5 %) and ethanol (70 %) alone or in combination with ultrasound (35 Hz) for 10 min at 4 °C and dipping in ethanol (70 %) was found to be effective treatment for pseudomonas, for Enterobacteriaceae and for lactic acid bacteria and moulds or yeasts. Samples can then be packaged with a microperforated film and stored at 4 °C for 28 days, during which time neither microbial counts nor sensory quality are affected (Rivera et al. 2010).

19.2.7 Irradiation

Food irradiation involves treatment with gamma and electron beams in order to enhance the shelf life and safety of food (Robertson and Hoy 2000). Consumers usually accept this processing technique, which they consider to be radiologically, microbiologically and toxicologically safe (Akram and Kwon 2010).

Generally, the radiation treatments are performed employing 1–10 kGy on truffles packed in impermeable films (Adamo et al. 2004). The recommended dose for extending the shelf life of mushrooms in different countries is up to 2.5 kGy (Jay et al. 2005) or 1–3 kGy. This method affects the microbial populations of *Nelumbo nucifera* samples, which were found to remain acceptable sensorially (Wen et al. 2006 and Khattak et al. 2009), while the pseudomonas populations and Enterobacteriaceae counts were reduced in irradiated Tuber. However, lactic acid bacteria and yeasts were less affected by the ionising radiation treatments. The carbon dioxide levels inside the packages containing irradiated truffles were lower than those of the non-irradiated ones, suggesting a decrease in the respiration rate of the treated ascocarps. The treatments did not negatively affect the sensory characteristics of truffles, but superficial yeast growth was visible in truffles irradiated with 1.5 kGy (Rivera et al. 2011). Sensory characteristics involve without softening and browning in the pulp and some anomalous odours (Rosnes et al. 2003).

Application of a radiation dose of 2 kGy proved sufficient to maintain the textural and sensorial quality and reduce the bioload of chopped carrots for 14 days at 5 °C (Chaudry et al. 2004). Doses of 2.5 kGy partially suppressed the growth of many bacteria and faecal coliforms (Nazzaro et al. 2007). This difference seems to be dependent on the lower degree of contamination of the raw product and justifies the assumption that lower gamma ray doses are more effective when truffles are previously cleaned, and improve if they are stored at −4 °C (Saltarelli et al. 2008). In 2.0 kGy samples, the protein profile was characterised by a 20 kDa polypeptide, which could be considered as a useful marker of the irradiation.

If fresh truffles are strongly contaminated, 5 and 10 kGy can be used although, while producing the highest microbial decontamination, these doses were seen to negatively influence sensorial parameters (Reale et al. 2009). A general decrease of microbial groups was evident immediately after treatment at different radiation doses. As expected, major doses provoked the highest decontamination of the product (Adamo et al. 2004).

Gamma radiation eliminates spoilage and pathogenic microorganisms by hitting the genetic material of microorganism using the high energy of these rays to ionise the water molecules (Andrews et al. 1998), depending upon its enzymatic DNA repair system and the given radiation dose (Molins 2001). Apparently, irradiation kills a certain fraction of the surface and sub-surface microflora at the time of application but has no residual effect to counter fresh infestation or regrowth during storage. In the pretreated samples, leftover antimicrobial agents like acetic acid and sodium oxalate may delay the resurgence of microbial growth, thereby enhancing the shelf life of the truffles. Under this premise, post-irradiation storage conditions are of greater relevance.

Gamma irradiation, also, prevents the enzymatic browning of mushrooms by inactivating polyphenol oxidase, although phenylalanine ammonia-lyase increases, and as a result, the total phenolic content significantly increases (Benoit et al. 2000). Irradiated mushrooms had equal or superior flavour and texture, but were somewhat discoloured (Kwon et al. 1990). However, irradiation alone is not sufficient to enhance the shelf life of truffles, and the influence of irradiation is greater when it is combined with antimicrobial treatment (Al-Rawi and Aldin 1979). As regards preservation methodology, a combination of different techniques is usually used to achieve the optimum results. There was no synergistic effect of sulphitation and irradiation on the prevention of browning (Vani et al. 2009). Another advantage of irradiation is that it replaces the use of toxic fumigants for eliminating microorganisms, insects or parasites capable of spoiling the food, while being economical and environmental friendly (Bhat et al. 2012).

The shelf life of truffles irradiated without pretreatment and stored at room temperature showed no significant differences in most cases from the 3 to 5 days observed for controls. However, after being pretreated with acid and alkali, the shelf life of irradiated samples increased up to 47 days (Al-Ruqaie 2009) (Table 19.5). For comparison purposes, similar sensory analyses as employed for assessing fruit body shelf life were carried out to determine the shelf life of rhizomes during storage. The appearance score of 1 kGy treated sample revealed a continuous decrease from 8.9 (day 0) to 6.3 (day 12), while scores of 7.6, 8.1 and 7.9 were recorded by 2 kGy, 4 kGy and 6 kGy treated samples, respectively, for appearance after 12 days of storage. As shown in Table 19.6, the flavour scores for control and radiated samples were found to be the same ($p > 0.05$) up to 6 days of storage.

Irradiation and pretreatment combined with storage at low temperature (7 °C) contributed towards delaying the development of microflora on the truffles. The synergistic effect of irradiation dose, packaging under vacuum and the storage temperature resulted in a direct effect on the microbial load, spoilage and shelf life. After irradiation, small variations in the intensity of some NMR resonances due to aromatic compounds and other unassigned compounds were observed. As confirmed by UV spectrophotometric data, the induced growth of soluble phenols

Table 19.6 Effect of gamma radiation on shelf life of desert truffles

Shelf life (days)				
	Storage conditions/treatment			
	Room temperature (25 °C)		Refrigerator (7 °C)	
Radiation dose (Gy)	Untreated	Pretreated	Untreated	Pretreated
Terfezia claveryi				
Control	3.33b	13.33c	10.67f	36.00[b]
150	3.33 b	43.67 a	22.33 e	41.00 [a]
250	3.00 b	13.67 c	26.33 c	41.00 [a]
500	3.00 b	13.33 c	38.67 a	40.67 [a]
1,000	4.00 a	13.33 c	30.33 b	41.00 [a]
2,500	4.00 a	17.33 b	23.33 de	41.00 [a]
3,000	3.33 b	13.00 c	25.33 dc	41.33 a
$LSD_{0.05}$	0.989	1.143	2.041	1.278
Tirmania nivea				
Control	4.00 ab	9.33 f	17.33 e	38.33 b
150	4.33 ab	24.67 c	33.33 b	41.33 a
250	4.67 a	21.33 d	29.00 c	41.33 a
500	5.00 a	47.33 a	41.33 a	41.00 a
1,000	3.33 b	26.33 b	32.33 b	41.00 a
2,500	4.67 a	18.00 e	28.33 c	41.33 a
3,000	3.33 b	18.00 e	26.00 d	41.00 a
$LSD_{0.05}$	1.616	1.363	1.512	1.452

Values in a column followed by the same superscripts are not significantly different at $LSD_{0.05}$
Values followed by the same letters are not significantly different by least significant difference $(LSD)_{0.05}$
Source: Al-Ruqaie (2009)

suggested that the 1.5 kGy dose can be considered as the radiation dose threshold beyond which clear chemical modifications appear in truffles (Adamo et al. 2004).

Massantini et al. (2002) maintained good texture and flavour in truffles by using a gas mixture containing 60 % CO_2 but only until the second or third week of storage. However the individuation of specific irradiation doses is essential to avoid negative effects on the quality of the product.

Finally, the extraction yield and phenolic contents increased with increasing radiation dose. Gamma radiation also enhanced DPPH scavenging. In general, no substantial change in proximate constituents (moisture, protein, fat and ash content) was observed in *Nelumbo nucifera*, while the fibre content was found to be slightly decreased at dose levels 4 and 6 kGy (Khattak et al. 2009).

The Joint Expert Committee of FAO/WHO/IAEA established that the irradiation of any food commodity up to an overall average dose of 10 kGy poses no toxicological hazards (CAC 1984, WHO 1994; Thakur and Singh 1995). The irradiation did not lead to the formation of mutagenic compounds on *Tuber aestivum* black truffles, but the level of anti-mutagenic activity was slightly decreased after the treatment (Fratianni et al. 2007b). It seems plausible that while irradiation is a safe and effective method for extending shelf life of truffles, it has to be augmented with other methods to achieve optimal results.

In 2.0 kGy samples, the protein profile was characterised by a 20 kDa polypeptide, which could be considered as a useful marker of the irradiation treatment and of the storage time of the product. MALDI-TOF mass spectrometry analysis did not permit a correct identification from tryptic peptides in databases, although the nano-ES/MS/MS analyses performed on the 10 kDa tryptic digest peptides showed an amino acidic sequence entirely contained in a protein of filamentous fungus *Neurospora crassa* (Filomena et al. 2007).

The main negative effects included the loss of structure and the presence of off-flavours, which seemed to be dependent on the increasing amounts of some volatile compounds such as sulphide dimethyl and disulphide dimethyl (Nazzaro et al. 2007). The production of these compounds in black truffles seem to be dependent on the radiation treatment, as demonstrated by the fact that they decreased during the storage of untreated truffles and increase in irradiated truffles. It was also observed that 2-methyl-butanol, one of the main alcohols influencing the aroma of truffles, maintained the same value in untreated samples until the end of the shelf life, whereas samples treated at 2 kGy showed an increase of this compound. According to these data, 1.5 kGy dose represented the maximum level of irradiation that did not influence the aroma features of truffles.

19.3 Cooking Methods

Besides the different methods of industrial processing, it is interesting to mention some cooking methods used by the catering industry, which reflect home-cooking conditions. The most common of these cooking treatments are summarised below:

19.3.1 Boiling

Boiling is one of most common home-cooking methods. The process, in which the food is added to the boiling water and cooked until tender, may modify the structure and composition of the foods, while the liquid is enriched in soluble substances. Pressure and a change in composition of the liquid may alter the boiling point of the liquid, which, in the case of water, is considered to be 100 °C (Ceserani et al. 1991).

This technique is the most certain way of killing all microorganisms in emergency situations. Boiling has several other advantages: it is safe and simple, is appropriate for large-scale cookery and improves food digestibility. However, its many disadvantages include changes in the nutritional values of foods and the loss of soluble vitamins to the water (if the water is discarded). Furthermore, some boiled foods may look unattractive (Ghazalas 1998). The treatment and the time will depend on the size of truffles and variety.

19.3.2 Pressure Cooking

The process of cooking food in a sealed vessel is known as pressure cooking. This cooking method allows food to be cooked with greater humidity and higher temperatures than are possible with conventional boiling or steaming methods. In a sealed pressure cooker, the boiling point of water increases as the pressure rises, resulting in superheated water. At a pressure of 15 psi (103 kPa) above atmospheric pressure, water in a pressure cooker can reach a temperature of up to 121 °C. Pressure is created initially by boiling water inside the closed pressure cooker. The trapped steam increases the internal pressure and temperature. The internal steam pressure from the boiling liquid causes saturated steam (or "wet steam") to permeate the food. This results in faster cooking times compared to conventional cooking methods because liquids and steam conduct heat more rapidly than dry air (Jimenez et al. 2009).

19.3.3 Frying

Frying involves cooking food in oil or another fat and was already used in Egypt around 2500 BC. Chemically, oils and fats are the same, differing only in their melting point, but the distinction is only made when needed. Foods can be fried in a frying pan in a variety of fats, including lard, animal and vegetable oil, and stirred until the sample becomes crisp-tender (Jimenezet al. 2009). To fry food in oil, especially in olive oil, is considered healthier than using lard, because the main fats in olive are unsaturated.

19.3.4 Microwave Oven

Microwaves are electromagnetic waves of high frequency of between 3,000 and 300 MHz. When absorbed by food, they are transformed into caloric energy. In this cooking method, the food is placed in a glass dish without additional water. The heat is generated by dielectric heating (food) and causes dipolar rotation in each water molecule with high-frequency vibration (Ceserani et al. 1995). Microwave ovens heat foods quickly, efficiently and uniformly.

19.3.5 Griddling

The food is cooked in a thick frying pan without oil and heat is transferred by direct conduction to the surface of the food, commonly from above or below (Ceserani et al. 1995).

19.3.6 Baking

This food cooking method uses prolonged dry heat acting by convection and not by thermal radiation, normally in an oven, but also by means of hot ashes or hot stones. When the desired temperature is reached within the heating instrument, the food is placed inside and baked for a certain amount of time.

19.3.7 Assessing the Best Cooking Method

Jimenez et al. (2009) have identified the best methods for cooking vegetables while retaining their radical-scavenging activity and antioxidant activity with health-related properties. Depending on the vegetables in question, griddling (35 %) and microwave (36 %) cooking produce the lowest losses, while pressure cooking (50 %) and boiling (61 %) lead to the greatest losses; in general, frying (43 %) occupies an intermediate position. In short, water is not the cook's best friend when it comes to preparing vegetables. However, there is no specific information about these methods for desert truffles.

19.4 Effects of Preservation on Biological Activities of *Terfezia* and *Picoa*

Taking into account that vegetables are often stored in a refrigerator, it is important to understand the possible losses that may occur during storage, but only data on the antioxidant properties of vegetables are available (Murcia et al. 2009). As Table 19.7 shows, Murcia et al. (2009) demonstrated a loss of 34 % inhibition (as measured by the scavenging of lipoperoxyl radical) by the seventh day with respect to the first day of analysis for samples rich in phenols. This method of preservation did not affect the hydroxyl radical scavenging of samples and losses up to 40 % were observed in the TEAC assay. Vegetables are rich in phenols, which seem to be more affected by storage factors such as temperature, atmosphere and light than carotenoids or vitamin C (Klimczak et al. 2007).

Table 19.7 Antioxidant activity of *Terfezia* and *Picoa* (raw and industrially processed) compared with the activity of common food antioxidants

		Lipid peroxidation	Deoxyribose assay			Peroxidase assay	Rancimat test		
Substance added to reaction mixtures	Processing	% inhibition	For RM + DR	% inhibition	Without ASC	(A_{436nm})	IP (h)	PF	TEAC assay
None (control)			1.226 ± 0.01	–	0.257	0.622 ± 0.02	7.58		–
Trolox (0.05 mM)									1.00 ± 0.0
Trolox (0.5 mM)									10.00 ± 0.0
Terfezia	Raw	95.7 ± 1	0.149 ± 0.03	87.8	0.110	0.162 ± 0.03	9.26	1.22 ± 0.1	4.77 ± 0.1
	Freezing	94.1 ± 2	0.179 ± 0.02	85.4	0.105	0.165 ± 0.05	8.00	1.05 ± 0.2	3.57 ± 0.01
	Canning	72.9 ± 2	0.371 ± 0.04	69.7	0.193	0.350 ± 0.01	2.35	0.31 ± 0.1	2.52 ± 0.1
Picoa	Raw	95.3 ± 1	0.070 ± 0.02	94.3	0.067	0.168 ± 0.04	8.75	1.15 ± 0.1	3.91 ± 0.1
	Freezing	94.5 ± 2	0.080 ± 0.02	93.5	0.061	0.321 ± 0.03	8.51	1.12 ± 0.1	2.57 ± 0.1
	Canning	74.1 ± 3	0.455 ± 0.01	62.9	0.174	0.400 ± 0.01	3.19	0.42 ± 0.2	0.56 ± 0.1
α-Tocopherol		15.3 ± 1	1.186 ± 0.03	3.2	0.240	0.711 ± 0.02	20.10	2.65 ± 0.2	1.16 ± 0.1
BHA		71.4 ± 1	0.914 ± 0.02	25.4	0.201	0.770 ± 0.03	8.68	1.14 ± 0.1	0.44 ± 0.1
BHT		22.3 ± 2	1.116 ± 0.05	8.9	0.559	0.700 ± 0.04	7.18	0.95 ± 0.2	0.26 ± 0.1
Propyl gallate		52.5 ± 1	2.070 ± 0.01	–	1.537	0.400 ± 0.02	14.21	1.87 ± 0.1	3.47 ± 0.1

Source: Murcia et al. (2009)

RM reaction mixtures, *DR* deoxyribose, Absorbance at 532 nm when hydroxyl radical was generated

ASC Absorbance at 532 nm when ascorbate was omitted from the reaction mixture

Rancimat tested at 120 °C, *IP* induced period, *PF* protection factor, PF = IP (olive oil + samples)/IP (olive oil)

TEAC value is the millimolar concentration of a Trolox solution showing the antioxidant capacity equivalent to that of the dilution of the substance under investigation. Propyl gallate dilution was selected to reduce the measurement within the appropriate part of the Trolox standard curve

(–) No inhibition detected

Murcia et al. (2002) investigated the biological activities of *Terfezia claveryi* and *Picoa juniperi* and the effects of freezing and canning processes on these properties using different analytical methods and then compared the results with those for common foods. Frozen truffles do not lose antioxidant activity, as measured by techniques which reveal the lipoperoxyl and hydroxyl radicals. Canning led to greater antioxidant losses (22 % for lipoperoxyl and 18–31 % for hydroxyl radical).

The Rancimat test is used in the food industry to determine oxidative stability. In this assay, the scavenger to be tested is added to a lipidic food and the degree of protection is evaluated. Frozen *Picoa* and *Terfezia* lost 2.7 % and 13.6 %, respectively, of their protection capacity in this system. In the case of canned truffles, the protection effect was reduced drastically. In the study of antioxidant stability during 30 days of storage, frozen truffles showed losses of 27 % (*Terfezia*) and 62 % (*Picoa*) (Murcia et al. 2002).

Finally, TEAC values can be assigned to all compound able to scavenge ABTS by comparing the scavenging capacities of these compounds with that of Trolox, a water-soluble vitamin E. The quantitative evaluation of antioxidant capacity based on TEAC can be used to provide an order of antioxidants (van den Berg et al. 1999). TEAC values fell by 25–33 % as a consequence of freezing and by 48–85 % in canned truffles.

References

Acton E, Morris GJ (1992) Method and apparatus for the control of solidification in Liquids. Int. Pat. WO 99/20420.

Adamo M, Capitani D, Mannina L, Cristinzio M, Ragni P, Tata A, Coppola R (2004) Truffles decontamination treatment by ionizing radiation. Rad Phys Chem 71:165–168

Ahmed AA, Mohamed MA, Hami MA (1981) Libyan truffles "*Terfezia boudieri* Chatin": chemical composition and toxicity. J Food Sci 11:927–929

Akram K, Kwon JH (2010) Food irradiation for mushrooms: a review. J Kr Soc Appl Biol Chem 53:257–265

Al-Rawi AM, Aldin MM (1979) New mycorhizal identification, truffle cultivation, and truffles radiation preservation. Rad Phys Chem 14:759–767

Al-Ruqaie IM (2006) Effect of different treatment processes and preservation methods on the quality of truffles: I. Conventional methods (drying/freezing). J Food Process Preserv 30: 335–351

Al-Ruqaie IM (2009) Effect of treatment process and preservation method on shelf life of truffles: II. Non-conventional methods (radiation). Int J Biol Chem 3(3):126–131

Allende A, Tomás-Barberán FA, Gil I (2006) Minimal processing for healthy traditional foods. Trends Food Sci Technol 17:513–519

Aletor VA (1995) Compositional studies on edible tropical species of mushrooms. Food Chem 54:265–268

Al-Laith AAA (2010) Antioxidant components and antioxidant/antiradical activities of desert truffle (*Tirmania nivea*) from various Middle Eastern origins. J Food Compos Anal 23:15–22

Andrews LS, Ahmedna M, Ahmedna GRM, Liuzzo JA, Murano PS, Murano EA (1998) Food preservation using ionizing radiation. Rev Environ Contam Toxicol 154:1–53

Benoit MA, D'Aprano G, Lacroix M (2000) Effect of y-irradiation on phenylalanine ammonia-lyase activity, total phenolic content, and respiration of mushroom (Agaricus bisporus). J Agric Food Chem 48:6312–6316

Bhat R, Karim AA, Paliyath G (2012) Use of electron beams in food preservation. In: Bhat R, Karim Alias A, Paliyath G (eds) Progress in food preservation. Wiley-Blackwell, Oxford. doi:10.1002/9781119962045.ch17

Bokhary HA, Parvez S (1993) Chemical composition of desert truffles *Terfezia claveryi*. J Food Compos Anal 6:285–293

Breene WM (1990) Nutritional and medicinal value of specially mushrooms. J Food Protect 53: 883–894

CAC (Codex Alimentarius Commission) (1984) Codex general standard for irradiated foods and recommended international code of practise for the operation of radiation facilities used for the treatment of foods, vol 15, 1st edn. CAC, Rome, Italy

Ceserani V, Kinton R, Foskett D (1991) Question and answers on practical cookery, 4th edn. Hodder&Stoughton, London

Ceserani V, Kinton R, Foskett D (1995) Practical cookery, 8th edn. Hodder&Stoughton, London

Chang ST, Buswell JA (1996) Mushroom nutriceuticals. World J Microbiol Biotechnol 12: 473–476

Chaudry MA, Bibi N, Khan M, Badshah A, Qureshi MJ (2004) Irradiation treatment of minimally processed carrots for ensuring microbial safety. Rad Phys Chem 71:171–175

Chemat F, Zill-e-Huma, Khan MK (2011) Applications of ultrasound in food technology: processing, preservation and extraction. Ultrason Sonochem 18:813–835

Chen X-H, Xia L-X, Zhou H-B, Qiu G-Z (2010) Chemical composition and antioxidant activities of *Russula griseocarnosa* sp. nov. J Agric Food Chem 58:6966–6971

Chow R, Blindt R, Chivers R, Povey M (2005) A study on the primary and secondary nucleation of ice by power ultrasound. Ultrasonics 43:227–230

Collins JL, Abdalla MO, Schoenemann DR (1994) pH and proximate composition of selected frozen, commercially processed vegetables. Tenn Farm Home Sci 170:32–38

Coskuner Y, Özdemir Y (1997) Effects of canning processes on the elements content of cultivated mushrooms (*Agaricus bisporus*). Food Chem 60:559–562

Danell E, Eaker D (1992) Amino acid and total protein content of edible mushroom *Cantharellus cibarius*. J Sci Food Agric 60:333–337

Delgado AE, Zheng L, Sun D-W (2008) Influence of ultrasound on freezing rate of immersion-frozen apples. Food Bioproc Technol 2:263–270

Desprez-Loustau ML, Robin C, Buée M, Courtecuisse R, Garbaye J, Suffert F, Sache I, Rizzo DM (2007) The fungal dimension of biological invasions. Trends Ecol Evol 22:472–480

Feofilova EP, Gornova IB, Memorskaia S, Garibova LV (1998) Lipid composition of fruiting bodies and submerged mycelium from *Lentinus edodes* (Berk.) Sing (*Lentinula edodes* (Berk.) Peglar). Mikrobiologia 67:655–659

Filomena N, Florinda F, Gianluca P, Raffaele C, Anna R, Aldo DL (2007) Evaluation of Mutagenic and antimutagenic properties of aqueous and ethanolic extracts from fresh and irradiated Tuber aestivum black truffle: a preliminary study. Food Chem 102:471–474

Fratianni F, Di Luccia A, Coppola R, Nazzaro F (2007a) Influence of gamma rays influence on some biochemical and microbiological aspects in black truffles. Food Chem 103(2):344–354

Fratianni F, Di Luccia A, Coppola R, Nazzaro F (2007b) Mutagenic and antimutagenic properties of aqueous and ethanolic extracts from fresh and irradiated *Tuber aestivum* black truffle: a preliminary study. Food Chem 102:471–474

Gazzani G, Papetti A, Daglia M, Bertè F, Gregotti C (1998a) Protective activity of water soluble components of some common diet vegetables on rat liver microsome and the effect of thermal treatment. J Agric Food Chem 46:4123–4127

Gazzani G, Papetti A, Massolini G, Daglia M (1998b) Anti- and prooxidant activity of water soluble components of some common diet vegetables and the effect of thermal treatment. J Agric Food Chem 46:4118–4122

Gebczynski P, Kmiecik W (2007) Effects of traditional and modified technology, in the production of frozen cauliflower, on the contents of selected antioxidative compounds. Food Chem 101: 229–235

Ghazalas S (1998) Sous vide and cook-chill processing for the food industry. Chapman&Hall, London

Heleno SA, Barros L, Sousa MJ, Martins A, Ferreira ICFR (2010) Tocopherols composition of Portuguese wild mushrooms with antioxidant capacity. Food Chem 119:1443–1450

Hokama Y, Hokama JLRY (1981) In vitro inhibition of platelet aggregation with low Dalton compounds from aqueous dialysates of edible fungi. Res Commun Chem Pathol Pharmacol 31:177–180

Honrubia M, Cano A, Molina-Niñirola C (1992) Hypogeous fungi from Southern Spanish semi-arid lands. Persoonia 14:647–653

Hussain G, Al-Ruqaie IM (1999) Occurrence in chemical composition, and nutritional value of truffles: overview. Pak J Biol Sci 2:510–514

Irtwange SV (2006) Application of modified atmosphere packaging and related technology in postharvest handling of fresh fruits and vegetables. Agr Eng Int 4:1–13

Isabelle M, Lee BL, Lim MT, Koh W-P, Huang D, Ong CN (2010) Antioxidant activity and profiles of common vegetables in Singapore. Food Chem 120:993–1003

Janakat S, Al-Fakhiri S, Sallal AK (2004) A promising peptide antibiotic from *Terfezia claveryi* aqueous extract against *Staphylococcus aureus* in vitro. Physiother Res 18:810–813

Janakat S, Al-Fakhiri S, Sallal AK (2005) Evaluation of antibacterial activity of aqueous and methanolic extracts of the truffle *Terfezia claveryi* against *Pseudomonas aeruginosa.* Saudi Med J 26:952–955

Janakat S, Nassar M (2010) Hepatoprotective activity of desert truffle (*Terfezia claveryi*) in comparison with the effect of *Nigella sativa* in the rat. Pak J Nutr 9:52–56

Jay JM, Loessner MJ, Golden DA (2005) Radiation protection of foods, and nature of microbial radiation resistance. In: Modern food microbiology, 7th edn. Springer Science and Business Media, New York, NY, pp 371–394

Jimenez AM, Garcia-Diz L, Martinez-Tome M, Mariscal M, Murcia MA (2009) Influence of cooking methods on antioxidant activity of vegetables. J Food Sci 74:97–103

Kalt W (2005) Effects of production and processing factors on major fruit and vegetable antioxidants. J Food Sci 70:11–19

Khattak KF, Simpson TJ, Ihasnullah (2009) Effect of gamma irradiation on the microbial load, nutrient composition and free radical scavenging activity of Nelumbonucifera rhizome. Rad Phys Chem 78:206–212

Kim KM, Ko JA, Lee JS, Park HJ, Hanna MA (2006) Effect of modified atmosphere packaging on the shelf-life of coated, whole and sliced mushrooms. Food Sci Technol 39:364–371

Klimczak I, Maleca M, Szlachta M, Gliszczynskaa-Swiglo A (2007) Effect of storage on the content of polyphenols, vitamin C and the antioxidant activity of orange juices. J Food Compos Anal 20:313–322

Kwon JH, Byun MW, Cho HO (1990) Browning and color characteristics in mushrooms (Agaricus bisporus) as influenced by ionizing energy. Kr J Food Sci Technol 22:509–513

Lakshmisha IP, Ravishankar CN, Ninan G, Mohan CO, Gopal TKS (2008) Effect of freezing time on the quality of Indian mackerel (*Rastrelliger kanagurta*) during frozen storage. J Food Sci 73:S345–S353

Li B, Sun D-W (2002a) Effect of power ultrasound on freezing rate during immersion freezing of potatoes. J Food Eng 55:277–282

Li B, Sun D-W (2002b) Novel methods for rapid freezing and thawing of foods – a review. J Food Eng 54:175–182

Longvah T, Deosthale YG (1998) Compositional and nutritional studies on edible wild mushroom from northeast India. Food Chem 63:331–334

Loo J (2009) Ecological impacts of non-indigenous invasive fungi as forest pathogens. Biol Invasions 11:81–96

Mandeel QA, Al-Laith AA (2007) Ethnomycological aspects of the desert truffle among native Bahraini and non-Bahraini peoples of the Kingdom of Bahrain. J Ethnopharmacol 110:118–129

Manzi P, Gambelli S, Marconi S, Vivanti V, Pizzoferrato L (1999) Nutrients in edible mushrooms: an inter-species comparative study. Food Chem 65:477–482

Manzi P, Pizzoferrato L (2000) Beta-glucans in edible mushrooms. Food Chem 68:315–318

Martinez-Tome M, Garcia-Carmona F, Murcia MA (2001) Comparison of the antioxidant and pro-oxidant activities of broccoli amino acids with dose of common food additives. J Sci Food Agric 81(10):1019–1026

Massantini R, Bruno M, Salcini C, Bellincontro A, Mencarelli F (2002) Conservazione in film plastico del tartufo fresco (Tuber aestivum). Ind Alim Novembre 41:123–126

Mattila P, Salo-Vaananen P, Konko K, Aro H, Jalava T (2002) Basic composition and amino acid contents of mushrooms cultivated in Finland. J Agric Food Chem 50:6419–6422

Molins RA (2001) In food irradiation: principles and applications. Wiley-Interscience, New York: NY

Murcia MA, Jiménez AM, Martinez-Tomé M (2009) Vegetables antioxidant losses during industrial processing and refrigerated storage. Food Res Int 42:1046–1052

Murcia MA, López-Ayerra B, García-Carmona F (1999) Effect of processing methods and different blanching times on broccoli: proximate composition and fatty acids. Lebensm Wiss Technol 32:238–243

Murcia MA, Martinez-Tomé M, Jiménez A, Vera A, Honrubia M, Parras P (2002) Antioxidant activity of edible fungi (truffles and mushrooms): losses during industrial processing. J Food Protect 65:1614–1622

Murcia MA, Martínez-Tomé M, Vera A, Morte A, Gutiérrez A, Honrubia M, Jiménez AM (2003) Effect of industrial processing on desert truffles *Terfezia claveryi* Chatin and *Picoa juniperi* Vittadini: proximate composition and fatty acids. J Sci Food Agric 83:535–541

Murcia MA, Vera A, García-Carmona F (1992) Effect of processing methods on spinach: proximate composition in fatty acids and soluble protein. J Sci Food Agric 59:473–476

Nazzaro F, Fratianni F, Picariello G, Coppola R, Reale A, Di Luccia A (2007) Evaluation of gamma rays influence on some biochemical and microbiological aspects in black truffles. Food Chem 103:344–354

Nicoli MC, Anese M, Parpinel MT, Franceschi S, Lerici CR (1997) Loss and/or formation of antioxidants during food processing and storage. Cancer Lett 114:71–74

Nöel-Suberville C, Cruz C, Guinberteau J, Montury M (1996) Correlation between fatty acid content and aromatic compound release in fresh blewit (*Lepista nuda*). J Agric Food Chem 44: 1180–1183

Norton T, Delgado A, Hogan E, Grace P, Sun D-W (2009) Simulation of high pressure freezing processes by enthalpy method. J Food Eng 91:260–268

Oms-Oliu G, Odriozola-Serrano I, Soliva-Fortuny R, Martín-Belloso O (2008) The role of peroxidases on the antioxidant potential of fresh-cut "Piel de Sapo" melon packaged under different modified atmospheres. Food Chem 106:1085–1092

Ooi VEC, Liu F (2000) Immunomodulation anti-cancer activity of polysaccharide-protein complexes. Curr Med Chem 7:715–729

Pegler DN (2002) Useful fungi of the World: the 'Poor man's truffles of Arabia' and Manna of the Israelites. Mycologist 16:8–9

Rawi AMA, Aldin MM (1979) New mycorhizal identification truffle cultivation and truffles radiation preservation. Rad Phys Chem 14:759–767

Reale A, Sorrentino E, Iacumin L, Tremonte P, Manzano M, Maiuro L, Comi G, Coppola R, Succi M (2009) Irradiation treatments to improve the shelf life of fresh black truffles (truffles preservation by gamma-rays). J Food Sci 74(4):196–200

Rickman JC, Bruhn CM, Barret D (2007a) Nutritional comparison of fresh, frozen and canned fruits and vegetables I. Vitamin A and carotenoids, vitamin C, B and phenolics compounds. J Sci Food Agric 87:390–944

Rickman JC, Bruhn CM, Barret D (2007b) Nutritional comparison of fresh, frozen and canned fruits and vegetables. II. Vitamin A and carotenoids, vitamin E, minerals and fiber. J Sci Food Agric 87:1185–1196

Rico D, Martín-Diana AB, Barat JM, Barry-Ryan C (2007) Extending and measuring the quality of fresh-cut fruit and vegetables: a review. Trends Food Sci Technol 18:373–386

Rivera CS, Blanco D, Marco P, Oria R, Venturini ME (2011) Effects of electron-beam irradiation on the shelf life, microbial populations and sensory characteristics of summer truffles (Tuber aestivum) packaged under modified atmospheres. Food Microbiol 28(1):141–148. doi:10.1016/j.fm.2010.09.008

Rivera CS, Blanco D, Salvador ML, Venturini ME (2010) Shelf-life extension of fresh Tuber aestivum and Tuber melanosporum truffles by modified atmosphere packaging with micro-perforated film. J Food Sci 75:225–233

Roberts RT (1993) High intensity ultrasonics in food processing. Chem Ind (London) 15:119–121

Robertson RE, Hoy J (2000) In food irradiation available research indicates that benefits outweigh risks. Report to congressional requesters, United States general accounting office. Washington, DC, USA

Rosnes JT, Sivertsvik M, Skåra T (2003) Combining MAP with other preservation techniques. In: Ahvenainem R (ed) Novel food packaging techniques. CRC Press, Finland, pp 287–304

Sagakami H, Aohi T, Simpson A, Tanuma S (1991) Induction of immunopotentiation activity by a protein-bound polysaccharide, PSK. Anticancer Res 11:993–1000

Saleh A (2006) Antimicrobial and antioxidant activity of two desert truffles, Tirmania and Terfezia. Mast Abst Int 46:109

Saltarelli R, Ceccaroli PP, Barbieri CE, Stocchi V (2008) Effect of storage on biochemical and microbiological parameters of edible truffle species. Food Chem 109:8–16

Sawaya WN, Al-Shalhat A, Al-Sogair A, Mohammad M (1985) Chemical composition and nutritive value of truffles of Saudi Arabia. J Food Sci 50:450–453

Song GS, Hu SQ, Li L, Chen P, Shen X (2009) Structural and physical changes in ultrasound-assisted frozen wet gluten. Cereal Chem 86:333–338

Stanikunaite R, Trappe JM, Khan SI, Ross SA (2007) Evaluation of therapeutic activity of *Hypogeous Ascomycetes* and *Basidiomycetes* from North America. Int J Med Mushrooms 9:7–14

Stone H, Siedel JL (1983) Sensory evaluation practices. Academic, New York, NY, pp 76–86

Sun D-W, Li B (2003) Microstructural change of potato tissues frozen by ultrasound-assisted immersion freezing. J Food Eng 57:337–345

Thakur BR, Singh RK (1995) Combination processes in food irradiation. Trends Food Sci Technol 6:7–11

Van den Berg R, Haenen GRMM, van den Berg H, Bast A (1999) Applicability o fan improved Trolox equivalent antioxidant capacity (TEAC) assay for evaluation of antioxidant capacity measurements of mixtures. Food Chem 66:511–517

Vani AM, Hussain PR, Meena RS, Dar MA, Mir MA (2009) Effect of gamma irradiation and sulphitation treatments on keeping quality of white button mushroom Agaricus bisporus. Int J Food Sci Technol 44:967–973

Wang S, Marcone MF (2011) The biochemistry and biological properties of the world's most expensive underground edible mushroom: truffles. Food Res Int 44:2567–2581

Wen HW, Chung HP, Chou FI, Lin IH, Hsieh PC (2006) Effect of gamma irradiation on microbial decontamination, and chemical and sensory characteristics of lyceum fruit. Rad Phys Chem 75:596–603

World Health Organization (1994) Safety and nutritional adequacy of irradiated food. Geneva, WHO

Chapter 20
The Medicinal Value of Desert Truffles

Elinoar Shavit and Efrat Shavit

20.1 Introduction

Fungi are extremely diverse and abundant and they grow all over the world, even in the most inhospitable places. Throughout human history, fungi have been used for nourishment, as functional foods to maintain good health and prevent ailments, and as a source of medicine to treat diseases (Wasser and Weis 1999; Wasser 2011; Badalyan 2012). Fungi have a prominent place in traditional medicines. Efforts have been made to document these traditional medicinal uses and contemporary research has validated many of their claims (Wasser 2011). Fungi are considered to be an enormous and mostly unexploited source of powerful new pharmaceutical products (Wasser 2011).

Desert truffles have a particularly long history of use (Shavit 2008; see Shavit in Chap. 15 of this volume). Almost everywhere where desert truffles have been used as food, records suggest that they have also been used medicinally (Trappe and Alsheikh 1983; Mandeel and Al-Laith 2007; Patel 2012; Kalotas 1996; Trappe et al. 2008a, b). Desert truffles (hypogeous Ascomycetes) are found in arid and semiarid areas on every continent other than Antarctica (Kagan-Zur 2001). However, they are best known from the Middle East and North Africa, the Mediterranean Basin, the African Kalahari, and the Australian Outback (Trappe et al. 2008a, b). Following the rainy season, they produce fruitbodies that ripen just beneath the surface of the sand, creating a pattern of cracks and bumps on the surface (Kagan-Zur 2001; see Fig. 20.1). People learn to gather the truffles by recognizing these telltale cracks and bumps on the sand (Trappe 1990; Mandeel and Al-Laith 2007; see Fig. 20.2).

E. Shavit (✉)
FUNGI Magazine, Concord, MA, USA
e-mail: elinoar.shavit@gmail.com

E. Shavit
Fletcher Allen Health Care, University of Vermont, Burlington, VT, USA

V. Kagan-Zur et al. (eds.), *Desert Truffles*, Soil Biology 38,
DOI 10.1007/978-3-642-40096-4_20, © Springer-Verlag Berlin Heidelberg 2014

Fig. 20.1 *Tirmania nivea* emerging from the sand. (**a**) A desert truffle "*bump*" on the "*skin*" of the sand. (**b**) Ripening *Tirmania nivea* emerging from the sand. Photographer: S. Al-Aseeri. Courtesy of Sultan Al-Aseeri and Ahmed Al-Aseeri, truffle researchers from Qatar.

Fig. 20.2 *Terfezia boudieri* whole and cut. Photographer: Elinoar Shavit.

Desert truffles are not a predictable crop and may not produce fruitbodies in dry years (Trappe et al. 2010).

Indigenous populations who gather desert truffles for food consider them an important source of sustenance. They appreciate the truffles not only for their nutritional value (believing them to be as nutritious as meat but also because they are often the only significant source of food available right after the rains, bridging the temporary food shortage created when stored provisions have been used up but new crops are still not ripe for the harvest (Shavit and Volk 2007; Gast 2000 in Volpato et al. 2013; De Roman 2010; Shavit 2008; Alsheikh 1994; Blunt and WSB 2007; see Shavit in Chap. 15 of this volume). Traditionally, desert truffles have been sliced and dried to preserve them for later use, but they are also preserved by pickling, salting, and in recent years also by freezing (Mandaville 2011; Patel 2012; Blunt and WSB 2007; see Martinez-Tome et al. in Chap. 17 of this volume).

The most popular desert truffles with the best recorded history of use are species of the genera *Terfezia* and *Tirmania* (Kagan-Zur and Roth-Bejerano 2008; Shavit 2008; see Figs. 20.3 and 20.4, respectively). Much of the information regarding the nutritional value and antioxidative and antimicrobial activities of desert truffles pertain primarily to these genera (Mandeel and Al-Laith 2007; Wang and Marcone 2011; Patel 2012; Volpato et al. 2013; see Martinez-Tome et al. in Chap. 17 and Al-Laith in Chap. 18 of this volume). The people with the longest and best recorded history of desert truffle use, for food and medicinal purposes, are the indigenous populations of the Middle East, North Africa, and the Mediterranean basin (Kagan-Zur and Roth-Bejerano 2008; Shavit 2008; see Shavit in Chap. 15 of this volume). The Aborigines of Central Australia and the San (!Kung San, Bushmen) of the African Kalahari have also used desert truffles for food and medicine since ancient times. Both cultures have oral traditions and their cultural knowledge is transferred from one generation to the next. The records of use of desert truffles by the Aborigines and the San are relatively recent, scant, and often anecdotal, and the determination of the identity of the truffles described may be challenging (Kalotas 1996; O'Connell et al. 1983; Trappe et al. 2008a, b). Nonetheless, there are surprising similarities in the ways in which desert truffles are used for food and medicinally by these populations, which are geographically separated on three continents and gather desert truffles of different genera (Trappe et al. 2008a, b).

20.2 Desert Truffles in the Traditional Medicines of the Middle East and North Africa

20.2.1 Desert Truffles in the Traditional Medicine of Islam

The juice of desert truffles, mainly of *Tirmania nivea*, *Terfezia claveryi*, and *Terfezia boudieri*, has been used in the Middle East and North Africa to treat eye diseases and skin lesions for centuries (Hussain and Al-Ruqaie 1999; Janakat et al. 2005; Shavit 2008). Such use was recommended by Islam's Prophet Muhammad (sixth to seventh centuries) (Sahih Muslim 2013; Sahih Al-Bukhari n.d.; Tirmidhi Hadith 2013). In a chapter called *Excellence of Truffles and Their Use as a Medicine for the Eyes* in Sahih Muslim (#s 5084–5088), it is written that Islam's prophet Muhammad said that desert truffles were the *manna* given by God to the Israelites and that their "*juice is a medicine for the eyes*" (Sahih Muslim # 5086).

Entry # 1194 in Tirmidhi Hadith provides a recipe for the medicinal preparation and application of desert truffle juice, and offers an example of its use on a patient. What is unique about this particular entry is that it also provides the context in which the Prophet Muhammad was inclined to emphasize the benefits of desert truffles and to point out their medicinal use. In particularly favorable desert truffle seasons, these truffles can produce numerous fruitbodies, creating a dense layer of bumps on the surface of the desert sand (Burckhardt 1831). People who observed this phenomenon thought that the truffle bumps (see Fig. 20.1) were the result of a

Fig. 20.3 *Tirmania nivea* on the cutting board. Photographer; John Feeney. Courtesy of Saudi Aramco World.

Fig. 20.4 *Terfezia boudieri*—whole, cut, and peelings. Photographer: Elinoar Shavit.

disease similar to skin diseases afflicting humans and would not eat them. It is written in Tirmidhi Hadith:

> *Narrated AbuHurayrah: When some of the companions of Allah's Messenger (peace be upon him) remarked to him that truffles were the smallpox of the earth, he replied, 'Truffles are a kind of 'manna,' and their juice is a remedy for the evil eye'. . . .AbuHurayrah said that he took three, five, or seven truffles, pressed them, put their juice in a bottle and applied it as an eye-lotion to a slave-girl of his who was bleary-eyed, and she recovered (*Tirmidhi Hadith, Number 1194; Alim CD-Rom Version).

The tenth-century Islamic Persian philosopher and physician Avicenna (Ibn Sina) recommended the use of desert truffles as a remedy for a number of ailments, such as weakness, vomiting, and wounds (Hall et al. 2007; Britannica.com; Shavit 2008). Avicenna, in his significant multivolume work, the *Qanun* (The Canon of Medicine), writes that "*The excellent physicians have recognized that [desert truffle] juice cleanses the eye,*" suggesting that the use of desert truffle juice to treat eye diseases was already a known treatment in the tenth-century Arabian traditional medicine (Wells 2011; Hall et al. 2007; Prioreschi 1991).

20.2.2 *Medicinal Use of Desert Truffles in the Traditional Medicine of the Middle East*

Sand storms are a problem in the desert, particularly in spring, often causing eye irritations and infections, and desert truffle juice is still a common treatment for these conditions among populations in the Middle East and North Africa (Trappe and Alsheikh 1983; "Ibn-Ar'ar" 2011; Volpato et al. 2013). The truffles that are dried in good desert truffle seasons are stored for future use and can easily be reconstituted, as needed, to be used for food or medicinally, by boiling the slices in clean water (Mandaville 2011; Blunt and WSB 2007; Volpato et al. 2013). Both the reduced boiling liquid and the rehydrated truffle slices are used to treat eye diseases and other ailments in people and livestock (Mandaville 2011; Volpato et al. 2013; Shavit 2008).

A 2011 Arabic language article, posted by "Ibn-Ar'ar" (an alias) on an online bulletin board associated with King Faisal University in the city of Al-Hufuf (Hofuf) in the Eastern Province of Saudi Arabia, entitled *Desert Truffles: Benefits and Information,*[1] offers the following information:

> *Truffles were much used in folk medicine, as has come down in old books on folk medicine. They said that anointing the eyes with it* [truffle juice] *is useful for dimness of vision and for stinging eye inflammation, and that its juice is the most effective medicine for the eye if it is mixed with antimony* [kohl] *and* [the eye is] *anointed with it. And it strengthens the eyelids and increases the visual function with respect to both power and acuity, and it protects it from accidents* ("Ibn-Ar'ar" 2011)

Studies have shown that the skin of desert truffles is particularly rich in nutrients and bioactive compounds (Patel 2012). The nutrient-rich peels have also been used medicinally, mostly to treat skin and nail conditions. *Desert Truffles: Benefits and Information* reports that

> ... [As for] *the rind* [or peelings] *of truffles, it is said that it will heal burns* [and it is used], *after drying the rind for a period of 10 days in the sun and then applying it daily to the burns. ...* [it is also used] *as a treatment for* [and to] *strengthen the finger and toe nails and to prevent their tendency to break or become brittle ... And as a treatment for cracked* [chapped] *lips* ("Ibn-Ar'ar" 2011) (see figure 4).

20.2.3 *The Medicinal Use of Desert Truffles by the Sahrawi of Western Sahara*

In their newly published ethnomycological study, Volpato et al. (2013) observed the culinary, medicinal, and veterinary uses of desert truffles among the nomadic

[1] These sections were retrieved, translated from the original Arabic, and interpreted courtesy of James P. Mandaville, author of Bedouin Ethnobotany—Plant Concepts and Uses in a Desert Pastoral World, referenced in this chapter.

Fig. 20.5 *Tirmania nivea* gathered by a nomadic Sahrawi refugee near Tifariti, Western Sahara. Photographer: Gabriele Volpato. Courtesy of Gabriele Volpato (referenced in this chapter).

Sahrawi refugees of the North-Western Sahara desert. Desert truffles, particularly *Tirmania spp.* (see Fig. 20.5), have always been an important staple of the diet of these desert nomads, who gather the truffles both for their own use and for sale and preserve them by drying. There are numerous similarities between the truffle traditions of the Sahrawi and the populations of the Arabian Peninsula (Volpato et al. 2013; Shavit 2008). Much like the Bedouins, the Sahrawi have long used desert truffles as a source of nutritious food. They consider desert truffles a complementary and/or emergency food resource in times of food scarcity, believing them to be a nutritious substitute for meat (Gast 2000 in Volpato et al. 2013; Shavit 2008; Mandaville 2011; see Shavit in this volume).

Volpato et al. (2013) report that the Sahrawi use the truffles as medicinal foods as well as in medicinal preparations to treat ailments in humans and livestock. As a medicinal food, they prepare a soup from desert truffles, which they use to treat colds and respiratory ailments. They also place compresses of boiled slices of desert truffles directly on limbs afflicted with arthritis or rheumatism, covering the compresses with a protective bandage (Volpato et al. 2013).

The most popular medicinal use of desert truffles among both Sahrawi and populations in the Arabian Peninsula is the use of truffle juice to treat eye ailments, particularly conjunctivitis and trachoma, which are common diseases in the Sahara desert. When truffles are in season, their juice is used by squeezing it directly into the afflicted eyes. A therapeutic liquid can also be prepared directly from the dry truffle slices by boiling the truffles in water for 30 min and using the reduced liquid like eye drops or squeezing the liquid out of a rehydrated slice directly into the eyes (Volpato et al. 2013). Volpato et al. (2013) noted that about 5 % of the households surveyed in the 2007 study had some quantity of dry desert truffles slices stored aside for the medicinal treatment of eye diseases (Volpato 2007 in Volpato et al. 2013).

The Sahrawi also use desert truffles in their veterinary medicine. Volpato et al. (2013) report that to treat metritis (inflammation of the wall of the uterus that often happens in livestock after an animal gives birth), the Sahrawi mash together boiled desert truffles, onions, and garlic and apply the mixture to the wall of the animal's uterus as a compress. External inflammations, like mastitis (inflammation of the udder often caused by excessive milking), are treated by washing the afflicted area with the cooking water of the truffles or by placing compresses of the boiled truffle slices directly on the afflicted area (Volpato et al. 2013).

20.3 The Medicinal Use of Desert Truffles in the Traditional Medicine of the Aborigines of the Australian Outback and the San of the African Kalahari

20.3.1 Medicinal and Cosmetic Use of Desert Truffles Among the Aborigines of Central Australia

Not all Aboriginal groups in the Australian Outback have used mushrooms and truffles. The Arunta of Central Australia, for example, would not eat any mushroom because they believe that all mushrooms are fallen stars and as such have properties of evil magic (Spencer and Gillen 1904 in Kalotas 1996). On the other hand, other aboriginal groups in the Australian Outback have cherished both mushrooms and truffles, gathering a variety of species and even using them "greedily" (Grey 1841 in Kalotas 1996). Kalotas (1996) reports that in 1841, Grey observed that the "Native Truffle," which he identified as *Choiromyces aboriginum* (presently *Elderia arenivaga*), was highly prized by Aboriginal groups, like the Alyawara. The Alyawara call this truffle *urkunga*, seek it out, and enjoy eating it both raw and baked (O'Connell et al. 1983; Latz 1982 in Kalotas 1996). Grey describes how the Alyawara took great pains to squeeze out all the juice until the truffle "ran dry" before eating it. The Warlpiri of Central Australia gather a desert truffle that they call *wilyiri*. They eat the truffle but also utilize its prolific, yellowish juice. Like the Alyawara, they do not consume the juice of the truffle (Kalotas 1996). Rather, the Warlpiri highly appreciate the juice for its medicinal and cosmetic applications. A Warlpiri informant described this cosmetic use as follows: "*Long time in bush—rub it in underarm to make the underarm hair fall out*" (Lofts 1997 in Trappe et al. 2008a). Valiquette (1993) in Kalotas (1996) explain that

> *Fluid from this fungus (known as Kumpu urine, on account of its dirty yellow colour) is squeezed into sore eyes and onto sores. Elderly people rub the fungus into their armpits like a deodorant stick. When rubbed into the hair, it prevents growth* (Kalotas 1996).

The identity of the desert truffle that the Warlpiri call *wilyiri* and use to treat eye diseases and heal sores is not clear. Different authors identify it as different truffles, including *Mycoclelandia bulundari* and *Elderia arenivaga* (Kalotas 1996; Trappe et al. 2008a).

In the absence of available studies of the bioactive properties of *Elderia arenivaga* or *Mycoclelandia bulundari*, Trappe et al. (2008a), referencing Stanikunaite et al. (2007), suggest that these Australian desert truffles may have anti-inflammatory or antibiotic properties, as has been determined for other species of desert truffles (Stanikunaite et al. 2007 in Trappe et al. 2008a).

20.3.2 The Medicinal Use of Desert Truffles Among the San of the African Kalahari

Although few records of the San's use of desert truffles are available, some of the San communities still practice their traditional medicine, of which only little is known outside of their communities (Dan et al. 2010). Further ethnomycological research in these communities may reveal more medicinal uses of desert truffles.

The San are nomadic hunter-gatherers. They have a long tradition of gathering Kalahari desert truffles for food and medicine (Trappe et al. 2008b; Chang and Mshigeni 2001). The main truffle gathered by the San is *Kalaharituber pfeilii*, which the San appreciate and eagerly gather for their own use as well as for sale locally and to outside distributors (Trappe et al. 2008b). Although information regarding the past use of desert truffles in the Kalahari is limited, it is known that some Kalahari San communities attribute particularly strong and even magical curative powers to these truffles (Tanaka 1980 in Trappe et al. 2008b). San hunters carry a dried piece of Kalahari truffle to eat as an antidote to the poison of their arrows, should they accidentally be cut by one (Tanaka 1980 in Trappe et al. 2008b).

Another medicinal use of the dry fruitbodies of the Kalahari desert truffles is reported from Botswana, where it is locally called *Mahupu* (Trappe et al. 2008b). Khonga and Mogotsi (2007) report that the fruitbodies of *Mahupu* are gathered and dried in season, and the powder is then used as needed to induce birth in both humans and livestock (Khonga and Mogotsi 2007).

20.4 Desert Truffles as Functional Foods

The concept of functional foods, summed up in Hippocrates' 2,500-year-old tenet, "Let food be thy medicine and medicine be thy food," has become extremely popular (Hasler 1998). Functional foods are foods that contain bioactive properties, such as dietary antioxidants, that provide health-enhancing and physiological benefits beyond that of meeting basic nutritional needs (Hasler 1998). Traditional medicines in the Middle East and North Africa have used desert truffles and their preparations both as functional foods and for their medicinal properties for centuries (Volpato et al. 2013; Trappe and Alsheikh 1983; Shavit 2008).

Scientific research has helped to substantiate traditional beliefs about the nutritional and medicinal properties of desert truffles. Desert truffles have been shown to provide a rich source of protein, fatty acids, minerals, and carbohydrates and to contain considerably higher values of fat, potassium, and phosphate than many commonly consumed vegetables (Al-Laith 2010; Bokhary et al. 1987, 1989; Nutrition Information Center, University of Stellenbosch, 2008, in Trappe et al. 2008b). For further information on the nutritional value of desert truffles, please see Martinez-Tome et al. in Chap. 17 and Al-Laith in Chap. 18 of this volume.

In recent years, particular attention has been focused on the important role of dietary antioxidants in health promotion and disease prevention (Al-Laith 2010). Cells use antioxidants to remove or inactivate toxic free radicals, which may contribute to the development of diseases such as cancer, rheumatoid arthritis, atherosclerosis, and the degenerative processes associated with aging (Halliwell and Gutteridge 1984; Al-Laith 2010). The same desert truffles that have been popular as functional foods in traditional medicine have been found to contain substances with powerful antioxidant activities, including ascorbic acid, anthocyanins, esterified phenols, free phenolics, flavonoids, and carotenoids (Patel 2012; Al-Laith 2010; Murcia et al. 2002; see Martinez-Tome et al. in Chap. 17 and Al-Laith in Chap. 18 of this volume).

Several studies have investigated the antioxidant activities of desert truffles and compared them to those of common food antioxidants. Murcia et al. (2002) compared the antioxidant properties of two desert truffles, *Terfezia claveryi* and *Picoa juniperi*, to four common food antioxidants, finding that both desert truffles exhibited higher percentages of oxidation inhibition than the food antioxidants. Dundar et al. (2012) evaluated the antioxidant activity of *Terfezia boudieri* and concluded that when compared with standard compounds, *T. boudieri* exhibited excellent antioxidant activity. This conclusion was corroborated by the results of a recent study by Dogan and Aydın (2013), which evaluated the antimicrobial effects, antioxidant activity, and phenolic content of *T. boudieri*. The extracts of *Tirmania nivea* and *Tirmania pinoyi* have also been found to have significant antioxidant activity (Al-Laith 2010; Stojkovic et al. 2013; see Al-Laith in Chap. 18 of this volume).

These studies suggest that desert truffles represent a good source of dietary antioxidants, validating the claims of traditional medicine and supporting the continued use of desert truffles as functional foods.

20.5 The Antimicrobial Properties of Desert Truffles

Desert truffles have been used for medicinal purposes for centuries. Traditional medicinal uses of desert truffles prompted modern scientific research into their bioactive properties (Mandeel and Al-Laith 2007; Patel 2012). As previously discussed, indigenous populations in the Middle East and North Africa have long used the extract of desert truffles as a remedy for eye diseases, including trachoma (Trappe and Alsheikh 1983; Patel 2012). In 1981, Al-Marzooky investigated the efficacy of several extracts of *Terfezia claveryi* on bacterial growth in vitro and found that all extracts showed antibacterial activity against a broad range of tested bacteria, including *Chlamydia Trachomatis*, which causes trachoma (Mandeel and Al-Laith 2007). Al-Marzooky performed a pilot study on patients infected with trachoma and found that treatment with a sterilized, aqueous extract of *T. claveryi* was a useful treatment, but took longer to work than standard antibiotics (Mandeel and Al-Laith 2007).

Since then, multiple studies have investigated the antibacterial activity of desert truffles. Several species of *Terfezia* and *Tirmania* have been evaluated against a wide variety of bacterial species, using multiple extraction methods and different solvent systems. The results varied (see Table 20.1). Janakat et al. (2004, 2005) investigated the efficacy of aqueous and methanolic extracts of *T. claveryi* as well as partially purified proteins from these extracts against *Staphylococcus aureus* and *Pseudomonas aeruginosa*. Both bacteria are known to cause eye infections. The 5 % aqueous extract inhibited the growth of *S. aureus* by 66.4 % and of *P. aeruginosa* by 40.9 %. The methanolic extracts from *T. claveryi* were found to be ineffective. However, the relative antimicrobial activities of the protein fractions were found to be superior to most of the reference antibiotics used for comparison against both *S. aureus* and *P. aeruginosa* (Janakat et al. 2004, 2005). Gouzi et al. (2011) evaluated aqueous extracts of three desert truffles, *T. claveryi, Terfezia leonis* (a synonym of *Terfezia arenaria*, see Chevalier in Chap. 9 of this volume), and *Tirmania nivea* against both *S. aureus* and *P. aeruginosa*. The results of this study were consistent with those of Janakat et al. (2004, 2005) in that extracts of both *T. claveryi* and *T. nivea* caused a significant inhibition in the growth of both *S. aureus* and *P. aeruginosa*. However, the aqueous extract of *T. leonis* did not show any antibacterial activity (Gouzi et al. 2011).

On the basis of such data, Aldebasi et al. (2012) undertook an in vivo study in rabbits to compare treatment with desert truffle extracts to a modern antibiotic (Vigamox). They evaluated several concentrations of a crude extract from *T. claveryi* as a treatment for induced corneal ulcers caused by a *S. aureus* ocular strain that was isolated from human corneal samples. Their results were noteworthy. Treatment with 1.5 and 3 % concentrations of *T. claveryi* extract eventually healed the ulcer, but more slowly than the Vigamox. However, treatment with the 5 % concentration of *T. claveryi* extract was toxic, caused improper epithelialization, and complicated the corneal ulcer (Aldebasi et al. 2012). This result is a sobering reminder that even though desert truffles have been a safe dietary staple for millennia, compounds derived from these truffles could potentially be harmful.

Other desert truffles have also demonstrated antimicrobial properties. Dogan and Aydın (2013) studied the antimicrobial effects of three different extracts of *Terfezia Boudieri* against four Gram-positive bacteria, five Gram-negative bacteria, and one yeast. All of the truffle extracts showed antimicrobial activity against all of the bacteria and the yeast, with the acetone extract showing high antimicrobial activity, the chloroform extract showing high to moderate activity, and the methanol extract showing moderate to weak antimicrobial activity. The strongest effect recorded was

Table 20.1 Summary of current research: the antimicrobial properties of desert truffles

Species	Microbe	Extraction method	Results and observations	Reference
Terfezia boudieri	*Bacillus subtilis, Staphylococcus aureus, Listeria monocytogenes, Streptococcus pyogenes, Escherichia coli, Klebsiella pneumoniae, Pseudomonas aeruginosa, Proteus vulgaris, Salmonella enteritidis, Candida albicans*	Chloroform, acetone, and methanol	All truffle extracts showed anti-microbial activity against all bacteria and yeast. The maximum inhibitory effect on the test microorganisms was observed with the acetone extract against *C. albicans* Overall, the acetone extract exhibited the maximum antimicrobial effect with values generally lower than 100 μg/mL (high antimicrobial activity) The effects of the chloroform extract against test microorganisms are of high and moderate antimicrobial activity The lowest antimicrobial effect was observed in methanol extracts, which demonstrated moderate (with a few weak) effects	Dogan and Aydin (2013)
Terfezia claveryi	*Staphylococcus aureus*	Aqueous, methanolic, partially purified proteins	5 % Aqueous extract inhibited the growth of *S. aureus* by 66.4 % Methanolic extract was ineffective The aqueous extract and two of the partially purified proteins were found to be superior to most reference antibiotics used for comparison	Janakat et al. (2004)

(continued)

Table 20.1 (continued)

Species	Microbe	Extraction method	Results and observations	Reference
Terfezia claveryi	*Pseudomonas aeruginosa*	Aqueous, methanolic, partially purified proteins	5 % aqueous extract inhibited the growth of *P. aeruginosa* by 40.9 % Methanolic extract was ineffective Relative antimicrobial activities of the protein fractions were found to be superior to most of the reference antibiotics used for comparison	Janakat et al. (2005)
Terfezia claveryI, Terfezia leonis, Tirmania nivea	*Pseudomonas aeruginosa, Staphylococcus aureus*	Aqueous	The aqueous extracts of *T. claveryi* and *T. nivea* in the growth medium of *S. aureus* caused a significant inhibition of *S. aureus* growth by 86.48 % and 99.09 %, respectively The aqueous extracts of *T. claveryi* and *T. nivea* were found to cause a significant inhibition of the growth of *P. aeruginosa* by 71.11 % and 100 %, respectively The aqueous extract of *Terfezia leonis* did not show any antibacterial activity	Gouzi et al. (2011)
Terfezia claveryi	*Staphylococcus aureus*	Crude extract prepared by homogenizing with sodium phosphate buffer solution, then diluted further when used	In treating induced corneal ulcers in rabbits, topical application of synthetic antibiotic Vigamox (0.5 %) improved the signs of corneal ulcer with marked healing within 3–5 days and left a transparent cornea 1.5 % of *Terfezia claveryi* significantly reduced the corneal	Aldebasi et al. (2012)

			ulcer within 9 days and healed as nebular type of central cornea opacity 3 % concentration of *T. claveryi* induced healing within 12–14 days The topical application of 5 % of *Terfezia claveryi* was toxic, the eye became dry, developed hypopyon and ultimately perforated	
Tirmania pinoyi	*Pseudomonas aeruginosa, Salmonella typhimurium, Salmonella enteritidis, Escherichia coli, Enterobacter cloacae, Listeria monocytogenes, Staphylococcus aureus*	Methanol in three solvents, water, methanol, and DMSO with and without potassium metabisulfite (E224)	Methanolic extract successfully inhibited the growth of *S. aureus* in chicken soup, at room temperature or in refrigerator, in a dose dependent manner Both streptomycin and ampicillin had higher antibacterial activity against all of the bacteria tested than any of the truffle extracts	Stojkovic et al. (2013)
Tirmania pinoyi	*Bacillus subtilis, Escherichia coli, Staphylococcus aureus, Pseudomonas aeruginosa, Enterococcus*	Ethyl acetate	Effective against *B. subtilis and S. aureus* Not effective against the three other bacteria tested GC-MS products included pyrazines, which have antibacterial activity in literature	Dib-Bellahouel and Fortas (2011)

(continued)

Table 20.1 (continued)

Species	Microbe	Extraction method	Results and observations	Reference
Tirmania pinoyi	*Staphylococcus aureus*, *Micrococcus flavus*, *Listeria monocytogenes*, *Pseudomonas aeruginosa*, *Enterobacter cloacae*, *Escherichia coli*, *Salmonella enteritidis*, *Salmonella typhimurium*	methanolic in three different solvent systems	The extract possessed an antibacterial effect The best solvent system for in vitro study seemed to be methanol: water (30:70) The best results were achieved for *E. coli* and *M. flavus* with MIC of 0.62 mg/mL and MBC of 1.25 mg/mL All results obtained were comparable to those obtained previously for standard antibiotics	Stojkovic et al. (2012)

the antifungal activity of the *T. Boudieri* acetone extract against *Candida albicans*, with a minimum inhibitory concentration (MIC) of 4.8 μg/mL (Dogan and Aydın 2013).

Several studies have also been done on the antimicrobial activity of *Tirmania pinoyi*. Dib-Bellahouel and Fortaz (2011) investigated the antibacterial activity of various fractions of an ethyl acetate extract of *T. pinoyi* against *E. coli, B. subtilis, S. aureus, P. aeruginosa,* and *Enterococcus*. The extracts were effective against *B. subtilis* and *S. aureus,* but not against the other 3 bacteria (Dib-Bellahouel and Fortaz 2011). Stojkovic et al. (2012, 2013) evaluated methanolic extracts of *T. pinoyi* against a variety of Gram-positive and Gram-negative bacteria. They concluded that the truffle extract had an antimicrobial effect (Stojkovic et al. 2012, 2013). Stojkovic et al. (2013) also examined the effect of the *T. pinoyi* extract on the control of *S. aureus* in a contaminated chicken soup under in situ conditions and found that the methanolic extract of *T. pinoyi* successfully inhibited the growth of *S. aureus* in chicken soup in a dose-dependent manner (Stojkovic et al. 2013).

This data reveals that desert truffles possess antimicrobial activity against a range of human pathogens and have the potential to be used as therapeutic agents. Further research needs to be done to identify, characterize, and purify the compounds responsible for this activity, so that this potential can be realized (Wang and Marcone 2011).

20.6 Conclusion

Much of the history of the medicinal use of desert truffles around the world has been lost. However, the continued use of desert truffles in traditional medicine, including in the treatment of eye infections, has led to modern scientific investigation of the medicinal properties of desert truffles. The research regarding the antioxidant properties of desert truffles demonstrates that desert truffles contain a variety of antioxidants and raises the question of whether desert truffles can be used as a functional food that could potentially play a role in health maintenance and disease prevention. The research into antibacterial substances found in desert truffles is quite encouraging as well. A significant body of evidence supports the claim that desert truffle extracts have antibacterial activity against common pathogens that cause disease in humans. The study of the medicinal properties of desert truffles, while still in its infancy, suggests that desert truffles could represent an untapped source of novel pharmaceuticals that could treat a wide variety of modern-day ailments.

References

"Ibn-Ar'ar" (2011) Desert Truffles: Benefits and [other] Information. Translation by JP Mandaville, author of Bedouin Ethnobotany, referenced in this chapter. Online bulletin board associated with King Faysal University in the city of al-Hufuf (Hofuf), in the Eastern Province of Saudi Arabia, posted on October 19th, 2011, at: http://kfuforums.kfu.edu.sa/

showthread.php?142749-%C7%E1%DD%DE%DA-(−%C7%E1%DF%E3%C3)-%DD%E6%C7%C6%CF-%E6%E3%DA%E1%E6%E3%C7%CA. Accessed 27 Apr 2013

Aldebasi YH, Nouh WG, Abdel-Atti NM, Salem-Bekhit MM, Qureshi MA, Aly SM (2012) Comparative pathological studies on the healing effect of natural (*Terfezia claveryi*) and synthetic (Vigamox) antimicrobial on corneal ulcers in rabbits. J Pharm Biomed Sci 2(6): 66–77

Al-Laith AAA (2010) Antioxidant components and antioxidant/antiradical activities of desert truffle (*Tirmania nivea*) from various Middle Eastern origins. J Food Comp Anal 23:15–22

Alsheikh AM (1994) Taxonomy and mycorrhizal ecology of the desert truffles in the genus *Terfezia*. A thesis submitted to Oregon State University toward the degree of Doctor of Philosophy, Corvallis, Oregon

Badalyan S (2012) Medicinal Aspects of Edible Ectomycorrhizal mushrooms. In: Zambonelli A, Bonito GM (eds) Edible ectomycorrhizal mushrooms: Current knowledge and future prospects, Chapter 18. Springer Publication, New York

Blunt A, WSB (2007) Bedouin tribes of the Euphrates. Kessinger Publishing's Rare Reprints, Kila, MT, USA

Bokhary HA, Suleiman AAA, Basalah MO, Parvez S (1987) Chemical composition of desert truffles from Saudi Arabia. C Inst F Sci Tech 20:336–341

Bokhary HA, Suleiman AAA, Basalah MO (1989) The fatty acid components of the desert truffle "Al Kamah" of Saudi Arabia. J Food Prot 52:668–669

Burckhardt JL (1831) Notes on the Bedouins and Wahábys: collected during his travels in the East, vol I, Henry Colburn and Richard Bentlev, London. http://archive.org/stream/notesonbedouins01burcgoog/notesonbedouins01burcgoog_djvu.txt. Accessed 26 Sep 2012

Chang ST, Mshigeni KE (2001) Mushrooms and their human health: their growing significance as potent dietary supplements. University of Namibia, Windhoek, pp 1–79

Dan V, Mchombu K, Mosimane A (2010) Indigenous medicinal knowledge of the San people: the case of farm Six, Northern Namibia. Information Development, Sage publication. Online at http://idv.sagepub.com/content/26/2/129.full.pdf+html. Accessed 5 Jun 2013

De Roman M (2010) The contribution of wild fungi to diet, income and health: a world review. In: Rai M, Kovics G (eds) Progress in Mycology. Springer and Scientific Publishers, India

Dib-Bellahouel S, Fortaz Z (2011) Antibacterial activity of various fractions of ethyl acetate extract from the desert truffle, *Tirmania pinoyi*, preliminary analyzed by gas chromatography–mass spectrometry (GC-MS). African J Biotech 10(47):9694–9699. doi:10.5897/AJB10.2687

Dogan HH, Aydın S (2013) Determination of antimicrobial effect, antioxidant activity and phenolic contents of desert truffles in Turkey. Afr J Tradit Complement Altern Med 10(4): 52–58, AJTCAM ISSN: 0189–6016

Dundar A, Faruk O, Acay H, Okumus V, Ozdemir S, Yildiz A (2012) Antioxidant properties, chemical composition and nutritional value of *Terfezia boudieri* (Chatin) from Turkey. Food Sci Technol Int 18:317–328

Gouzi H, Belyagoubi L, Abdelali KN, Khalifi A (2011) *In vitro* antibacterial activities of aqueous extracts from Algerian desert truffles (*Terfezia* and *Tirmania*, Ascomycetes) against *Pseudomonas aeruginosa* and *Staphylococcus* aureus. Int J Med Mushrooms 13:553–558

Hall IR, Brown GT, Zambonelli A (2007) Taming the truffle. Timber Press, Portland, Oregon

Halliwell B, Gutteridge JMC (1984) Oxygen toxicity, oxygen radicals, transition metals and disease: review article. Biochem J 219:1–14

Hasler CM (1998) Functional foods: their role in disease prevention and health promotion. Food Technology 52(2):57–62

Hussain G, Al-Ruqaie MA (1999) Occurrence, chemical composition, and nutritional value of truffles: an overview. Pakistan J Bio Sci 2(2):510–514

Janakat S, Al-Fakhiri S, Sallal AK (2004) A promising peptide antibiotic from *Terfezia claveryi* aqueous extract against *Staphylococcus aureus in vitro*. Phytother Res 18:810–813

Janakat SM, Al-Fakhiri SM, Sallal AK (2005) Evaluation of antibacterial activity of aqueous and methanolic extracts of the truffle *Terfezia claveryi* against *Pseudomonas aeruginosa*. Saudi Med J 26(6):952–955

Kagan-Zur V (2001) Terfezias, a family of mycorrhizal edible mushrooms for arid zones. In: Pasternak D, Schlissel A (eds) Combating deforestation with Plants. Plenum Publishers, New York

Kagan-Zur V, Roth-Bejerano N (2008) Desert truffles. Fungi Magazine 1(3):32–37

Kalotas AC (1996) Aboriginal knowledge and use of fungi. In: Mallet K, Gregurinovic G (eds) Fungi of Australia, vol IB, CSIRO Publishing. Canberra, Australia, pp 269–295

Khonga EB, Mogotsi KK (2007) Indigenous knowledge on edible, poisonous, and medicinal macrofungi. Botswana Int. J. Med. Mushrooms, Beagle House, In

Mandaville JP (2011) Bedouin ethnobotany: plant concepts and uses in a desert pastoral world. University of Arizona Press, Tucson, USA

Mandeel QA, Al-Laith AA (2007) Ethnomycological aspects of the desert truffle among native Bahraini and non-Bahraini people of the Kingdom of Bahrain. J Ethn 110:118–129

Murcia MA, Martinez-Tome M, Jimenez A, Vera A, Honrubia M, Parras P (2002) Antioxidant activity of edible fungi (truffles and mushrooms): losses during industrial processing. J Food Protect 65:1614–1622

O'Connell JF, Latz PK, Barnett P (1983) Traditional and modern plant use among the Alyawara of Central Australia. Econ Bot 37(1):80–109

Patel S (2012) Food, health and agriculture importance of truffles: a review of current scientific literature. Curr Trends Biotech Pharm 6(1):15–27

Prioreschi P (1991) The history of medicine. Edwin Mellen Press, New York

Sahih Al-Bukhari, Volume 7, Translated by MM Khan, PhD, Darussalam publishers, Riyadh, Saudi Arabia. Volume 7, Book 76: The Book of Medicine, chapter 20—Al-Mann heals eye disease, hadith 5708.p 337

Sahih Muslim, translated by AH Siddiqui, volume: The book of drinks (Kitab al-Ashriba), chapter: excellence of truffles and their use as a medicine for the eyes. *Ahadith* 5084–5088. Published online at http://www.theonlyquran.com/hadith/Sahih-Muslim/?volume=23&chapter=26. Accessed 23 Apr 2013

Shavit E (2008) Truffles roasting in the evening fires: pages from the history of desert truffles. Fungi Magazine 1(3):18–23

Shavit E, Volk TJ (2007) *Terfezia* and *Tirmania*, desert truffles (*terfez, kama, p/faqa*): delicacies in the sand or manna from heaven? Fungus of the Month for January 2007. http://botit.botany.wisc.edu/toms_fungi/jan2007.html. Accessed 2 Dec 2012

Stojkovic D, Reis F, Glamoclija J, Ciric A, Nikolic M, Barros L, Ferreira Isabel, Giweli A and Sokovic M (2012) Nutritional value and antibacterial activity of desert truffle *Tirmania pinoyi* (Maire) Malencon (1973) growing wild in Lybia. Proceedings of the International Conference Biological Food Safety and Quality. Belgrade, at http://hdl.handle.net/10198/7845. Accessed 2 May 2013

Stojkovic D, Reiss FS, Ferreira ICFR, Barros L, Glamoclija J, Ciric A, Nikolic M, Stevic T, Giveli A, Sokovic M (2013) *Tirmania pinoyi*: chemical composition, *in vitro* antioxidant and antibacterial activities and in situ control of staphylococcus aureus in chicken soup. Food Res Int 53:56–62

Tirmidhi Hadith, Number 1194; Alim CD-Rom Version, at: http://www.answering-islam.org/Shamoun/muhammad_poison.htm. Accessed 2 May 2013

Trappe JM (1990) Use of truffles and false truffles around the world. In: Bencivenga M, Granetti B (eds) Proceedings, Atti del Secondo Congresso Internazionale sul Tartufo. Communita Montana dei Monti Martini e del Serano, Spoleto, Italy, pp 19–30

Trappe JM, Alsheikh AM (1983) Desert truffles: the genus *Tirmania*. Trans Br Mycol Soc 81(1): 83–90

Trappe JM, Claridge AW, Liddle L (2008a) Desert truffles of the Australian outback: ecology, ethnomycology and taxonomy. Econ Bot 62(3):497–506

Trappe JM, Claridge AW, Arora D, Smit WA (2008b) Desert truffles of the African Kalahari: ecology, ethnomycology and taxonomy. Econ Bot 62(3):521–529
Trappe JM, Kovacs GM, Claridge AW (2010) Comparative taxonomy of desert truffles of the Australian Outback and the African Kalahari. Mycol Prog 9:131–143. doi:10.1007/s11557-009-0621-6
Volpato G, Rossi D, Dentoni D (2013) A reward for patience and suffering: Ethnomycology and commodification of desert truffles among Sahrawi refugees and nomads of Western Sahara. Econ Bot XX(X):1–14, Published online by The New York Botanical Garden Press, Bronx, NY 10458–5126, USA, on June 6th, 2013. Accessed 8 Jun 2013
Wang S, Marcone MF (2011) The biochemical and biological properties of the world's most expensive underground edible mushroom: truffles. Food Res Int 44:2567–2581
Wasser SP (2011) Current findings, future trends, and unsolved problems in studies of medicinal mushrooms. Appl Microbial Biotechnol 890:1323–1332
Wasser SP, Weis AL (1999) Medicinal properties of substances occurring in higher Basidiomycetes mushrooms: current perspectives (review). Int J Med Mushrooms 1:31–62
Wells P (2011) Simply truffles: recipes and stories that capture the essence of the black truffles. Harper Collins Publishing, New York

Part V
Cultivation

Chapter 21
Domestication: Preparation of Mycorrhizal Seedlings

Asunción Morte and Alberto Andrino

21.1 Introduction

There are many mycorrhizal fungi with a great potential as edible fungi, but only some species have been cultivated (Zambonelli and Bonito 2012). Among these fungal species, two species of desert truffle have been successfully cultivated and reported, *Terfezia claveryi* Chatin in Spain (Honrubia et al. 2001; Morte et al. 2008, 2009, 2010, 2012) and *Terfezia boudieri* Chatin in Tunisia (Slama et al. 2010) and Israel (Khagan-Zur, pers. com.).

Since the first plantation of *Terfezia* mycorrhizal plants was established in 1999 in Murcia (south-eastern Spain), the increasing demand for this crop, not only in Spain but also in other countries, has prompted research into new strategies to help the passage from experimental scale to medium-large-scale cultivation. The first step in this process is the selection and production of suitable mycorrhizal seedlings of quality and adapted to different cultivation sites. The present chapter describes our experience and the experiments carried out to improve this first step.

21.2 Host Plant Selection and Propagation

The election of a suitable host plant species is a very important factor in the production of mycorrhizal plants. The wide edaphic tolerance of *Helianthemum* as host allows them to share desert truffle symbionts (Díez et al. 2002), although bioclimatic conditions are probably the most important factor for choosing a host plant for a specific area.

A. Morte (✉) • A. Andrino
Dpto. Biología Vegetal (Botánica), Facultad de Biología, Universidad de Murcia, Campus de Excelencia Internacional Regional "Campus Mare Nostrum", Campus de Espinardo, 30100 Murcia, Spain
e-mail: amorte@um.es

V. Kagan-Zur et al. (eds.), *Desert Truffles*, Soil Biology 38,
DOI 10.1007/978-3-642-40096-4_21,

Table 21.1 Different mycorrhizal associations, obtained under controlled conditions, between species of desert truffles and different *Helianthemum* species

Fungal species	*Helianthemum* species	References
Terfezia claveryi	*H. salicifolium*	Awameh et al. (1979)
		Dexheimer et al. (1985)
	H. guttatum	Fortas and Chevalier (1992)
	H. almeriense	Morte et al. (1994, 2008)
	H. ledifolium	Gutiérrez (2001)
	H. violaceum	Morte et al. (2009)
	H. hirtum	Torrente et al. (2009)
	H. canariense	Andrino et al. (2012)
Terfezia boudieri	*H. salicifolium*	Awameh et al. (1979)
	H. sessiliflorum	Slama et al. (2010)
	H. lippii	Pers. com.
Terfezia leonis (redefined as *Terfezia boudieri)*	*H. sessiliflorum*	Roth-Bejerano et al. (1990)
Terfezia leptoderma	*H. salicifolium*	Dexheimer et al. (1985)
	H. guttatum	Fortas and Chevalier (1992)
Terfezia arenaria	*H. guttatum*	Fortas and Chevalier (1992)
Terfezia terfezioides (redefined as *Mattirolomyces terfezioides)*	*H. ovatum*	Kovács et al. (2003)
Tirmania nivea	*H. salicifolium*	Awameh et al. (1979)
	H. lippii	Pers. com.
Tirmania pinoyi	*H. salicifolium*	Awameh et al. (1979)

Among the plant families cited in the literature that contain some species which form mycorrhiza with desert truffles are *Cistaceae* (Alsheikh 1984; Awameh et al. 1979; Morte et al. 1994; Roth-Bejerano et al. 1990; Gutiérrez et al. 2003; Zaretsky et al. 2006), *Fagaceae* (Díez et al. 2002), *Pinaceae* (Díez et al. 2002; Honrubia et al. 2007), *Fabaceae* (Kovács et al. 2003) and even *Cyperaceae* (Ammarellou and Saremi 2008; Jamali and Banihashemi 2012). However, most of plant species reported as host plants for experimental desert truffle mycorrhization are perennial and annual species from *Helianthemum* genus, belonging to the Cistaceae (Table 21.1). We have chosen to employ *Helianthemum almeriense* in our experiments because we have a lot of previous information about its culture, which help us to continue improving its domestication.

21.2.1 Photoautotrophic (PA) Versus Photomixoautotrophic (PM) Cultivation Methods

Many *Helianthemum* species show erratic seed germination, and seed scarification is necessary to increase germination rates (Pérez-García and González-Benito 2006). Moreover, high mortality of the germinated seedlings is common during the first 2 months after germination in nursery conditions (Morte et al. 2012).

Micropropagation techniques have been used for plant production since they improve seed germination and plant survival (Morte and Honrubia 2009; Morte et al. 2008). The in vitro micropropagation protocols of the *Helianthemum* species studied are straightforward and rapid because plant multiplication, elongation and rooting occur in the same subculture. Consequently, they are also cheap because only small amounts of plant growth regulators and little labour are required (Morte and Honrubia 2009). However the plants that have been grown under in vitro conditions generally present non-functional stomata, weak root systems and poor development of waxes and cuticle (Mathur et al. 2010). The acclimation process helps to develop the necessary morphological and metabolic adaptations before plant goes to ex vitro conditions (Pospóšilová et al. 1999), although this does not always occur. To solve this problem, we have developed a photoautotrophic *Helianthemum* micropropagation system (Andrino et al. 2012) based on the methodology described by Kozai (1991). The photoautotrophic micropropagation technique which overcomes these problems can be defined as micropropagation without sugar in the culture medium, the growth or accumulation of carbohydrates by cultures being fully dependent on photosynthesis and inorganic nutrient uptake (Kozai 1991; Zobayed et al. 2004). Photoautotrophic (PA) micropropagation has many advantages over conventional or photomixoautotrophic (PM) micropropagation with including improved plantlet physiology (biological aspect) and operation/management of the production process (engineering aspect), although it also has some disadvantages (Xiao et al. 2011).

Helianthemum almeriense Pau has been successfully micropropagated by PM method (Morte and Honrubia 1992, 1997), and the same plant was used as a model to improve *Helianthemum* propagation by PA. When cultured in the absence of sucrose, this plant increased its survival rate during acclimation using a PA system (Andrino et al. 2009). At the same time, exchanging agar for perlite, as physical support, contributed to a significant reduction in plant losses during acclimation. In addition, the absence of sucrose reduced the presence of microbial contamination during the cultivation vessel phase. In the light of the above, our objectives were to ascertain why these differences in the survival rates existed between *H. almeriense* plants growth under PA and PM conditions. Also, we looked for any parameter that could help determine the quality of the seedling before the acclimation phase. For this purpose, *H. almeriense* was cultivated inside vessels in the following conditions: 21–23 °C, 60–70 % relative humidity, 350–360 ppm CO_2, 4,000–4,300 lux, 140–160 μmol/m^2 s, photoperiod of 16 h light under PM (salts and vitamins of Murashige and Skoog 1962) medium with 3 % sucrose and 8 g/l agar) or PA (salts and vitamins of MS medium without sucrose and with horticultural perlite). The plants were maintained for 60 days in these conditions, after which seedlings were randomly harvested for different physical (leaf area and stem and root lengths) and photosynthetic (photosynthetic pigments and the chlorophyll metre SPAD-502-Konica Minolta, Japan-value) parameters to be analysed.

The results showed significant differences ($p \leq 0.05$) between the PA and PM treatments (Fig. 21.1). Leaf area (Fig. 21.1a) was approximately 50 % greater in PA than in PM. The leaf area is a parameter that defines the quality of the crop

(Miyashita et al. 1996): greater leaf areas course with higher initial photosynthetic rates and, therefore, higher growth rates. Likewise, low relative humidity in the PA cultivation vessels enhances the development of leaf area and the length of the upper stems (Afreen 2005; Nguyen and Kozai 2005). According to Jackson et al. (1991), the ethylene content inside a PM cultivation container can play an important role, inhibiting stem growth, explant development and leaf expansion. This was reflected in the *H. almeriense* plants, which showed leaf areas and stem lengths that were significantly higher ($p \leq 0.05$) in the PA than in the PM treatment (Fig. 21.1a, b). Thanks to greater stem length and larger overall plant size, the cultivation period is shortened by 2 weeks in the PA treatment, a saving of time that is crucial for medium-large-scale mycorrhizal plant production.

H. almeriense plants showed significantly higher root development in PA than in PM conditions (Fig. 21.1c). Kozai and Kubota (2005) claimed that replacing conventional agar gel by porous materials, like perlite, significantly affects the environment of roots and, therefore, their anatomical characteristics. This improvement in the root environment in PA is based on experience gained from hydroponics, a technique that normally works without sugar in the culture medium and by irrigation with liquid medium on porous substrates. In our case, perlite was selected as substrate because it is widely used in the hydroponic cultivation of numerous horticultural species, due to its good aeration properties and water holding capacity (Urrestarazu-Gavilán 2004). A well-developed root system permits the plant to transport water and nutrients efficiently, promoting proper growth of the seedlings and improving their health. This translates into higher rates of survival during transplanting and in the final environmental conditions (Afreen et al. 1999). In other woody species, correct root development has been correlated with a high survival rate in ex vitro conditions (Kirdmanee et al. 1995).

The photosynthetic pigment concentrations (total chlorophyll, chlorophylls a and b, xanthophyll and carotene) were significantly ($p \leq 0.05$) higher in PA of *H. almeriense* plants (Fig. 21.1d–g), as has been observed in many other crops (Kozai 2005). In a PA cultivation system, photosynthesis is the only source for carbohydrate production and accumulation. In general, PM cultivation systems show poor photosynthetic function, non-functional stomata, poor development of epicuticular waxes and hyper-hydrated tissues, resulting in a low survival rate of the plant material in ex vitro conditions (Kozai 2005). Chlorophyll *a* is usually found to be 3 times more concentrated than chlorophyll *b* (Afreen 2005), as occurred in the *H. almeriense* plants (Fig. 21.1d, e). Moreover, it is well known that the air exchange rate is the key factor for increasing the photosynthetic pigment concentration, favouring suitable CO_2 levels for each phase and preventing the accumulation of harmful ethylene (Cui et al. 2000; Jo et al. 2002; Park et al. 2004; Righetti 1996; Zobayed et al. 2000). The PA cultivation system favoured gas exchange with the external atmosphere of cultivation, causing a significant increase in photosynthetic pigments in *H. almeriense* plants (Fig. 21.1d–h).

The quantity and quality of light and the absence of sucrose play an important role in the *H. almeriense* cultivation (Andrino et al. 2009, 2012). In the case of the xanthophylls and total carotenoids, there were significant differences ($p \leq 0.05$)

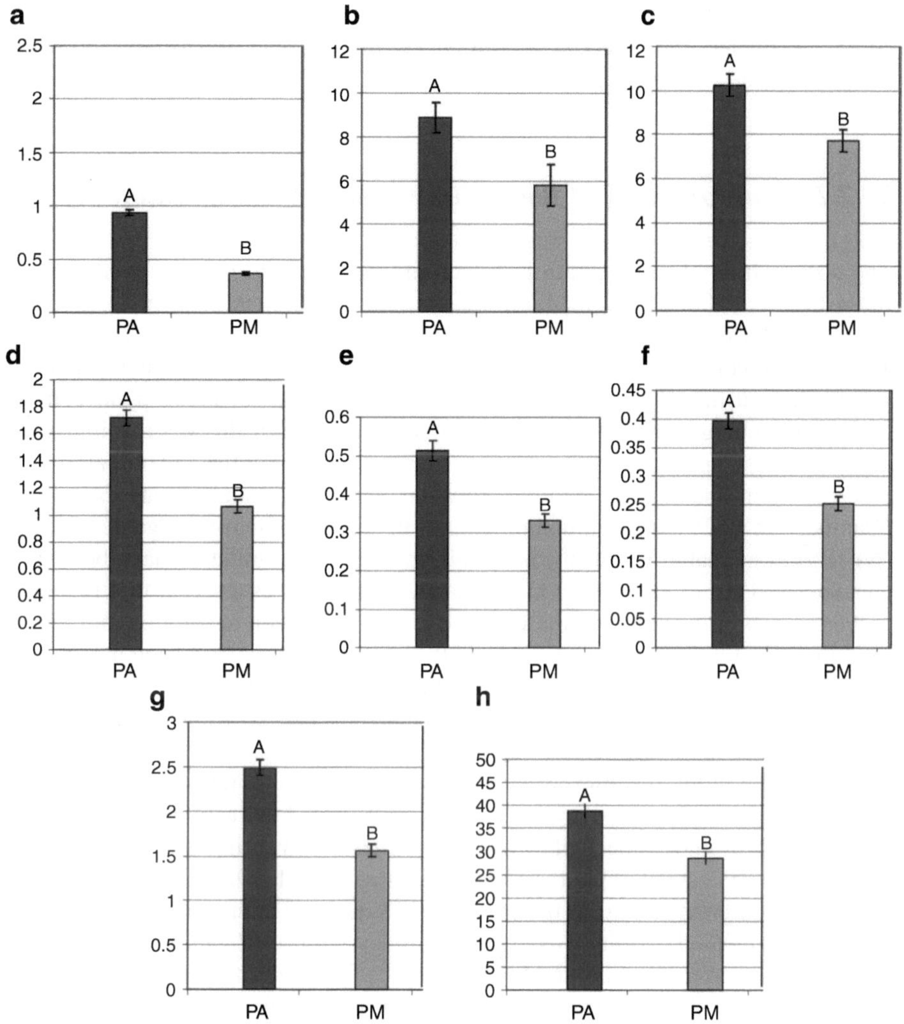

Fig. 21.1 Effect of PA and PM cultivation systems on leaf area (**a**, cm^2), stem (**b**, cm) and root (**c**, cm) lengths, chlorophyll a (**d**, mg/g leaf), chlorophyll b (**e**, mg/g leaf), xanthophylls and carotenoids (**f**, mg/g leaf), total chlorophylls (**g**, mg/g leaf), and SPAD-502 value (**h**) of *H. almeriense* seedlings. Each bar is the mean of fourteen measurements and its standard deviation value. The parameters, for each PA or PM treatment, that do not share the same letter are significantly different according to ANOVA test ($p = 0.05$)

between the PA and PM treatments (Fig. 21.1). These pigments provide mechanisms for the dissipation and extinction of light energy, as may also be antenna pigments that work between 450 and 500 nm (Rivas 2008). These protection pigments increase proportionally with the chlorophyll content of plants grown in vitro during the adaptation to a more intense light (ex vitro environment) (Hofman et al. 2002). The data obtained for *H. almeriense* indicated that the plants grown under the PA cultivation system had higher values for these pigments, which could be related with better adaptation to higher light intensities following transplantation.

21.2.2 Improvement of the Culture Medium Composition

The composition of the medium is a determining factor for in vitro plant growth. Murashige and Skoog (MS) medium was used for *Helianthemum* species micropropagation (Morte and Honrubia 1992, 1997; Morte et al. 2009; Torrente et al. 2009). The medium contains the correct amounts and proportion of inorganic nutrients to satisfy the nutritional as well as the physiological needs of many plant cells in culture (Gamborg et al. 1976). A distinguishing feature of the MS medium is its high content of nitrate (NO_3^-), potassium and ammonium (NH_4^+) compared with other nutrient media, the relation between ammonium and nitrate being 1:2 (Gamborg et al. 1976). Ammonium and nitrate are the primary inorganic nitrogen (N) sources available for plants, and their uptake and assimilation have been well characterised (Li et al. 2006). On one hand, nitrate acts as a signal that regulates carbon metabolism, inducing the required genes for carbon absorption and reduction, the assimilation of ammonium and the synthesis of the carbonated skeletons necessary for amino acid synthesis. On the other hand, nitrate inhibits the genes involved in carbohydrate biosynthesis (Maldonado et al. 2008). The lack of sufficient quantity and quality of light, in combination with the presence of carbohydrates in the medium, produces low photosynthetic rates, high activity of the PEPC (phosphoenolpyruvate carboxylase) and low activity of RuBisCo (ribulose-1,5-bisphosphate carboxylase oxygenase), favouring the respiration of the plant tissue, instead of photosynthesis (Cristea et al. 1999; Genoud et al. 2001).

PA cultivation method uses the complete MS medium in the absence of a carbon source. This method permitted us to grow a large volume of *H. almeriense* seedlings with germination rates of around 80–90 % and very satisfactory results. However, some seedlings stopped growing and turned yellowish green during the first 2 months after germination before finally dying. The problem was attributed to the possible excess ammonium concentration in the MS medium. Theoretically ammonium should be the preferred N source because it consumes less energy than nitrate in the plant metabolism (Britto and Kronzucker 2002). However, excess ammonium nutrition usually has deleterious effects on growth and deprives cells of osmotic adjustment. The reported symptoms of NH_4^+ toxicity range widely and generally appear at external NH_4^+ concentrations above 0.1–0.5 mmol/L in sensitive species like barley (Britto and Kronzucker 2002). Two of the most dramatic of these symptoms are leaf chlorosis and the total suppression of growth. Yield reduction among sensitive species can range from 15 to 60 % and may even result in death. Other symptoms often include a decrease in the root:shoot ratio although the reverse effect has been observed for some species (Britto and Kronzucker 2002). Symptoms not so readily visible, but equally important, may include a decline in mycorrhizal associations. Moreover, seed germination and seedling establishment can be inhibited by NH_4^+ toxicity, a feature of high ecological significance (Britto and Kronzucker 2002).

In aerated soils, the major form of inorganic N is nitrate; in flooded wetland or acidic soils, the major form is ammonium. In the rhizosphere, the root can release oxygen and exudates that greatly influence the local redox potential and the density and activity of microbial populations (Xu et al. 2012). The relative abundance of NH_4^+ compared with NO_3^- in the soil solution is determined by a number of factors, of which the accumulation of organic matter, soil pH, soil temperature, the presence of allelopathic chemicals and soil oxygenation status are the most important (Britto and Kronzucker 2002). Typically, low pH, low temperature, the accumulation of phenolic-based allelopathic compounds and poor oxygen supply inhibit many nitrifying microorganisms, resulting in higher rates of net ammonification than net nitrification. Soils exhibiting these conditions tend to be late successional, while NO_3^--rich soils tend to be early successional (Britto and Kronzucker 2002). The soils where *H. almeriense* mycorrhizal plants live are normally open places, dry rocky soils, limestone, loamy soils or gypsum soils; sometimes also in volcanic terrain and even in sandy soils, which are also rich in carbonates, with low/no organic matter and high soil temperature. The pH of the soils goes from slightly to moderately alkaline (7.5–8.5) (Honrubia et al. 2007). Carbonates influence the availability of plant nutrients such as phosphorus, molybdenum, iron, boron, zinc and manganese (Navarro-Blaya and Navarro-García 2003). Plant symbioses play an important role in the ability to take up these scarce nutrients. Our hypothesis is that the ecosystem in which *H. almeriense* develops does not present the necessary characteristics for ammonification, and it is assumed that the vegetation that grows naturally develops in its absence.

To assess this hypothesis, several culture media with different NH_4^+/NO_3^- ratios and different light conditions were tested. The complete formulation of MS medium was used as control treatment. This medium presents a $NH4^+/NO_3^-$ ratio of 0.52. The idea was to reduce the presence of ammonium ion but not the total concentration of nitrate ion. For the ratios in which the presence of ammonium was diminished, the lack of nitrogen ion was compensated by calcium nitrate salt. Moreover, for in vitro plant multiplication, plants are normally grown without daylight, and the light source selected has a higher proportion of blue to promote stem elongation and multiplication. The experimental design was made to study the effect of light source and NH_4^+/NO_3^- ratio on plant survival (Table 21.2). One month after sowing the seeds inside the vessels, the results pointed to the high toxicity of different ammonium/nitrate ratios, especially under high-intensity light conditions.

Statistically, under low-intensity light conditions (Fig. 21.2a), the 0 and 0.52 ratios (0 and 20.6 mM NH_4^+) presented the same low survival rates, the only difference being that the control treatment (ratio 0.52) showed a very high standard deviation value, which explains the erratic behaviour of some plant lots during seedling production. The best results were obtained in the 0.13–0.39 range (5.15–15.45 mM NH_4^+), in which *H. almeriense* seedlings were able to grow and showed very good survival rates of up to 80 %.

Table 21.2 Different treatments carried out to evaluate possible toxicity of NH_4^+/NO_3^- ratios

Treatment						
1. Different light sources	2. Different NH_4^+/NO_3^- ratios					
	NH_4NO_3 (mM)	KNO_3 (mM)	$Ca(NO_3)_2$ (mM)	Total NH_4^+ (mM)	Total NO_3^- (mM)	Ratio NH_4^+/NO_3^-
Fluorescent tube	0	18.8	10.3	0	39.4	0.00
Daylight 18 W (Sylvania)	5.15	18.8	7.7	5.15	39.4	0.13
PAR: 70–90 μmol/m^2 s	10.3	18.8	5.15	10.3	39.4	0.26
Sodium vapour bulb	15.45	18.8	2.6	15.45	39.4	0.39
Sodium vapour 400 W (Gavita)	Control: 20.6	18.8	0	20.6	39.4	0.52
PAR: 140–200 μmol/m^2 s						

The MS media composition has two nitrate forms, ammonium nitrate and potassium nitrate. The objective of this trial is to test the possible ammonium toxicity, for this purpose there has been established decreasing concentrations of ammonium nitrate. To restore MS initial nitrate concentration, but decreasing the total N, calcium nitrate has been used as substitute

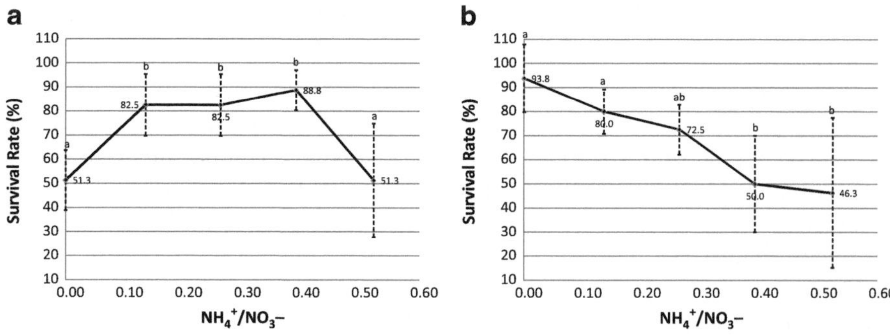

Fig. 21.2 Survival rates of *H. almeriense* in the tests carried out with different NH_4^+/NO_3^- ratios under fluorescent light (**a**) and sodium vapour light (**b**). Each point is the mean of eight measurements and its standard deviation value. The survival rates for each NH_4^+/NO_3^- ratio which do not share the same letter are significantly different in ANOVA test ($p = 0.05$)

Under high-intensity light conditions (Fig. 21.2b), seedlings did not show statistical differences in survival rates at NH_4^+/NO_3^- ratios ranging from 0 to 0.26 (0–10.3 mM NH_4^+), in which the survival of *H. almeriense* plants was not negatively affected. The 0 ratio (absence of ammonium) provided the best survival rate (up to 90 %) The survival rate changed as the ratio increased from 0.39 to 0.52 (15.4–20.6 mM NH_4^+) when *H. almeriense* started to suffer the deleterious effects of high ammonium concentrations, the survival rates being low and very variable (Fig. 21.2). Therefore, the presence of NH_4^+ starts to be harmful at a ratio of 0.26 (10.3 mM NH_4^+) (Fig. 21.2b). In the light of these findings, the nitrogen source of MS medium should be modified in order to improve *H. almeriense* seedling survival for commercial production.

Many studies have reported that ammonium toxicity is frequently more pronounced at high light intensity (Magalhaes and Wilcox 1984; Zhu et al. 2000; Britto and Kronzucker 2002), which agrees with the results obtained for *H. almeriense* (Fig. 21.2). It may be that the light optimum with NH_4^+ (relative to NO_3) nutrition is shifted to a higher intensity to compensate for increased carbon utilisation for respiration and amino acid production (Britto and Kronzucker 2002).

Plants susceptible to NH_4^+ toxicity are typically afflicted by reduced rates of net photosynthesis and enhanced photorespiratory rates (Blasco et al. 2010). Photorespiration is a possible means of alleviating light stress; it can produce 20-fold more NH_4^+ than is generated by the reduction of NO_3^- and is considered the major source of this cation, especially in C3 plants. Around 25 % of the fixed CO_2 is released during this metabolism; however, the suppression of photorespiration has negative effects on plants, producing a decrease in the CO_2 assimilation rate, poor vegetable growth and alterations in the chloroplast structure (Blasco et al. 2010). Further physiological analyses should be carried out in order to determine the importance of the photorespiration process under different light conditions with different NH_4^+ concentrations.

21.2.3 Plant Quality Control Before the Acclimation Phase

During seedling production, it is very important to know the most adequate moment for transplantation. The main objective is to know when the best moment is to start the transplantation from vessel to pot. For this purpose, a non-destructive determination of total chlorophyll was carried out using a SPAD-502. A linear relation was observed between SPAD-502 values and total chlorophyll in *H. almeriense* for each treatment (Fig. 21.3).

During the last decade, the use of this type of device for non-destructive measurements has increased in agricultural (Uddling et al. 2007) and forestry (Hawkins et al. 2009) field research, being associated with the estimation of chlorophyll and leaf nitrogen concentration with time. However, these measurements must take into account the different species cultivated, growth conditions, season of the year, stage of the crop, crop variety, light reflection, light dispersion and even the unequal distribution of leaf chloroplasts (Nauš et al. 2010). For these reasons, any measurement in *H. almeriense* should take into consideration its growth conditions.

Good linear correlations were observed between the total chlorophyll obtained by pigment extraction and the SPAD-502 measurement from the same leaves (Table 21.3), according to the Pearson correlation coefficient for both the PA and PM treatments. Once the linear correlation was established, the efforts were centred on modelling this relation. The desired relation is linear because it would permit to interpolate SPAD-502 values to total chlorophyll once as an indicator of plant health. Three different simulations were carried out, the first with only the PM data (total chlorophyll vs. SPAD-502 measurement, $n = 14$), the second with the PA data and the third with both PM and PA data ($n = 28$). For the first two simulations, the R^2 were very low, 0.693 for PA and 0.624 for PM, compared with the results of other studies, which ranged from 0.8 to 0.9 (Hawkins et al. 2009; Nauš et al. 2010; Uddling et al. 2007). The slope shown by the PA and PM linear functions was identical (0.04). However, the R^2 value reached 0.804 when PA and PM data were considered together, and the mathematical relation between total chlorophyll and SPAD-502 measurement could be expressed as a linear function (Fig. 21.3).

Once the linear relation had been described between both quality parameters, a logistic regression was carried out. The logistic regression is useful in those cases in which it is necessary to predict the presence or absence of a particular feature or result, according to the values of a set of predictor variables (independent), which may be quantitative or qualitative (Ferran-Aranaz 2001). A pseudo-R^2 value (Nagelkerke R^2) measures the strength of association of the model, whose values are between 0 and 1. Nagelkerke R^2 is considered a good indicator of the association strength (Dominguez-Rojas et al. 1993).

The dependent and dichotomous parameter is live (value = 1) or dead (value = 0), after the acclimation process. As co-variable, the set of physical parameters (stem and root lengths, leaf area) and photosynthetic parameters (SPAD-502, C_a, C_b, C_{x+c}, C_{total}) were used. This statistical test has two different purposes:

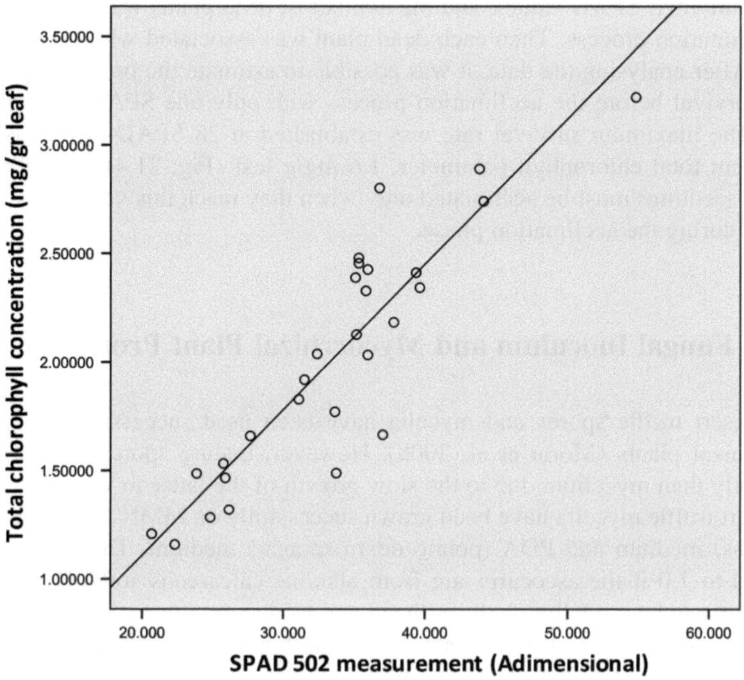

Fig. 21.3 Graphic representation of total chlorophyll and the SPAD-502 value ($R^2 = 0.804$) of *H. almeriense* seedlings before acclimation. The linear equation: total chlorophyll = −0.232 + (0.067 × SPAD 502). Data obtained from PM and PA treatments together ($n = 28$)

Table 21.3 Correlations between total chlorophyll and the SPAD-502 measurement

Treatment	Correlation coefficient	SPAD-502	C_{total}
PA	Pearson correlation test	1.000	0.832
	Significance		0.000
PM	Pearson correlation test	1.000	0.790
	Significance		0.001

The correlation was significant at $\alpha = 0.01$ level

(1) to identify which parameter (physical or photosynthetic) is able to determine whether a pre-acclimated plant is going to survive (or not) after the acclimation process and (2) to establish the interval value of the parameter between the live or dead *H. almeriense* seedlings.

The SPAD-502 parameter obtained the best Nagelkerke R^2 value, 0.71, which indicates a moderately strong relationship between prediction (live or dead) and the SPAD-502 measurement in 71 % of cases. Therefore, the SPAD-502 seems to be an easy and robust measurement to estimate the pre-acclimation condition in *H. almeriense*. The second objective was to identify the SPAD-502 interval value between live and dead plants. For this purpose, 68 acclimated plants were selected to measure their SPAD values, and the number of dead plants was registered after the acclimation process. Then each dead plant was associated with its SPAD-502 value. After analysing the data, it was possible to estimate the probability (0.1) of plant survival before the acclimation process with only one SPAD-502 measurement. The maximum survival rate was established at 28 SPAD-502 units, or its equivalent total chlorophyll parameter, 1.6 mg/g leaf (Fig. 21.4), and, therefore, in vitro seedlings must be acclimated only when they reach this value to be able to survive during the acclimation phase.

21.3 Fungal Inoculum and Mycorrhizal Plant Productions

Both desert truffle spores and mycelia have been used successfully to produce mycorrhizal plants (Morte et al. 2008). However, mature spores are used more frequently than mycelium due to the slow growth of the latter in vitro.

Desert truffle mycelia have been grown successfully on MMN (modified Melin-Norkrans) medium and PDA (potato dextrose agar) medium. The pH should be adjusted to 7.0 if the ascocarps are from alkaline calcareous soils. Desert truffle mycelium can be used directly from the plates as inoculum for in vitro mycorrhizal synthesis (Morte et al. 1994; Morte and Honrubia 1995, 1997) and from liquid fermentation for both in vitro and in vivo inoculation trials (Morte and Honrubia 2009; Morte et al. 2008). However, only fungal strains well adapted to in vitro conditions should be used to produce mycelium by liquid fermentation in a bioreactor. A study on in vitro mycelium cultures of two mycorrhizal desert truffles in conditions of water stress demonstrated that *Terfezia* mycelium (strain TcS2) grows better under slight water stress (−0.45 MPa), which could improve the production of this mycelial inoculum in a bioreactor (Navarro-Ródenas et al. 2011).

The spore suspension is made taking into account the maturity of the spores. The spore suspension from mature ascocarps consists of 6 g of dried and crushed ascocarps per litre of distilled water. This spore solution is shaken overnight (12 h), and, then, instead of inoculating the plants directly, the spore solution was added to the perlite, allowing spore adhesion to the pores and cavities within (Andrino et al. 2012; Morte et al. 2012). Using such a technique allowed the quantity of spores per litre to be reduced from 10 (Morte et al. 2008) to 6 g (which means 3.5–4.5 × 10^5 mature spores/plant). The percentage of inoculum per plant represents 5 % of the final container volume.

For the production of desert truffle mycorrhizal plants, four ex vitro and two in vitro inoculation options were designed, the time required for each of them ranging between 4 and 9 months, depending on the type of plant propagation system and inoculum source used (Table 21.4). The photoautotrophic (PA) *Helianthemum* micropropagation system (7 months) allowed this time to be reduced to 3 months with respect to PM system (4 months) since fungal inoculation is carried out at the moment plants are transferred from in vitro to ex vitro conditions, so that plant acclimation and mycorrhization occur at the same time. Moreover, this last way has

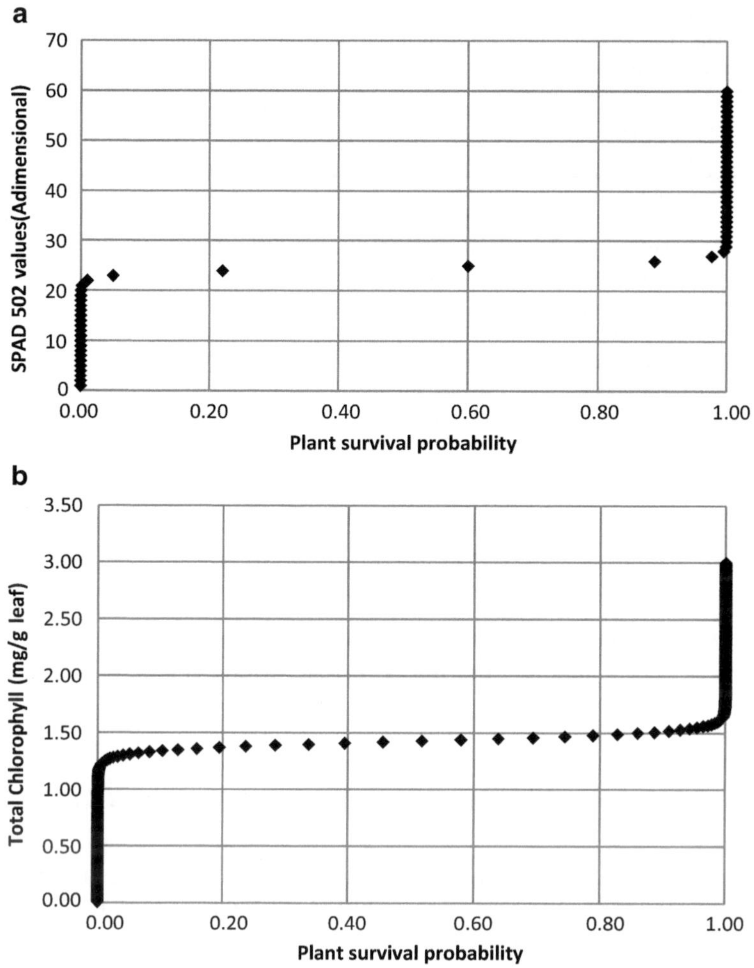

Fig. 21.4 Plant survival probability graphs, where 0 value means dead plant and 1 alive plant after acclimation process. (**a**) Representation of the survival interval value of SPAD-502 for pre-acclimated *H. almeriense* plant. (**b**) Representation of the survival interval value of total chlorophyll (mg/g leaf) for pre-acclimated *H. almeriense* plant

other advantages, including (1) a saving of fungal inoculum, (2) high survival percentage/smooth transition to ex vitro environment, (3) elimination of plant physiological disorders, (4) increased annual productivity per floor area, (5) reduced labour costs, (6) simplified micropropagation system and (7) no unwanted contamination due to the absence of sugar in the medium (Andrino et al. 2012; Morte et al. 2012).

After the acclimation phase, seedling irrigation is established to maintain the pot water potential between −15 and −30 kPa in nursery conditions. Approximately 30–40 days after transplanting, it is necessary to make a mycorrhization quality

Table 21.4 Options designed by in pot and in vitro methods to produce desert truffle mycorrhizal plants and the time required for each, depending on the type of plant propagation system and inoculum source used

Inoculation	Plant material (*Helianthemum*)	Fungal material	Time for plant production (months)	Time for plant mycorrhization (months)	Total time (months)
In pots	Seedlings directly germinated In pots	Mature spores	6	3	9
		Mycelial suspension	6	1–2	7.5
	Acclimated PM micropropagated plants	Mature spores	4	3	7
	Micropropagated PA plants	Mature spores or mycelium in perlite	3	1	4
In vitro	Micropropagated PM plants	Pieces of agar with mycelium	3	2	5
		Mycelial suspension	3	2	5

control. With all these production systems, mycorrhization rates range between 75 and 85 % after 2 months (Andrino et al. 2012; Morte et al. 2012).

21.3.1 Use of Fertilisation of Mycorrhizal Nursery Plants

Given that the objective is to obtain quality mycorrhizal plants, nursery maintenance tasks are crucial to ensure the quality of the final product. Using the desert truffle mycorrhizal plant production method described by Andrino et al. (2012), mycorrhization can reach values of around 70–80 %. Nine-month-old plants are normally taken to the field for plantations.

If the mycorrhizal plant is maintained longer than 12 months in the pot in nursery conditions, the mycorrhization percentage starts to decrease due to limited pot space and the absence of nutrients, which are periodically leached by the irrigation. For this reason, it is necessary to establish a viable fertilisation protocol of mycorrhizal *H. almeriense* plants that permit to maintain alive the symbiosis during potting.

One-year-old *H. almeriense* mycorrhizal plants were selected ($n = 4$ per treatment) to determine the amount of fertiliser necessary to support the desert truffle mycorrhizal plants (Table 21.5). Granular forms of fertiliser were spread over the soil surface around the potted plants. Two different fertilisers were used, a solid controlled release fertiliser and a solid soluble fertiliser. Irrigation was programmed at 16 mm/week and the experiment lasted 3 months, from14/02 to 14/05, when the average temperature was 16.4 °C (Fig. 21.5a). At the end of the experiment, all

Table 21.5 Fertilisation treatments carried out in 1-year-old desert truffle *H. almeriense* mycorrhizal plants in nursery conditions

TREATMENTS

Solid controlled release fertilizer: PLANTACOTE PLUSS 6M NPK-[Mg]

Brand: AGLUKON (Germany)
Composition: 14-9-15-[2]
Description and use: the nutrients are released as needed during the 6 months following deposition depending only on soil moisture and temperature:
15/16 °C -> longevity: 6-7 months
20/21 °C -> longevity: 5-6 months
26/27 °C -> longevity: 4-5 months

Dose:
AGLUKON recommended dose: 2 - 6.5 Kg/m^3.

Test	Fertilizer dose Kg/m^3	Fertilizer for 220 cc pot (g).
Control	0	0
L1	1	0.22
L2	2	0.44
L3	3	0.66
L4	4	0.88
L5	5	1.1
L6	6	1.32
L7	7	1.54

Solid soluble fertilizer: COMBIFERT NPK-[Mg]

Brand: FUENTES FERTILIZERS (Spain)
Composition: 12-12-17-[2]
Description and use: the nutrients are released after deposition and depending only on the irrigation.

Dose:
Fertilizer doses were calculated in terms of electrical conductivity measured from a solution made between irrigation water and solid fertilizer (Figure 21.5 b).

Test	Fertilizer conductivity (dS/m)	Fertilizer for 220 cc pot (g).
Control	1 (tap water)	0
R1	2	0.14
R2	3	0.42
R3	4	0.69
R4	5	0.97
R5	6	1.25
R6	7	1.53

roots from each fertiliser treatment were analysed, and the mycorrhization % was estimated following the methodology established by Phillips and Hayman (1970).

The controlled release fertiliser was applied following the recommended doses (from 2 to 6.5 kg/m^3). A total of seven fertilisation treatments were established (Table 21.5). For the solid soluble fertiliser, whose use depends on plant species, a calibration curve was made using the electrical conductivity (EC) values from 30 fertiliser solutions (Fig. 21.5b). The model curve describes a positive lineal regression ($R^2 = 0.98$) between EC and fertiliser weight. Six EC levels were established to apply different amounts of solid fertiliser.

After 3 months of fertiliser application, the best results were obtained with the controlled release fertiliser (Fig. 21.5c). The fertilisation levels from L1 to L4 (0.22–0.88 g) showed significant positive differences compared with the non-fertilised control samples. Mycorrhization values were three times higher for the L1–L4 interval (60.5–73.5 %) compared with the control (22.75 %). The

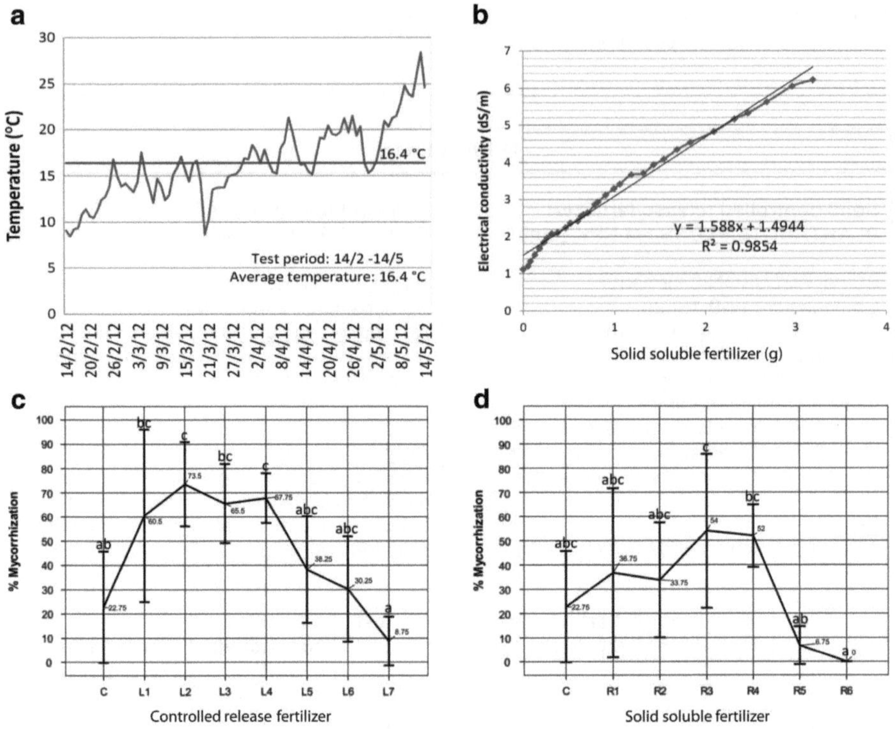

Fig. 21.5 Results of *H. almeriense* mycorrhizal plants under different fertilisation regimes. (**a**) Registry of cultivation temperatures (°C) period that goes from 14/02 to 14/05. (**b**) Electrical conductivity behaviour of increasing solid soluble fertiliser solutions in tap water. The graph shows a linear correlation explained by the equation: $y = 1.588x + 1.4944$ ($R^2 = 0.9854$). (**c**) Results for the controlled release fertiliser effects on *H. almeriense* mycorrhizal plants. Each point is the mean of four measurements and its standard deviation value. The mycorrhization % value belonging to each fertilisation step and that does not share the same letter is significantly different in ANOVA test ($p = 0.1$). (**d**) Results for the solid soluble fertiliser effects on *H. almeriense* mycorrhizal plants. Each point is the mean of four measurements and its standard deviation value. The mycorrhization % value belonging to each fertilisation step and that does not share the same letter is significantly different in ANOVA test ($p = 0.1$)

mycorrhization percentage fell significantly in the L5 to L6 interval (1.1–1.32 g) compared with the L1–L4 interval but was still significantly greater than in the control. The L7 (1.54 g) level shared the same mycorrhization value as the control (Fig. 21.5c).

The results were not better for the solid soluble fertiliser than for the controlled release fertiliser (Fig. 21.5d). Mycorrhization percentages from R1 to R2 (2–3 dS/m) did not show significant differences with regard to the control. The best results were obtained with the R3 and R4 values (4–5 dS/m) although these percentages showed very high standard deviations, especially R3 (±31.75 %). The results were statistically worse than the control for R5 and R6 (6–7 dS/m) treatments; moreover, all the plants died during the first month in R6 treatment.

Most studies on the fertilisation of nursery mycorrhizal plants have been made with ectomycorrhizal plants (Rudawska et al. 1994; Väre 1990; Danielson et al. 1984; Quoreshi 2003; Browning and Whitney 1992; Shaw Iii et al. 1982; Khasa et al. 2001; Gagnon et al. 1987, 1988, 1995; Quoreshi and Timmer 1998, 2000; Chakravarty and Chatarpaul 1990; Ek 1997) and only few studies look at ericoid mycorrhizal plants (Caporn et al. 1995; Johansson 2000). In general, fertiliser application has a positive response in these mycorrhizal associations, but it is necessary to establish the optimal values for each plant symbiosis. Neutral or negative results normally occur for high fertiliser concentrations. For mycorrhizal *H. almeriense* nursery seedlings, it has been possible to adjust the correct concentration using solid controlled release fertiliser. *H. almeriense* mycorrhizal seedlings respond to very low fertiliser concentrations (L1; 0.22 g.) and are able to withstand very high EC values (6 dS/m).

21.3.2 Certification of Desert Truffle Mycorrhizal Plants

Characterisation of the mycorrhiza formed in the *Helianthemum* root systems by the different desert truffle species is extremely important to ensure the high quality of mycorrhizal plants (Morte and Honrubia 2009). For this reason, a morphological and/or molecular analysis of the mycorrhiza should be carried out before planting. Such characterisation is also important to evaluate the permanence of the mycorrhiza under field conditions.

The morphological evaluation process consists of examining the entire root system by binocular microscope, observing the abundance and condition of mycorrhizal root morphotypes (Gutiérrez et al. 2003). The analyst should examine any root tips of doubtful identification by staining the roots (with 5 % blue ink in acetic acid or 0.01 % acid fuchsine solution). *T. claveryi* with *H. almeriense* forms an endomycorrhiza under natural field conditions, an ecto- and ectendomycorrhiza without a sheath in pot cultures, and an ectomycorrhiza with a characteristic sheath and Hartig net in vitro (Gutiérrez et al. 2003; Morte et al. 2008). Therefore, culture conditions can induce changes in mycorrhiza morphology, and there is no clear cut-off between these two main types of mycorrhiza organisation in *Helianthemum* species (Gutiérrez et al. 2003). A recent study has demonstrated that this colonisation varies from ecto- to endomycorrhiza according to the water available in the soil, the scarcer the water, the more intracellular the colonisation (Navarro-Ródenas et al. 2012). Therefore, this symbiosis should be considered as an *ectendomycorrhizal continuum* (Navarro-Ródenas et al. 2012), and water availability modifies the relative amount between intra- and intercellular hyphae in this *continuum*.

Certification of plant lots for colonisation by desert truffle mycorrhizae is a destructive and laborious process, but it is important to sample a minimum number of plants to statistically test the percentage of mycorrhizal plants. We suggest examining 12 plants for each lot of 1,000 plants and consider a mycorrhization percentage exceeding 33 % of the root system as good (Morte et al. 2012). *Terfezia* mycorrhiza has no problems with other contaminant mycorrhizal fungi due to its host specificity.

Moreover, molecular identification of the desert truffle mycorrhiza is very useful for evaluating the permanence of the mycorrhiza in field conditions. Due to the number of ITS sequences from different desert truffles currently available in molecular data bases, it has been possible to design specific primers for this purpose (Kovács et al. 2008).

Acknowledgments This work was supported by project CGL2011-29816 (MINECO-FEDER, Spain). Authors thank the staff of SEAF-SAI from the University of Murcia, with special emphasis to J. A. Sánchez-Saorin for his help in the greenhouse.

References

Afreen F (2005) Physiological and anatomical characteristics of in vitro photoautotrophic plants. In: Kozai T, Afreen F, Zobayed SMA (eds) Photoautotrophic (sugar-free medium) micropropagation as a new micropropagation and transplant production system. Springer, Netherlands, pp 61–90. doi:10.1007/1-4020-3126-2_6

Afreen F, Zobayed SMA, Kubota C, Kozai T, Hasegawa O (1999) Supporting material affects the growth and development of in vitro sweet potato plantlets cultured photoautotrophically. Vitro Cell Dev Biol Plant 35(6):470–474

Alsheikh A (1984) Mycorrhizae of annual *Helianthemum* species formed with desert truffles. In: 6th North American Conference on Mycorrhizae, Bend, Oregon (USA), 25–29 Jun 1984. Oregon State University. Forest Research Laboratory

Ammarellou A, Saremi H (2008) Mycorrhiza between *Kobresia bellardii* (All) Degel and *Terfezia boudieri* Chatin. Turk J Bot 32:17–23

Andrino A, Morte A, Honrubia M (2009) Mejora en la aclimatación de dos especies del género Helianthemum sp. Paper presented at the Octava reunión de la Sociedad Española de Cultivo In Vitro de Tejidos Vegetales, Murcia, España

Andrino A, Morte A, Honrubia M (2012) Method for producing plants of the *Cistaceae* family that establish mycorrhiza with different desert truffle species. Spain Patent ES2386990

Awameh M, Alsheikh A, Al-Ghawas S (1979) Mycorrhizal synthesis between *Helianthemum ledifolium*, *H. salicifolium* and four species of the genera *Terfezia* and *Tirmania* using ascospores and mycelial cultures obtained from ascospore germination. In: Proceedings of the 4th North American Conference on Mycorrhizae, Colorado State University, Fort Collins, Colorado, pp 23–23

Blasco B, Rios JJ, Cervilla LM, Sánchez-Rodríguez E, Rubio-Wilhelmi MM, Rosales MA, Ruiz JM, Romero L (2010) Photorespiration process and nitrogen metabolism in lettuce plants (*Lactuca sativa* L.): induced changes in response to iodine biofortification. J Plant Growth Regul 29(4):477–486

Britto DT, Kronzucker HJ (2002) NH_4 toxicity in higher plants: a critical review. J Plant Physiol 159(6):567–584

Browning MHR, Whitney RD (1992) The influence of phosphorus concentration and frequency of fertilization on ectomycorrhizal development in containerized black spruce and jack pine seedlings. Can J Forest Res 22(9):1263–1270. doi:10.1139/x92-168

Caporn SJM, Song W, Read DJ, Lee JA (1995) The effect of repeated nitrogen fertilization on mycorrhizal infection in heather [*Calluna vulgaris* (L.) Hull]. New Phytologist 129 (4):605–609. doi:10.1111/j.1469-8137.1995.tb03028.x

Cristea V, Dalla Vecchia F, La Rocca N (1999) Developmental and photosynthetic characteristics of a photoautotrophic *Chrysanthemum* culture. Photosynthetica 37(1):53–59. doi:10.1023/a:1007066812957

Cui Y-Y, Hahn E-J, Kozai T, Paek K-Y (2000) Number of air exchanges, sucrose concentration, photosynthetic photon flux, and differences in photoperiod and dark period temperatures affect growth of *Rehmannia glutinosa* plantlets in vitro. Plant Cell Tissue Organ Cult 62(3):219–226. doi:10.1023/a:1006412321864

Chakravarty P, Chatarpaul L (1990) Effect of fertilization on seedling growth, ectomycorrhizal symbiosis, and nutrient uptake in *Larix laricina*. Can J Forest Res 20(2):245–248. doi:10.1139/x90-034

Danielson RM, Griffiths CL, Parkinson D (1984) Effects of fertilization on the growth and mycorrhizal development of container-grown Jack pine seedlings. Forest Sci 30(3):828–835

Dexheimer J, Gerard J, Leduc J-P, Chevalier G (1985) Étude ultrastructurale comparée des associations symbiotiques mycorhiziennes *Helianthemum salicifolium—Terfezia claveryi* et *Helianthemum salicifolium—Terfezia leptoderma*. Can J Bot 63(3):582–591. doi:10.1139/b85-073

Díez J, Manjón JL, Martin F (2002) Molecular phylogeny of the mycorrhizal desert truffles (*Terfezia* and *Tirmania*), host specificity and edaphic tolerance. Mycologia 94(2):247–259

Dominguez-Rojas V, Astasio-Arbiza P, Ortega-Molina P, Gordillo-Florencio E, Garcia-Nunez J, Bascones-Martinez A (1993) Analysis of several risks factors involved in dental caries through multiple logistic regression. Int Dent J 43(2):149

Ek H (1997) The influence of nitrogen fertilization on the carbon economy of *Paxillus involutus* in ectomycorrhizal association with *Betula pendula*. New Phytol 135(1):133–142. doi:10.1046/j.1469-8137.1997.00621.x

Ferrán-Aranaz M (2001) SPSS para Windows. Análisis estadístico. McGraw Hill, España

Fortas Z, Chevalier G (1992) Effects des conditions de culture sur la mycorrhization de l´ *Helianthemum guttatum* par trois espèces de terfez des genres *Terfezia* et *Tirmania* d´Algerie. Can J Bot 70:2453–2460

Gagnon J, Langlois CG, Bouchard D, Tacon FL (1995) Growth and ectomycorrhizal formation of container-grown Douglas-fir seedlings inoculated with *Laccaria bicolor* under four levels of nitrogen fertilization. Can J Forest Res 25(12):1953–1961. doi:10.1139/x95-210

Gagnon J, Langlois CG, Fortin JA (1987) Growth of containerized jack pine seedlings inoculated with different ectomycorrhizal fungi under a controlled fertilization schedule. Can J Forest Res 17(8):840–845. doi:10.1139/x87-133

Gagnon J, Langlois CG, Fortin JA (1988) Growth and ectomycorrhiza formation of containerized black spruce seedlings as affected by nitrogen fertilization, Inoculum type, and symbiont. Can J Forest Res 18(7):922–929. doi:10.1139/x88-140

Gamborg OL, Murashige T, Thorpe TA, Vasil IK (1976) Plant tissue culture media. Vitro Cell Dev Biol Plant 12(7):473–478. doi:10.1007/bf02796489

Genoud C, Sallanon H, Hitmi A, Maziere Y, Coudret A (2001) Growth, stomatal conductance, photosynthetic rate, ribulose-1,5-bisphosphate carboxylase/oxygenase and phosphoenolpyruvate carboxylase activities during rooting and acclimatisation of *Rosa hybrida* plantlets. Photosynthetica 38(4):629–634. doi:10.1023/a:1012425927121

Gutiérrez A (2001) Caracterización, identificación y cultivo en campo de las trufas de desierto. Murcia University, Murcia

Gutiérrez A, Morte A, Honrubia M (2003) Morphological characterization of the mycorrhiza formed by *Helianthemum almeriense* Pau with *Terfezia claveryi* Chatin and *Picoa lefebvrei* (Pat.) Maire. Mycorrhiza 13(6):299–307. doi:10.1007/s00572-003-0236-7

Hawkins TS, Gardiner ES, Comer GS (2009) Modeling the relationship between extractable chlorophyll and SPAD-502 readings for endangered plant species research. J Nat Conserv 17(2):123–127. doi:10.1016/j.jnc.2008.12.007

Hofman P, Haisel D, Komenda J, Vágner M, Tichá I, Schäfer C, Čapková V (2002) Impact of in vitro cultivation conditions on stress responses and on changes in thylakoid membrane proteins and pigments of tobacco during ex vitro acclimation. Biol Plantarum 45(2):189–195. doi:10.1023/a:1015180219628

Honrubia M, Gutiérrez A, Morte A (2001) Desert truffle plantation from south-east Spain. In: Edible mycorrhizal mushrooms and their cultivation: proceedings of the second international conference on edible mycorrhizal mushrooms. Christchurch, New Zealand, pp 3–5

Honrubia M, Morte A, Gutiérrez A (2007) Las Terfezias. Un cultivo para el desarrollo rural en regiones áridas y semi-áridas. In: Reyna Domenech S (ed) Truficultura, fundamentos y técnicas, vol 10. Mundi-Prensa, España, pp 365–397

Jackson M, Abbott A, Belcher A, Hall K, Butler R, Cameron J (1991) Ventilation in plant tissue cultures and effects of poor aeration on ethylene and carbon dioxide accumulation, oxygen depletion and explant development. Ann Bot 67(3):229–237

Jamali S, Banihashemi Z (2012) Host and distribution of desert truffles in Iran, based on morphological and molecular criteria. J Agri Sci Tech 14:1379–1396

Jo MH, Ham I, Lee M, Lee E, Song N, Han G, Woo I (2002) Effects of sealing materials and photosynthetic photon flux of culture vessel on growth and vitrification in carnation plantlets in vitro. J Kor Soc Hort Sci 43(2):133–136

Johansson M (2000) The influence of ammonium nitrate on the root growth and ericoid mycorrhizal colonization of *Calluna vulgaris* (L.) Hull from a Danish heathland. Oecologia 123 (3):418–424. doi:10.1007/s004420051029

Khasa PD, Sigler L, Chakravarty P, Dancik BP, Erickson L, Mc Curdy D (2001) Effect of fertilization on growth and ectomycorrhizal development of container-grown and bare-root nursery conifer seedlings. New Forests 22(3):179–197. doi:10.1023/a:1015674921878

Kirdmanee C, Kitaya Y, Kozai T (1995) Effects of CO_2 enrichment and supporting material in vitro on photoautotrophic growth of *Eucalyptus* plantlets in vitro and ex vitro. Vitro Cell Dev Biol Plant 31(3):144–149. doi:10.1007/bf02632010

Kovács GM, Trappe JM, Alsheikh AM, Bóka K, Elliott TF (2008) *Imaia*, a new truffle genus to accommodate *Terfezia gigantea*. Mycologia 100(6):930–939

Kovács GM, Vágvölgyi C, Oberwinkler F (2003) In Vitro interaction of the truffle *Terfezia terfezioides* with *Robinia pseudoacacia* and *Helianthemum ovatum*. Folia Microbiol 48 (3):369–378. doi:10.1007/bf02931369

Kozai T (1991) Photoautotrophic micropropagation. Vitro Cell Dev Biol Plant 27(2):47–51. doi:10.1007/bf02632127

Kozai T (2005) Introduction. In: Kozai T, Afreen F, Zobayed SMA (eds) Photoautotrophic (sugar-free medium) micropropagation as a new micropropagation and transplant production system. Springer, The Netherlands, pp 1–5. doi:10.1007/1-4020-3126-2_6

Kozai T, Kubota C (2005) In vitro root zone environments and their effects on growth and development of plants. In: Kozai T, Afreen F, Zobayed SMA (eds) Photoautotrophic (sugar-free medium) micropropagation as a new micropropagation and transplant production system. Springer, The Netherlands, pp 53–60. doi:10.1007/1-4020-3126-2_5

Li B, Xin W, Sun S, Shen Q, Xu G (2006) Physiological and molecular responses of nitrogen-starved rice plants to re-supply of different nitrogen sources. Plant Soil 287(1):145–159

Magalhaes JR, Wilcox G (1984) Ammonium toxicity development in tomato plants relative to nitrogen form and light intensity. J Plant Nutr 7(10):1477–1496

Maldonado J, Agüera E, Pérez Vicente R (2008) Asimilación del nitrógeno y del azufre. In: Azcon-Bieto J, Talon M (eds) Fisiología y bioquímica vegetal. McGraw-Hill, Madrid, España, pp 215–236

Mathur A, Mathur AK, Verma P, Yadav S, Gupta ML, Darokar MP (2010) Biological hardening and genetic fidelity testing of micro-cloned progeny of *Chlorophytum borivilianum* Sant. et Fernand. Afr J Biotechnol 7(8):1046–1053

Miyashita Y, Kitaya Y, Kubota C, Kozai T (1996) Photoautotrophic growth of potato plantlets as affected by explant leaf area, fresh weight and stem length. Scientia Horticulturae 65 (2–3):199–202. doi:10.1016/0304-4238(96)00877-1

Morte A, Andrino A, Honrubia M, Navarro-Ródenas A (2012) *Terfezia* cultivation in arid and semiarid soils. In: Zambonelli A, Bonito GM (eds) Edible ectomycorrhizal mushrooms. Soil biology, vol 34. Springer, Heidelberg, Berlin. doi:10.1007/978-3-642-33823-6_14

Morte A, Cano A, Honrubia M, Torres P (1994) In vitro mycorrhization of micropropagated *Helianthemum almeriense* plantlets with *Terfezia claveryi* (desert truffle). Agr Food Sci Finland 3(3):309–314

Morte A, Honrubia M (1992) In vitro propagation of *Helianthemum almeriense* Pau (*Cistaceae*). Agronomie 12(10):807–809

Morte A, Honrubia M (1995) Improvement of mycorrhizal synthesis between micropropagated *Helianthemum almeriense* plantlets with *Terfezia claveryi* (desert truffle). In: Elliot TJ (ed) Science and Cultivation of Edible Fungi, vol 2, Balkema, Rotterdam pp 863–868

Morte A, Honrubia M (1997) Micropropagation of *Helianthemum almeriense*. In: Bajaj YPS (ed) High-tech and micropropagation VI, vol 40. Springer, India, p 163

Morte A, Honrubia M (2009) Biotechnology for the industrial production of ectomycorrhizal inoculum and mycorrhizal plants. In: Chauhan AK, Varma A (eds) A textbook of molecular biotechnology. IK International Pvt Ltd, India, pp 691–704

Morte A, Honrubia M, Gutiérrez A (2008) Biotechnology and cultivation of desert truffles. In: Varma A (ed) Mycorrhiza: state of the art genetics and molecular biology, eco-function, biotechnology, eco-physiology, structure and systematics. Springer, Heidelberg, Berlin, pp 467–483

Morte A, Navarro-Ródenas A, Nicolás E (2010) Physiological parameters of desert truffle mycorrhizal *Helianthemum almeriense* plants cultivated in orchards under water deficit conditions. Symbiosis 52(2–3):133–139. doi:10.1007/s13199-010-0080-4

Morte A, Zamora M, Gutiérrez A, Honrubia M (2009) Desert truffle cultivation in Semiarid Mediterranean areas. In: Azcón-Aguilar C, Barea JM, Gianinazzi S, Gianinazzi-Pearson V (eds) Mycorrhizas—functional processes and ecological impact. Springer, Heidelberg, Berlin, pp 221–233. doi:10.1007/978-3-540-87978-7_15

Murashige T, Skoog F (1962) A revised medium for rapid growth and bio assays with tobacco tissue cultures. Physiol Plantarum 15(3):473–497

Nauš J, Prokopová J, Řebíček J, Špundová M (2010) SPAD chlorophyll meter reading can be pronouncedly affected by chloroplast movement. Photosynth Res 105(3):265–271. doi:10.1007/s11120-010-9587-z

Navarro-Blaya S, Navarro-García G (2003) Química agrícola: El suelo y los elementos químicos esenciales para la vida vegetal. Mundi-Prensa Libros Chapter 6, pp 99–118, ISBN 84-8476-155-X

Navarro-Ródenas A, Lozano-Carrillo MC, Pérez-Gilabert M, Morte A (2011) Effect of water stress on in vitro mycelium cultures of two mycorrhizal desert truffles. Mycorrhiza 21(4): 247–253. doi:10.1007/s00572-010-0329-z

Navarro-Ródenas A, Pérez-Gilabert M, Torrente P, Morte A (2012) The role of phosphorus in the ectendomycorrhiza *continuum* of desert truffle mycorrhizal plants. Mycorrhiza 22(7):565–575. doi:10.1007/s00572-012-0434-2

Nguyen QT, Kozai T (2005) Photoautotrophic micro-propagation of woody species. In: Kozai T, Afreen F, Zobayed SMA (eds) Photoautotrophic (sugar-free medium) micropropagation as a new micropropagation and transplant production system. Springer, The Netherlands, pp 123–146. doi:10.1007/1-4020-3126-2_8

Park SW, Jeon JH, Kim HS, Park YM, Aswath C, Joung H (2004) Effect of sealed and vented gaseous microenvironments on the hyperhydricity of potato shoots in vitro. Scientia Horticulturae 99(2):199–205. doi:10.1016/s0304-4238(03)00097-9

Pérez-García F, González-Benito ME (2006) Seed germination of five *Helianthemum* species: effect of temperature and pre-sowing treatments. J Arid Environ 65(4):688–693. doi:10.1016/j.jaridenv.2005.10.008

Phillips J, Hayman D (1970) Improved procedures for clearing roots and staining parasitic and vesicular-arbuscular mycorrhizal fungi for rapid assessment of infection. Trans Br Mycol Soc 55(1):158–161

Pospóšilová J, Tichá I, Kadleček P, Haisel D, Plzáková Š (1999) Acclimatization of micropropagated plants to ex vitro conditions. Biol Plantarum 42(4):481–497. doi:10.1023/a:1002688208758

Quoreshi AM (2003) Nutritional preconditioning and ectomycorrhizal formation of *Picea mariana* (Mill.) BSP seedlings. Euras J For Res-Hokkaido University 6:1–63

Quoreshi AM, Timmer VR (1998) Exponential fertilization increases nutrient uptake and ectomycorrhizal development of black spruce seedlings. Can J Forest Res 28(5):674–682. doi:10.1139/x98-024

Quoreshi M, Timmer VR (2000) Growth, nutrient dynamics, and ectomycorrhizal development of container-grown *Picea mariana* seedlings in response to exponential nutrient loading. Can J Forest Res 30(2):191–201. doi:10.1139/x99-208

Righetti B (1996) Chlorophyll, ethylene and biomass determination in *Prunus avium* cv. Victoria cultivated in vitro under different atmospheric conditions. J Hort Sci 71(2):249–255

Rivas J (2008) La luz y el aparato fotosintético. In: Azcon-Bieto J, Talon M (eds) Fisiología y bioquímica vegetal. McGraw Hill, Madrid, España, pp 165–190

Roth-Bejerano N, Livne D, Kagan-Zur V (1990) *Helianthemum-Terfezia* relations in different growth media. New Phytol 114(2):235–238. doi:10.1111/j.1469-8137.1990.tb00395.x

Rudawska M, Kieliszewska-Rokicka B, Debaud J-C, Lewandowski A, Gay G (1994) Enzymes of ammonium metabolism in ectendomycorrhizal and ectomycorrhizal symbionts of pine. Physiol Plantarum 92(2):279–285. doi:10.1111/j.1399-3054.1994.tb05338.x

Shaw Iii CG, Molina R, Walden J (1982) Development of ectomycorrhizae following inoculation of containerized *Sitka* and white spruce seedlings. Can J Forest Res 12(2):191–195. doi:10.1139/x82-029

Slama A, Fortas Z, Boudabous A, Neffati M (2010) Cultivation of an edible desert truffle (*Terfezia boudieri* Chatin). Afr J Microbiol Res 4(22):2350–2356

Torrente P, Navarro A, Gutiérrez A, Morte A (2009) Micropropagación de *Helianthemum hirtum* y micorrización in vitro con micelio de *Terfezia claveryi*. Paper presented at the Octava reunión de la Sociedad Española de Cultivo In Vitro de Tejidos Vegetales, Murcia, España

Uddling J, Gelang-Alfredsson J, Piikki K, Pleijel H (2007) Evaluating the relationship between leaf chlorophyll concentration and SPAD-502 chlorophyll meter readings. Photosynth Res 91 (1):37–46. doi:10.1007/s11120-006-9077-5

Urrestarazu-Gavilán M (2004) Tratado de cultivo sin suelo, 3rd edn. Mundi-Prensa, Madrid

Väre H (1990) Effects of soil fertility on root colonization and plant growth of *Pinus sylvestris* nursery seedlings inoculated with different ectomycorrhizal fungi. Scand J Forest Res 5 (1–4):493–499. doi:10.1080/02827589009382632

Xiao Y, Niu G, Kozai T (2011) Development and application of photoautotrophic micropropagation plant system. Plant Cell Tiss Organ Cult 105(2):149–158. doi:10.1007/s11240-010-9863-9

Xu G, Fan X, Miller AJ (2012) Plant nitrogen assimilation and use efficiency. Ann Rev Plant Biol 63:153–182

Zambonelli A, Bonito GM (2012) Edible ectomycorrhizal mushrooms. Soil biology, vol 34. Springer, Heidelberg, Berlin. doi:10.1007/978-3-642-33823-6_14

Zaretsky M, Kagan-Zur V, Mills D, Roth-Bejerano N (2006) Analysis of mycorrhizal associations formed by *Cistus incanus* transformed root clones with *Terfezia boudieri* isolates. Plant Cell Rep 25(1):62–70. doi:10.1007/s00299-005-0035-z

Zhu Z, Gerendás J, Bendixen R, Schinner K, Tabrizi H, Sattelmacher B, Hansen UP (2000) Different tolerance to light stress in NO_3^- and NH_4^+ grown *Phaseolus vulgaris* L. Plant Biol 2 (5):558–570. doi:10.1055/s-2000-7498

Zobayed SMA, Afreen F, Kubota C, Kozai T (2000) Water control and survival of *Ipomoea batatas* grown photoautotrophically under forced ventilation and photomixotrophically under natural ventilation. Ann Bot 86(3):603–610. doi:10.1006/anbo.2000.1225

Zobayed SMA, Afreen F, Xiao Y, Kozai T (2004) Recent advancement in research on photoautotrophic micropropagation using large culture vessels with forced ventilation. Vitro Cell Dev Biol Plant 40(5):450–458. doi:10.1079/ivp2004558

Chapter 22
Preparation and Maintenance of Both Man-Planted and Wild Plots

Mario Honrubia, Alberto Andrino, and Asunción Morte

22.1 Introduction

The first *Terfezia claveryi* orchard was established in 1999 in the province of Murcia (Spain), and since then more than 20 plantations have been established in mainland Spain and the Canary Islands (Morte et al. 2008, 2009, 2012). More recently, experimental results have been obtained in Tunisia with *Terfezia boudieri*, using *Helianthemum sessiliflorum* as a host plant (Slama et al. 2010), in Israel with the same symbionts (Kagan-Zur, pers. com.) and in Argentina with *T. claveryi* (Cortes 2012, pers. com.).

Desert truffle fructification should occur one to three years after plantation, depending on mycorrhized seedling quality, site suitability, plantation frame and management practices, which are a critical factor for the regularity of future truffle production. In the first plantation in Murcia, over the last twelve years, carpophores have fructified yearly and production has increased as a result of the land management techniques used and irrigation (Morte et al. 2008, 2009, 2012). The application of suitable plantation management techniques is necessary to maintain desert truffle production, because, without them, plantations would lose their productivity after two or three years (Morte et al. 2008). However, even so, desert truffle production fluctuates from one year to another in the same orchard. Among the factors which most influence the start of desert truffle production are as follows: soil characteristics, planting season, the frame of the plantation, water availability (irrigation) and weed management. Irrigation control and cutting a quantity of plants to open up the orchard ecosystem are also very important for maintaining high productivity. Soil characteristics (acid or basic) determine the group of desert truffle species and corresponding host plans that can be selected to establish the

M. Honrubia (✉) • A. Andrino • A. Morte
Dpto. Biología Vegetal (Botánica), Facultad de Biología, Universidad de Murcia, Campus de Excelencia Internacional Regional "Campus Mare Nostrum," Campus de Espinardo, 30100 Murcia, Spain
e-mail: honrubia@um.es

V. Kagan-Zur et al. (eds.), *Desert Truffles*, Soil Biology 38,
DOI 10.1007/978-3-642-40096-4_22,

orchards. For example, *Terfezia arenaria* and *T. leptoderma* grow in acid soils while *T. claveryi, T. boudieri* and *Tirmania nivea* grow in basic soils (Honrubia et al. 1992).

There is a growing interest in establishing desert truffle plantations, especially in countries of the Mediterranean Basin, Middle East, Iran, the Arabian Peninsula, Persian Gulf, Southern Africa and South American countries such as Chile and Argentina, where desert ecosystems cover large areas, and the cultivation of desert truffles could increase the development and wealth of rural areas.

The present chapter describes our experience in improving desert truffle production in basic/calcareous soils.

22.2 Establishment and Management of Desert Truffle Plots

The first step in the establishment of a desert truffle plot is to choose suitable host plants and fungal species that are well adapted to the environmental conditions and soil characteristics. Moreover, high-quality mycorrhizal plants, with certified mycorrhization levels, should be selected (see Morte and Adrino, Chapter 21 this volume).

The second step is to take into account the characteristics described below (Sects. 22.2.1–22.2.4).

22.2.1 Soil Characteristics and Soil Preparation

The characteristics of the soils in which most of the plantations in Spain have been established are shown in Chapter 4 (Bonifacio and Morte, this volume). In general, the soils used for plantations are poor and are characterised by low values of electrical conductivity, organic carbon and C/N ratio. These characteristics, especially pH and texture, change from site to site, and depending on these changes, a particular *Terfezia* species should be used—as mentioned above, some grow in acid soils and others in basic soils. Soil texture can influence ascomata shape (Malençon 1973) but is not a determinant factor for truffle formation, since *T. claveryi* ascomata have been produced in loamy clay soils (Murcia) as well as in sandy soils (Lanzarote, Canary Islands) (Honrubia et al. 2007; Morte et al. 2012). Similar results have been obtained for *T. boudieri* ascocarps in plantations in loamy sand and gypsum soils (Slama et al. 2010).

Soil fertilisation in Murcia has not been necessary to date. Fertilisation has never been applied even to the oldest plantation (12 years old), and mycorrhizal *Helianthemum* plants are still producing *T. claveryi* truffles yearly (Morte et al. 2012).

The cultivation of *Terfezia* spp. and *Tirmania* spp. is compatible with woody crops like almond, cherry tree and olive trees, among others, to optimise land use and irrigation facilities if installed. A double crop is possible in the same plot (one tree-borne the other hypogeous—truffles). This is possible because there is no competition for host plants between the arbuscular mycorrhizal fungi from almond and olive trees, etc., and the different desert truffle species, which are mostly associated in symbiosis with some species of the genera *Helianthemum* in the Mediterranean areas.

Before the mycorrhized seedlings are planted in field plots, the plot should be left fallow and totally tilled for at least one year. This is to prevent soil structure and microbe destruction in depth and so as not to stimulate weed seed germination. Small holes of about 15–20 cm in depth and diameter are sufficient for planting the seedlings. The use of herbicides is unadvisable at any time (pre- or post-plantation). The elimination of weeds should be carried out mechanically by trimmer.

22.2.2 Season and Frame of Plantation

In our latitudes, spring (from April to May) is the best time for planting due to moderate temperatures, the abundance of precipitation and the long photoperiod (Morte et al. 2008). Moreover, planting in spring is essential if the first ascocarps are to be obtained the following spring (11–12 months after plantation). Autumn is also a suitable season for planting, although, when *Terfezia* orchards have been established in September or October, desert truffle production did not start in the following spring but the subsequent one (19–20 months after plantation).

Different frames of plantation were tested: (0.5 × 0.5, 1 × 1, 3 × 3, 3 × 2 and 4 × 2 m), distributed alternately in rows, ridges and groups (9–12 plants). According to our experience, although a wide frame (4 × 2 m, in rows) facilitates mechanical soil tilling and plant physiological measurements, a narrow frame (0.5 × 0.5 m, in groups) is advisable to obtain desert truffles the first year after plantation (Morte et al. 2012). This is because the introduced mycorrhized root systems will come into contact immediately, and hyphal anastomosis will take place and more mycelium will be produced because of a synergistic effect. An unknown amount of mycelial biomass should be sufficient to start the production of small "truffettes". The exploration type of the studied desert truffles, *Terfezia claveryi* and *Picoa lefebvrei*, is between "contact exploration type" and "short-distance exploration type" according to Agerer (2001). We observed some emanating hyphae, but rhizomorphs have never been observed. This is why we consider it more worthwhile to establish narrow frames of plantation, for example, in groups separated by broad rows and ridges.

Distribution in ridges much facilitates mycorrhiza sampling and truffle collection. This is also true for wild animals so that a fence is necessary to keep out undesirable desert truffle hunters.

Fig. 22.1 (**a**) Aspect of a plantation of *H. almeriense* vs *Terfezia claveryi*, flowering 16 months after plantation in Caniles (Granada, Spain). The frame of plantation was 1 × 1 m and every two rows separate 2 m. (**b**) The first truffles were produced in April 2013

Figure 22.1 shows an orchard of *H. almeriense* with *Terfezia claveryi*, established in autumn of 2010, at Caniles (Granada, Spain). The plantation frame was 1 × 1 m in rows separated by 2 m. The first produced ascocarps were collected in April 2013 in this orchard. Other examples of productive plantation frames are shown in Fig. 22.2.

22.2.3 Irrigation

The importance of rainfall is well known for fungal fructification in general. So, irrigation is one of the most important factors for maintaining cultivation since desert truffle fruiting depends directly on rainfall (Morte et al. 2008). Therefore, to guarantee truffle production, even in dry years, an irrigation system is needed. Watering by (micro) sprinklers or by trickle irrigation are two common and efficient systems that can be installed in orchards, as shown in Fig. 22.3.

Estimated desert truffle production in wild areas varies between 50 and 170 kg ha^{-1} in the province of Murcia after rainfall of between 350 and 400 mm during the year (Honrubia et al. 2001, 2007). An irrigation system is not necessary when rainfall is guaranteed because the mycorrhizal association is well adapted to arid and semiarid climates (Morte et al. 2000). However, natural desert truffle production dramatically decreases or even disappears when rainfall is less than 150 mm, and so irrigation is necessary in dry years.

After following *T. claveryi* production for 10 years in an orchard established in 1999, we observed a statistical correlation between the amount of precipitation during autumn (September, October and November) and *T. claveryi* truffle production the following year (Morte et al. 2012). This finding will help maintain desert truffle production after dry years by enabling us to adjust soil water potential to the plant physiological parameters necessary to keep the mycorrhizal symbiosis productive.

The main question is what level of water stress can mycorrhizal plants sustain without affecting the production of desert truffles.

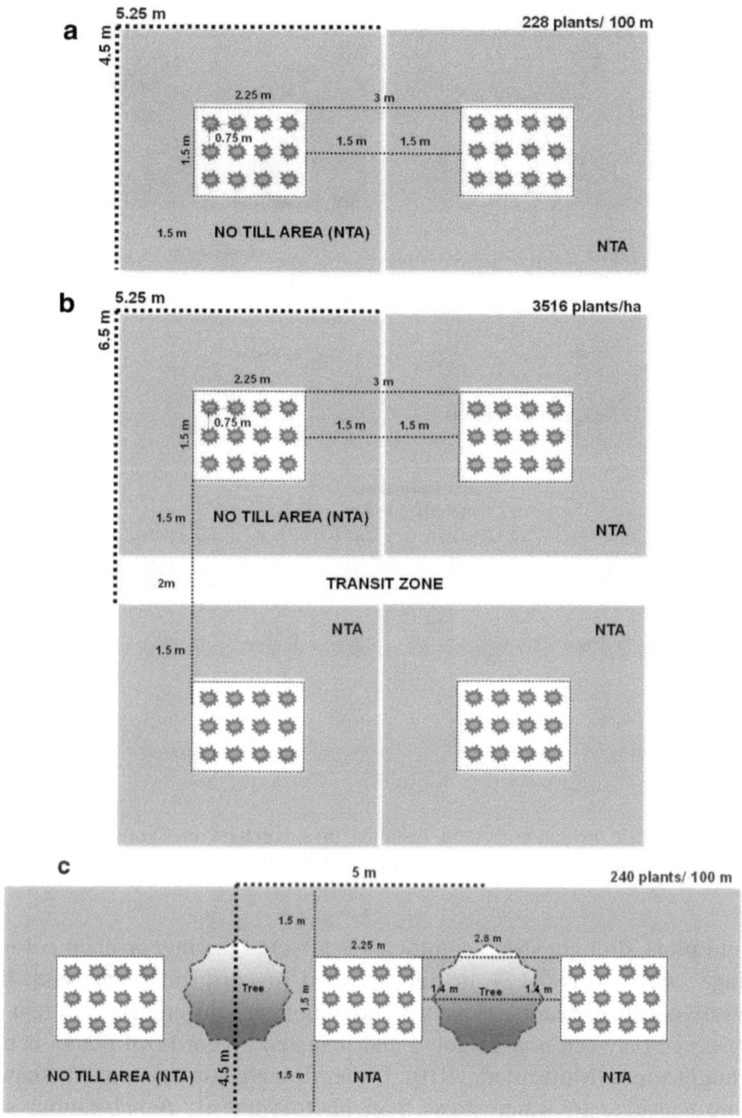

Fig. 22.2 Different plantation plots for desert truffle cultivation. The recommended plots contain 9–12 plants. (**a**) One-line distribution. (**b**) Two-dimension plant distribution. (**c**) Distribution between trees, for example, olive or almond tree

It is known that desert truffle mycelia are able to resist moderate water stress in in vitro conditions (Navarro-Ródenas et al. 2011). The effect of drought stress on the growth and water relations of the mycorrhizal association of *H. almeriense* with *T. claveryi* has been studied under both nursery (Morte et al. 2000) and field (Morte et al. 2010) conditions.

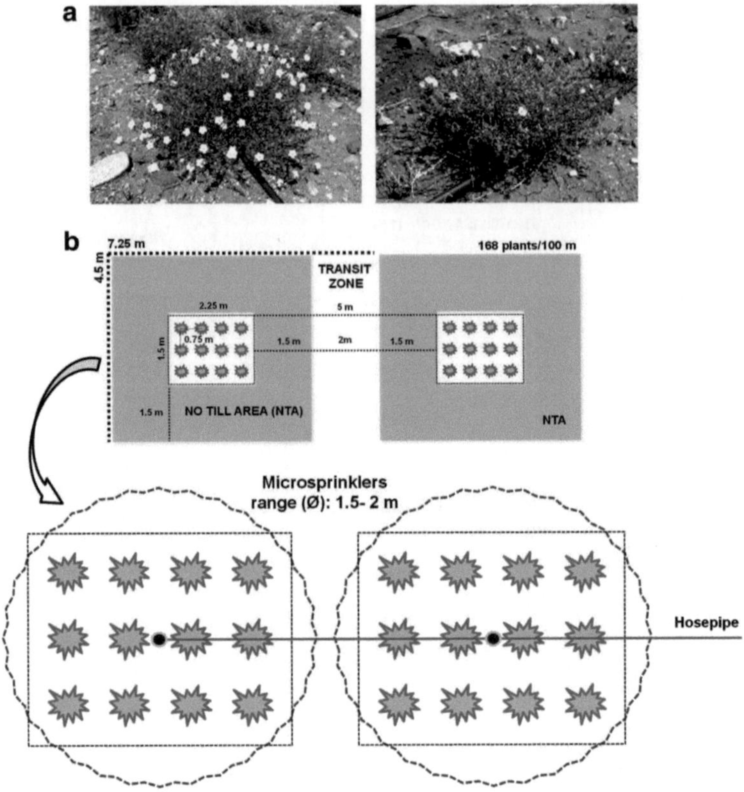

Fig. 22.3 (**a**) Trickle irrigation system installed in an orchard in Caniles (Granada, Spain). (**b**) Schematic location of microsprinklers along the desert truffle plantation

In field plots, drought stress significantly affected the mycorrhizal colonisation percentage, which was 70 % in non-irrigated mycorrhizal plants and 48 % in irrigated mycorrhizal plants. However, no significant differences in plant growth were observed between non-irrigated and irrigated mycorrhizal plants before and after drought stress (Morte et al. 2010). Under drought stress, stomatal conductance was more sensitive to water stress than photosynthesis. A substantial stomatal closure took place under water-deficit and low-radiation conditions, which improved water-use efficiency in the plants grown under drought conditions (Morte et al. 2010).

H. almeriense, like many other Mediterranean perennial plants, has a conservative water-use strategy, based mainly on the avoidance of drought: the lower the soil water potential, the drier the atmospheric conditions transpiration rate and stomatal conductance are reduced (Morte et al. 2010). The results show that mycorrhizal *H. almeriense* plants are able to maintain good physiological parameters at low soil matric potentials (around −120 KPa) in field conditions, making them a suitable alternative agricultural crop in arid and semiarid areas.

In Murcia (Spain), 2 years after plantation, only the *H. almeriense* plants subjected to drought produced *T. claveryi* truffles (three ascocarps in May 2011). Plants irrigated to field water capacity (−30 KPa) did not produce any ascocarps. This means that an excess of irrigation water inhibits truffle production in the field and a certain level of water deficit may actually stimulate the development of *T. claveryi* mycorrhizae, while elimination of the water deficit reduces that stimulus.

22.2.4 Weed Elimination

Competition from weeds has been shown to reduce desert truffle production (Morte et al. 2009). To reduce the impact of herbaceous competition, weeding is necessary, at least once a year, in order to avoid plant competition for water and to maintain the open and sunny desert truffle ecosystem.

Two different weed control methods have been used in Murcia: mechanical tilling between rows with cultivator tines set at 5–8 cm depth and the application of a commercial systemic glyphosate-based herbicide at half the recommended dose. In field conditions, the application of glyphosate has not been shown to cause any inhibition of short root formation or mycorrhizal colonisation (Chakravarty and Chatarpaul 1990). However, mechanical soil tilling is preferred to herbicide application to keep *Terfezia* cultivation as an organic crop (Morte et al. 2012).

Due to the difficulty of controlling weeds in the La Garrobera plantation (Murcia, Spain), which was established in April 2008 with 3,000 *Helianthemum* mycorrhizal plants distributed in 3 ha, the first *T. claveryi* truffles were obtained the third year after plantation (May 2011) rather than in the first or second year. A combination of the two different weed control methods in early winter and late summer (never during or close to the fruiting season) allowed such weed control (Morte et al. 2012). However, it is advisable to study the long-term effects of weed control methods for desert truffle cultivation under specific field conditions.

22.2.5 Collection of Desert Truffles

The collection of desert truffles is a manual task for well-trained people who are able to recognise the crack formed by the fungus in the soil near the host plant. Dogs or any olfactory animals are not necessary if the desert truffle hunter knows how to recognise the host plant and the crack in the soil (Morte et al. 2012). Figure 22.4 shows a crack (left) and collected truffle (right) in field.

Fig. 22.4 *Terfezia claveryi* fructification under *H. almeriense* host plant

22.3 Desert Truffle Production in the Wild

To understand the possibility of producing truffles in a given geographic area, we must first know the potential zones where desert truffles can grow. For that, a technological tool, based on a geographic information system (GIS), was developed, as described below.

Knowing the chorology and potential productive areas for desert truffles, it is possible to establish a new plantation with an almost certain guarantee of success and to manage these new orchards, in a process known as "**turmiculture**" (from "*turmas*", Spanish name for desert truffles). Alternatively, suitable ecosystems in open forests can be managed in order to maintain the productivity of these areas (**desert truffle silviculture**).

22.3.1 *Chorology and Desert Truffle Potential Zones in Murcia Province: A GIS Approach*

A geographic information system (GIS) is a technological tool for understanding geography and making intelligent decisions. GIS organises geographical data in such a way that the data necessary for a specific project or can be selected. We have used the methodology to establish a model of potential desert truffle areas in the province of Murcia. These potential areas were delimited in accordance with the habitat selection descriptions made in the Interpretation Manual of natural and semi-natural habitats for the province of Murcia (2008) and the cartography of the habitats (2007), both documents result from the Habitats Directive 92/43/EEC.

The European Habitats Directive 92/43/EEC is a community legislative instrument in the field of nature conservation that establishes a common framework for the conservation of wild animal and plant species and natural habitats of community importance; it provides for the creation of a network of special areas of conservation, called Natura 2000. The Habitats Directive divides the EU into 9 biogeographical regions—the Atlantic, Continental, Alpine (which includes the Pyrenees, the Alps, the Carpathian mountains and parts of Scandinavia), Mediterranean, Boreal (Finland, Sweden, Estonia, Latvia and part of Lithuania), the Macaronesian (Madeira, Azores and Canary Islands), the Pannonian (essentially

Table 22.1 List of the different layers that have been used for the thematic cartography elaboration

Cartography layer	Source	Scale	Format
Province of Murcia digital terrain model	National Geographic Institute: www.ign.es	1:200,000	Raster
Spain forest map (Region of Murcia province)	MAGRAMA: http://www.magrama.gob.es	1:50.000	Vectorial
Region of Murcia grass map	IMIDA: http://www.imida.es	1:200,000	Vectorial
Soil map of Region of Murcia	IMIDA: http://www.imida.es	1:100,000	Vectorial
Habitats from the Region of Murcia	Dirección General del Medio Natural www.murcianatural.carm.es/	1:50,000	Vectorial

Hungary and parts of the Czech Republic, Romania and Slovakia) and the Steppic and the Black Sea region (parts of Bulgaria and Romania). Natura 2000 sites are selected according to each biogeographical region. Working on this level makes it easier to conserve species and habitat types under similar natural conditions across country boundaries.

The distribution of the different types of habitats and plant associations in the province of Murcia is far from random. Murcia presents a Mediterranean climate that is characterised by a minimum rainfall period during the summer accompanied by the highest temperatures. The rains are distributed throughout the rest of the year, often with two peaks, one in autumn or winter and another in spring. These climatic characteristics of the province of Murcia are determined by its geographical situation and by its orography. These factors determine the existence of different bioclimatic zones and the presence of different species in the same type of habitat with different plant associations. The climatic situation is determined by the proximity to the Azores anticyclone which is responsible for the summer drought, the low cloud cover and, therefore, the high level of radiation. It also affects the alternation of maritime subtropical and continental Saharan air flows, while the relative proximity to the Mediterranean is a determinant of the most important territorial rains. On the other hand, the orography modifies the climate due to the Föehn effect (Alcaraz Ariza et al. 2008).

In a first stage, it was necessary to establish the aspects that will define the potential presence of desert truffles. This selection was based on the main characteristics that define desert truffle mycorrhizal establishment: selection of the host plant habitats, the delimitation of soil characteristics and characterisation of the bioclimatic zones. To classify all these features, the different layers, detailed in Table 22.1, were used.

Using the habitat cartography as base map, six mycorrhizal *Helianthemum* species that are naturally present in the province of Murcia were selected (Table 22.1). Each species may be present in one or more type of habitat, and each specific habitat is described by a numerical code (e.g. 5330; thermo-Mediterranean and pre-steppe scrub). Once the habitats of the six *Helianthemum* species had been selected, these selections were combined with the soil map restrictions. All the calcareous soils that are contained in calcic cambisols petrocalcic cambisols, calcaric regosols and calcic and gypsic xerosols in the province of Murcia were selected.

The third restriction applied to the habitat area was made with the forest cartography from the province of Murcia. All the potential desert truffle areas are in places lacking dense woodland, which is why the selected zones are scrublands, hills with sparse trees, hills with scattered trees and treeless hills.

To characterise the average annual precipitation for each *Helianthemum* species, 90 weather stations from Murcia and the surrounding provinces (Albacete, Almería, Granada, Jaén and Alicante) were used. Twenty-year historic precipitation data were used to elaborate the precipitation layer, using the ordinary co-kriging technique. Co-kriging is a geostatistical tool that is applied to estimate the precipitation of an area, using the sampled meteorological data. The technique defines a spatial continuity of the variable and can be extrapolated to estimate the precipitation value from non-sampled places.

Finally, the bioclimatic zones were extracted in an isolated vector layer from the province of Murcia grass map. The bioclimatic zones are the result of the thermicity index which is the sum of annual mean temperature (T), mean daily maximum temperature of the coldest month (M) and mean daily minimum temperature of the coldest month (m) multiplied by 10; f., $I_t = 10(T + M + m)$.

The second stage consisted of developing of thematic layers and their integration into a GIS software. The mapping was carried out through the overlapping of vector and raster layer types that codified the selected aspects in the first stage. The result was a group of thematic maps: a bioclimatic zone map (Fig 22.5a) and a precipitation map (Fig. 22.5b); six maps for each *Helianthemum* species grouped and presented in Figs. 22.6 and 22.7, a combined map for all the *Helianthemum* species (Fig. 22.8).

This thematic cartography aims to provide a prediction tool for the presence of desert truffle fungi. For the purpose of future mycosilvicolous management of the lands, knowledge of fungal diversity and functionality of the species in their respective ecosystems (Table 22.2). Some of the future uses may include:

1. Construction of data bases of the real presence contrasted with potential maps
2. Mapping, classification and counting fungal resources in each locality
3. Edition of local distribution maps
4. Calculation of new maps of density and diversity
5. Search for areas to protect and manage
6. Implementation of projects by enterprises and institutions: creating tourist routes and interactive presentation programmes (in museums, interpretation centres, natural parks)
7. Application in environmental impact studies: risk maps (flooding, erosion, mapping of the pollution and its effects on fauna and flora), managing the impact of new roads and search for the least traumatic solutions, impact of human activities on plant and animal populations and the diversity of a given areas

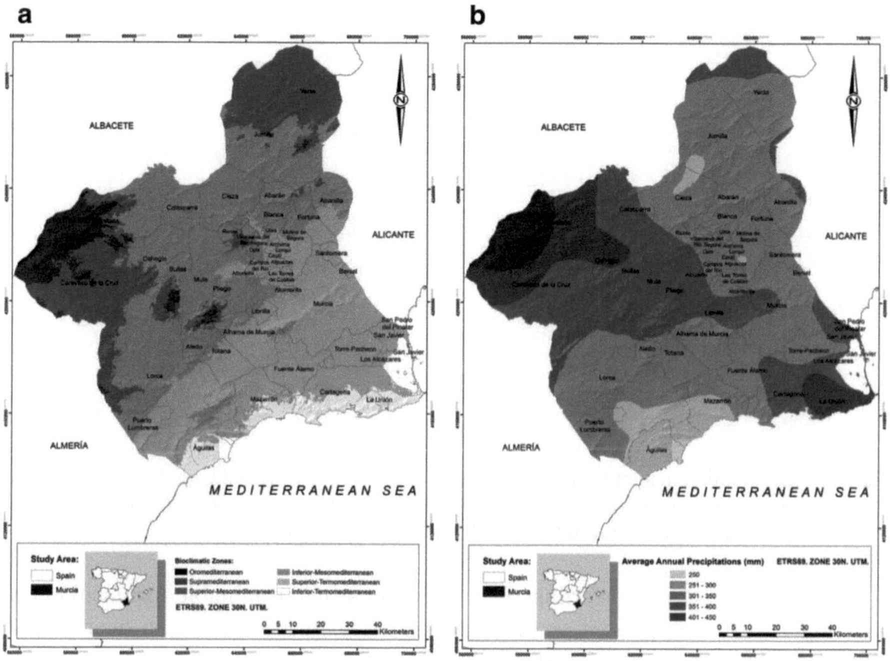

Fig. 22.5 Climate attributes of Murcia. (**a**) Distribution map of the principal bioclimatic zones from Murcia. (**b**) Distribution map of the average annual precipitations from Murcia. The precipitations are expressed in 50 mm/rainfall intervals

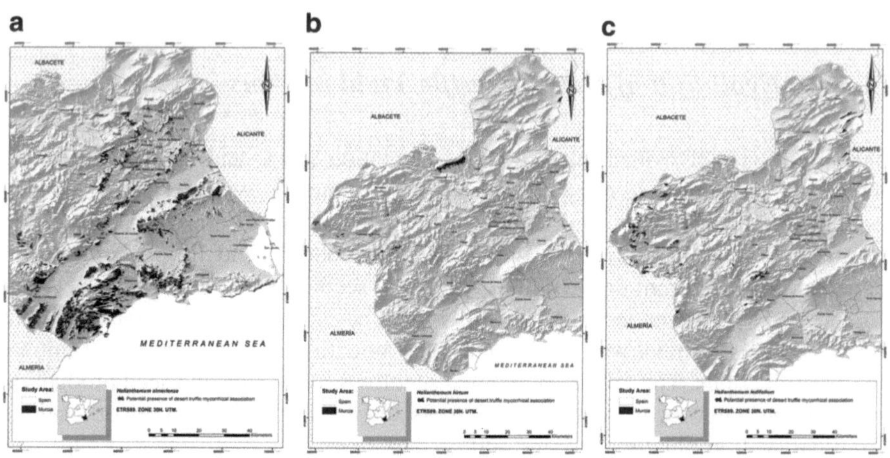

Fig. 22.6 Distribution map of the desert truffle potential areas associated with the presence of the following: (**a**) *Helianthemum almeriense*, (**b**) *Helianthemum hirtum*, (**c**) *Helianthemum ledifolium*. These maps are the result of a GIS multivariate analysis

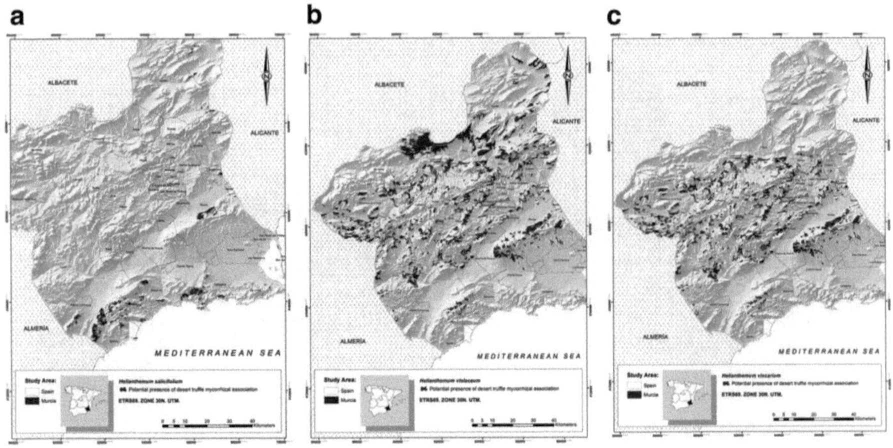

Fig. 22.7 Distribution map of the desert truffle potential areas associated with the presence of the following: (**a**) *Helianthemum salicifolium,* (**b**) *Helianthemum violaceum,* (**c**) *Helianthemum viscarium.* These maps are the result of a GIS multivariate analysis

This mapping offers a starting point for addressing sustainable management of the fungal resources present in the areas of study. It also offers an overview of species richness and their presence in the Region of Murcia.

The principal aim is to enrich documentation concerning the digital cartographic mycological potential value and provide additional information to facilitate the task of decision-making at territorial planning level.

22.3.2 Evolution of Desert Truffle Yield in Plots

Depending on mycorrhizal plant quality, plantation frame and management after plantation, seedlings will grow and produce seeds once or twice per year. The new seedlings will be rapidly and directly mycorrhized in the soil in a population dynamic process. Root systems will come into contact, and the mycelium will come from the different original plants and new seedlings. A complicated network formed by mycorrhized roots and mycelium will be established until a basic amount of mycelium is formed, which will start the production of sexual cells and initiate first fructifications. The growth of planted seedlings, flowering and seed production will make up a new ecosystem in which roots and mycelium form a network and the aerial part will gradually cover the soil surface.

It is only when the soil characteristics are established and reach equilibrium and when the new seedlings arising from the flowering of the introduced host plants reach a soil cover of over 60 % that the plantation will start to produce truffles. Truffle production should start between 12 and 24 months after plantation depending on water availability, irrigation, etc. The soil should not be tilled for

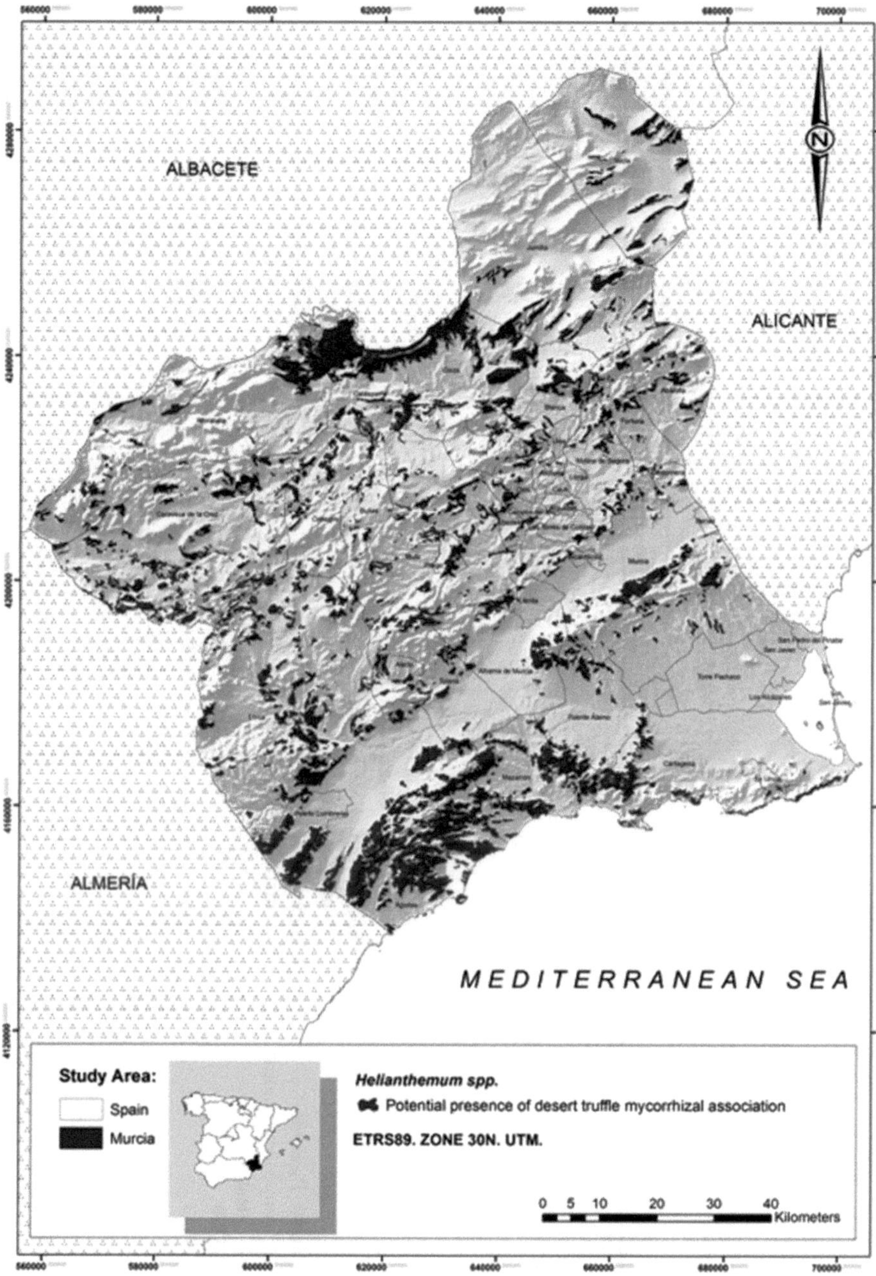

Fig. 22.8 Distribution map of the desert truffle potential areas associated with *Helianthemum* sp. presence. This map is the result of a GIS multivariate analysis

Table 22.2 List of *Helianthemum* host species for desert truffles. The table presents a data summary from the GIS thematic cartography

Host	European habitat code	Bioclimatic zones	Average annual precipitation (mm)	Area (ha)	% Total *Helianthemum* spp. area (122,694 ha)
H. almeriense	**5330:** Thermo-Mediterranean and pre-steppe scrub	Inferior-meso-Mediterranean to inferior-thermo-Mediterranean	250–400	61,827	50.4
H. hirtum	**1430:** Halo-nitrophilous shrubs **2260:** Dune sclerophyllous scrubs	Supra-Mediterranean to inferior-meso-Mediterranean	300–450	3,104	2.5
H. ledifolium	**6220:** Pseudo-steppe with grasses and annuals	Oro-Mediterranean to superior-thermo-Mediterranean	300–450	6,119	5.0
H. salicifolium	**6220**	Superior-meso-Mediterranean to inferior-thermo-Mediterranean	250–400	9,087	7.4
H. violaceum	**4090:** Endemic oro-Mediterranean heaths **5330**	Supra-Mediterranean to superior-thermo-Mediterranean	250–450	75,372	61.4
H. viscarium	**5330**	Supra-Mediterranean to superior-thermo-Mediterranean	250–450	46,048	37.5

2–3 years after plantation, to prevent the wild seed bank of the soil plantation becoming active when the soil is moved.

Truffle production increases during the first four to five years until the soil shrub cover reaches 80 % or more. At this moment tilling of at least 50 % of the cover should be made to open up the ecosystem and recover orchard productivity. After the first time (2–3 years after plantation), tilling must be done every 5–6 years to optimise desert truffle production, depending on how the cover develops.

After following *T. claveryi* production for 10 years in an orchard established in 1999, where the appropriate management was done, statistical correlation was observed between the amount of precipitations during autumn (September, October and November) of a year and the *T. claveryi* truffle production the following year (Morte et al. 2012). This new finding will help maintain desert truffle production after dry years, by enabling us to adjust soil water potential to the plant physiological parameters necessary to keep the mycorrhizal symbiosis productive. The average desert truffle production in this orchard was about 400 kg/Ha, but this production fluctuated from one year to another, the minimum being 2 kg/Ha, the maximum 1,050 kg/Ha along these 10 years (Morte et al. 2012).

22.3.3 *Desert Truffle Silviculture (Turmiculture) to Improve Wild Production: Proposal for Sustainable Management*

Silviculture is the science and associated technology that allows the management and use/exploitation of a forest, hill or brushwood (in the widest sense of a Mediterranean forest environment), to provide a stable and continuous supply of forest, while desert truffle silviculture can be considered as the science and associated technology that allows the management and use/exploitation of a land to provide a stable and continuous supply of wild desert truffles. Turmiculture refers to the cultivation of desert truffles in orchards or plantations.

22.3.3.1 Diagnosis of the Ecosystem

Most species of the genus *Helianthemum* (Cistaceae), perennial or annual, are the natural host plants of desert truffles around the world (see Morte and Andrino in this volume).

Pastures of annuals *Helianthemum* spp. need different silvicultural treatments from brushwood of perennial species for sustainable turmiculture.

In Murcia (Spain), we mainly work with *Helianthemum* spp. shrubs growing in calcareous soils, where *H. almeriense*, *H. viscarium* and *H. violaceum* have the widest distribution in our province.

The strength of the symbiotic mycorrhizal association between host plants and desert truffles permits silvicultural procedures to be practiced in stands inside the forest or scrubland to help the development of plant populations and so improve the growth and fructification of desert truffles.

These species of *Helianthemum* are heliophilous plants that form dense populations, similar to caespitous populations in more or less wide stands, where desert truffles fructify abundantly. These are open and sunny spaces, where water stands in small streambeds. In these conditions, wild desert truffle productivity is usually more productive than in other less dense, more open *Helianthemum* ecosystems similar to those formed by *Thymus* spp. populations.

The root systems of *Helianthemum* shrubs are very branched. The thin fraction of the root systems is usually very abundant, which is why mycorrhization is high and, in the field, endomycorrhiza type dominates in the context of an *ectendomycorrhizal continuum* (Navarro-Ródenas et al. 2012). The rhizospheric surface is usually bigger than the cover area of the individual *Helianthemum* shrubs; the root system and mycelium of desert truffles also grow horizontally (not only in depth) as they look for the cover of other plants of the ecosystem, such as grass of *Stipa tenacissima, Thymus, Sideritis, Genista, Artemisia* and others. These conditions are more favourable as regards water availability, because the cover provides protection against evapotranspiration and the mycelia of desert truffles could fructify. All these companion plants of *Helianthemum* in the ecosystems establish arbuscular mycorrhiza with *Glomeromycota* fungi and have no negative interactions with desert truffle symbiosis since there is no direct competition for the host plants.

The *Helianthemum* spp. annuals form part of the meadow populations where plantaginaceae, grasses, small legumes, etc., are present but do not affect the mycorrhizal symbiosis between desert truffle mycelia and the *Helianthemum* spp.

22.3.3.2 Management of the Desert Truffle Ecosystem for a Sustainable Use

Any preparation of the ground should not be too traumatic and, however it is carried out, be separated in time, since we are dealing with scrublands or meadows of perennial or annual *Helianthemum* spp., respectively—open spaces with no tree cover whose preservation to avoid soil erosion is closely linked to the presence and density of the different plant species that make up part of the ecosystem in question. Any uncontrolled intervention may damage not only the resource per se but also the ecosystem as a whole, not least from a landscape point of view.

Given the germinative capacity of *Helianthemum* spp. seeds, the pioneering nature of both the host plant and the symbiont fungus—both good colonisers of open and sunny edaphic systems—silvicolous interventions should be aimed at maintaining the conditions in which both symbiotic elements remain competitive in the ecosystem.

Pruning in perennial plants can be carried out in overgrown shrubs which cover a large area (more than 0.75 m^2), by cutting the young parts, without cutting the stem or of the more woody parts. Below the shrubs, organic matter (litter) accumulates and germinated seedlings at different ages appear. The shaded micro-system gradually formed with the passage of time is no longer suited to truffle fruiting, which requires a more aerated and lighter surface soil. Therefore, along with the light pruning, a shallow tilling (10–15 cm) is recommended to eliminate competitive, opportunist adventitious vegetation. At the same time soil sponginess is recovered, favouring aeration and oxygenation, improving the water holding capacity and rain infiltration and preventing surface runoff and the loss of water through evaporation since the surface capillaries are broken. This process could be repeated every 6–8 years after harvesting (end of April or May) and before summer sets in. The land should be tilled following the contours of sloping areas in lines at least 4 m apart to avoid erosion. The combination of both treatments should be sufficient to maintain zones of desert truffle production over several decades. As mentioned above, the hosts of the truffles are usually annual plants and small-sized perennial plants, so that simply tilling the soil in this way should be sufficient to ensure the sustainability of truffle production.

However, to favour the spread of the truffle areas we have seen that subsoiling (breaking up of subsoils, usually because they are compacted, without inverting them. This is usually performed with a chisel-like device that is pulled through the soil) are beneficial and long lasting. The depth of the subsoiling, which does not involve turning the soil horizons, leads to greater penetration and rainwater infiltration, although it initially eliminates much of the vegetation. The temporary risk of erosion is minimised in part by the invasive capacity of *Helianthemum* species, pioneering colonisers with a large germinative capacity. Indeed, their populations may be favoured by this type of action. The landscape effect is noticeable during the first few years, although the scrubland quickly regenerates and within a couple of seasons, the cover will have exceeded the original cover. At the same time, it is possible to encourage the expansion of *Helianthemum* by sowing seeds, perhaps of several species and helped by sporal inoculation. Protection against erosion can be carried out as above, by subsoiling in lines of 4–5 m apart. This, together with tilling, sowing and inoculation should help expand the area of truffle production, which will begin a few years after the intervention.

Irrigation can be considered an optional (and effective) practice in the case of potentially very productive areas to maximise production or to ensure the same during particularly dry years. Irrigation should be provided at the end of summer/in autumn and, if the dry conditions continue, at the beginning of the fructification season—an extra 50–80 l/m^2 may greatly improve the yield, whether applied by a drip or a sprinkler system. However, it cannot be denied that installing an irrigation system in the field is sometimes a difficult task.

Other treatments, such as fertilisation, the use of phytosanitary products, clearing and weeding have not even been attempted in Murcia for natural truffle production because they are considered unnecessary, which obviously keeps costs down. Only in irrigated plots is weeding recommended every 3 or 4 years.

The presence of other plants near *Helianthemum* may represent microsites in which fructiferous fungus primordia can develop, which is why carpophores sometimes appear below other shrubs such as thyme or esparto grass, as mentioned above.

The presence of carpophores in a desert truffle plantation is evident from the slight hump in the ground caused by the truffle as it matures, ending in a small crack. The truffle can be extracted using a screwdriver or other hard implement of small diameter and measuring about 20 cm in length. The soils tend to be compacted or hardened so that it is advisable to break the earth to facilitate extraction. The fruiting season begins with the new year and lasts well into spring, depending, obviously, on rainfall and the weather in general during the growing period. In "normal" years most truffles are harvested at the end of February to mid-March.

However, the carpophores take a long time to mature (March–April) so that most truffles harvested earlier are immature, which represents a problem for propagule regeneration. But if the ecological conditions are not drastically altered the mycelium of desert truffles is well adapted and may last for many years in the subsoil, ensuring fructification. The mycelium in the field may suffer alterations, especially in prolonged droughts, when they, presumably, survive by forming microsclerotia, chlamydospores, root-stromes or other resistant structures, as do most fungi under stress conditions (Jennings and Rayner 1986).

Despite the above, it would seem advisable to maintain a minimum number of truffles (no less than 10 %) in the ecosystem to ensure the formation of spores that will permit the autoconservation of propagules. This can be done by limiting the permitted harvesting season, with special attention to the end date of the gathering season. In dry years or when the rainfall is close to the mean inter-annual level of the truffle season, harvesting can be concentrated from the beginning of the year to mid/end of March and be prolonged in especially rainy years until mid-April. In this way the sustainability of truffle production can be ensured.

22.3.3.3 Costs Estimates for Wild Sites Maintenance

As already mentioned, desert truffles are a natural resource that require little care, thus minimising the intervention needed to ensure its sustainability. Whatever tasks are necessary in this respect need only be carried out every 6–8 years, and, even then, little labour is required. Only the pruning demands an increased workforce, while weeding and tilling are totally mechanised operations.

Neither is excessive or continuous irrigation necessary, so that the cost of water is not great, although whatever outlay is involved should be repaid in increased production. No other costs are involved.

Given that the desert truffle is not an expensive crop to cultivate, the earnings attainable in a medium-low quality fruiting season may well be: 50 kg/ha/year $\times$ 30 euros/kg $=$ 1,500 euros/ha/year. Indeed, this is a conservative estimate. It is true that there may be no production during periods of drought if no irrigation is provided, but given the little expense involved in maintaining the crop, the potential for promoting harvesting as a tourist attraction, the landscape and ecological benefits, the soil-fixing capacity, not to mention its role as a wild fauna niche, desert truffle plantations would seem to be a promising alternative in areas where other crops would be difficult to cultivate.

22.3.3.4 Conservation, Uses and Future Possibilities in Spain

In the previous section we mentioned three aspects of gradually increasing importance in the rural environment, in which new hierarchies have been established for evaluating territorial resources. Biodiversity is understood as a resource, as is the landscape and the sense of wilderness that an increasingly urbanised population may feel when confronted with an open space. The problem is how such resources can be made economically productive while being protected.

Until recently, practically the only forestry resource considered to be of value was the wood—and even this was difficult in south-eastern Spain because of the biogeographical peculiarities of the region. Hunting, fishing and other secondary practices (aromatic plants, apiculture and honey, firewood, fruits, mushrooms and even truffles) were introduced in accordance with firmly controlled regulations. At present there is a growing demand for leisure pursuits and open spaces, the economic value of which sometimes exceeds the traditional economic functions of the countryside.

In this context, desert truffles can be considered a natural resource with a promising economic future if managed in an ordered and sustainable way. Truffle "hunting" can be considered as an economic activity or, for most people, a leisure-sociocultural activity.

Quite another matter are the collectors, who can be regarded as more or less professional and whose ability to find truffles is at times surprising. However, such people tend to be respectful of the resources from which they obtain benefits (not always declared!). Neither do they interfere with small animal hunting. Furthermore, both sets of hunters tend to have the same interest in protecting the countryside. Both activities can even be combined because the respective seasons overlap.

The landscape/leisure activity aspect should not be overlooked especially in the flowering season of *Helianthemum*, which tends to coincide with the desert truffle harvesting season (early spring). At this time of the year, a walk in the countryside is regarded by many as the perfect antidote to winter lethargy, especially for people of a certain age and condition (pensioners, retirees, etc.) and not only those drawn from a rural background but, increasingly, representing an urban population.

The richness and value, whether or not considered from a monetary point of view, of these scrublands is high, and so they should be conserved through rational and controlled use. This, in turn, should be able to count on the agreement and compliance of the rural population because only in this way will such populations be sustained and their quality of life improved.

The sustainability of the ecosystems in question implies a compromise between exploiting all the resources they harbour and respecting all the interests involved and those which may arise as social demands change. Only by producing (through the exploitation of resources), conserving (following criteria of sustainability) and improving (as regards the biodiversity and multifunctionality of the space in question) will ecosystem management offer guarantees for the future development of rural areas.

Acknowledgments This work has been supported by projects CGL2011-29816 (MINECO-FEDER) and 08812/PI/08 (Fundación Séneca, Region of Murcia, Spain).

References

Agerer R (2001) Exploration types of ectomycorrhizae. A proposal to classify ectomycorrhizal mycelial systems according to their patterns of differentiation and putative ecological importance. Mycorrhiza 11:107–114

Alcaraz Ariza F, Barreña Cayuela JA, Clemente Díaz M, González Garnés AJ, López Bernal J, Rivera Núñez D, Ríos Ruiz S (2008) Manual of interpretation of natural and semi-natural habitats from the Region of Murcia Dirección General del Medio Natural Consejería de Desarrollo Sostenible y Ordenación del Territorio, Murcia. Spain

Chakravarty P, Chatarpaul L (1990) Effect of fertilization on seedling growth, ectomycorrhizal symbiosis, and nutrient uptake in *Larix laricina*. Can J Forest Res 20:245–248. doi:10.1139/x90-034

DGMN (2007) Community interest regional inventory habitats for the region of Murcia Dirección general del Medio natural, Murcia

Honrubia M, Cano A, Molina-Niñirola C (1992) Hypogeous fungi from Southern Spanish semi-arid lands. Persoonia 14(4):647–653

Honrubia M, Gutiérrez A, Morte A (2001) Desert truffle plantation from south-east Spain, edible mycorrhizal mushrooms and their cultivation. Proceedings of the 2nd international conference on edible mycorrhizal mushrooms. Christchurch, New Zealand, p 3–5

Honrubia M, Morte A, Gutiérrez A (2007) Las terfezias. Un cultivo para el desarrollo rural en regiones áridas y semi-áridas. In: Reyna Domenech S (ed) Truficultura, Fundamentos y Técnicas, Mundi-Prensa, España. p 365–397, ISBN 10: 84-8476-305-6 ISBN 13: 978-84-8476-305-5

Jennings DH, Rayner AD (1986) The ecology and physiology of the fungal mycelium. Cambridge University Press, British Mycological Society, Cambridge. ISBN 0 521 254132

Malençon G (1973) Champignons hypogés du Nord de l'Afrique. I Ascomycètes Persoonia 7:261–288

Morte A, Lovisolo C, Schubert A (2000) Effect of drought stress on growth and water relations of the mycorrhizal association *Helianthemum almeriense-Terfezia claveryi*. Mycorrhiza 10:115–119. doi:10.1007/s005720000066

Morte A, Honrubia M, Gutiérrez A (2008) Biotechnology and cultivation of desert truffles. In: Varma A (ed) Mycorrhiza: state of the art genetics and molecular biology, eco-function, biotechnology, eco-physiology, structure and systematics. Springer, Heidelberg, pp 467–483

Morte A, Navarro-Ródenas A, Nicolás E (2010) Physiological parameters of desert truffle mycorrhizal *Helianthemum almeriense* plants cultivated in orchards under water deficit conditions. Symbiosis 52:133–139. doi:10.1007/s13199-010-0080-4

Morte A, Zamora M, Gutiérrez A, Honrubia M (2009) Desert truffle cultivation in semiarid mediterranean areas. In: Azcón-Aguilar C et al (eds) Mycorrhizas—functional processes and ecological impact. Springer, Heidelberg, pp 221–233

Morte A, Andrino A, Honrubia M, Navarro-Ródenas A (2012) *Terfezia* cultivation in arid and semiarid soils. In: Zambonelli A, Bonito GM (eds) Edible ectomycorrhizal mushrooms. Soil biology. Springer, Heidelberg

Navarro-Ródenas A, Lozano-Carrillo MC, Pérez-Gilabert M, Morte A (2011) Effect of water stress on in vitro mycelium cultures of two mycorrhizal desert truffles. Mycorrhiza 21:247–253. doi:10.1007/s00572-010-0329-z

Navarro-Ródenas A, Pérez-Gilabert M, Torrente P, Morte A (2012) The role of phosphorus in the ectendomycorrhiza *continuum* of desert truffle mycorrhizal plants. Mycorrhiza 22:565–575. doi:10.1007/s00572-012-0434-2

Slama A, Fortas Z, Boudabous A, Neffati M (2010) Cultivation of an edible desert truffle (*Terfezia boudieri* Chatin). Afr J Microbiol Res 4:2350–2356

Index

V. Kagan-Zur et al. (eds.), *Desert Truffles*, Soil Biology 38,
DOI 10.1007/978-3-642-40096-4, © Springer-Verlag Berlin Heidelberg 2014

O

P

W

MIX
Papier aus verantwortungsvollen Quellen
Paper from responsible sources
FSC® C105338

If you have any concerns about our products,
you can contact us on
ProductSafety@springernature.com

In case Publisher is established outside the EU,
the EU authorized representative is:
Springer Nature Customer Service Center GmbH
Europaplatz 3, 69115 Heidelberg, Germany

Printed by Libri Plureos GmbH
in Hamburg, Germany